Ecological Applications

Ecological Applications: toward a sustainable world

Colin R. Townsend
Department of Zoology, University of Otago, Dunedin, New Zealand

Blackwell
Publishing

BLACKWELL PUBLISHING
350 Main Street, Malden, MA 02148–5020, USA
9600 Garsington Road, Oxford OX4 2DQ, UK
550 Swanston Street, Carlton, Victoria 3053, Australia

First published 2008 by Blackwell Publishing Ltd

1   2008

Library of Congress Cataloging-in-Publication Data

Townsend, Colin R.
    Ecological applications : toward a sustainable world / Colin R. Townsend.
        p.   cm.
    Includes bibliographical references and index.
    ISBN-13: 978-1-4051-3698-3 (pbk. : alk. paper)
    ISBN-10: 1-4051-3698-7 (pbk. : alk. paper)   1. Applied ecology.   2. Nature–
Effect of human beings on.   3. Sustainable development.   I. Title.

    QH541.29.T69 2008
    639.9–dc22

                                                        2006100267

A catalogue record for this title is available from the British Library.

Set in 9.5 on 12 pt Sabon
by SNP Best-set Typesetter Ltd., Hong Kong
Printed and bound in Singapore
by Fabulous Printers Pte Ltd

The publisher's policy is to use permanent paper from mills that operate a
sustainable forestry policy, and which has been manufactured from pulp
processed using acid-free and elementary chlorine-free practices. Furthermore,
the publisher ensures that the text paper and cover board used have met
acceptable environmental accreditation standards.

For further information on
Blackwell Publishing, visit our website:
www.blackwellpublishing.com

Contents

List of plates xii
List of boxes xiii
Preface xiv
Acknowledgments xvi

1 Introduction – humans, nature and human nature 1
   1.1 *Homo* not-so-*sapiens*? 2
      1.1.1 *Homo sapiens* – just another species? 3
      1.1.2 Human population density and technology underlie environmental impact 3
   1.2 A biodiversity crisis 4
      1.2.1 The scale of the biodiversity problem 6
      1.2.2 Biodiversity, ecosystem function and ecosystem services 7
      1.2.3 Drivers of biodiversity loss – the extinction vortex 11
      1.2.4 Habitat loss – driven from house and home 12
      1.2.5 Invaders – unwanted biodiversity 13
      1.2.6 Overexploitation – too much of a good thing 14
      1.2.7 Habitat degradation – laying waste 17
      1.2.8 Global climate change – life in the greenhouse 18
   1.3 Toward a sustainable future? 20
      1.3.1 Ecological applications – to conserve, restore and sustain biodiversity 22
      1.3.2 From an economic perspective – putting a value on nature 28
      1.3.3 The sociopolitical dimension 29

Part 1: Ecological applications at the level of individual organisms

2 Ecological applications of niche theory 36
   2.1 Introduction 37
   2.2 Unwanted aliens – lessons from niche theory 41
      2.2.1 Ecological niche modeling – predicting where invaders will succeed 42
      2.2.2 Are we modeling fundamental or realized niches? 44
      2.2.3 When humans disrupt ecosystems and make it easy for invaders 44

2.3   Conservation of endangered species – each to its own niche   46
    2.3.1   Monarch's winter palace under siege   46
    2.3.2   A species off the rails – translocation of the takahe   48
2.4   Restoration of habitats impacted by human activities   49
    2.4.1   Land reclamation – prospecting for species to restore mined sites   49
    2.4.2   Agricultural intensification – risks to biodiversity   51
    2.4.3   How much does it cost to restore a species?   52
    2.4.4   River restoration – going with the flow   53

3   Life-history theory and management   59
3.1   Introduction – using life-history traits to make management decisions   60
3.2   Species traits as predictors for effective restoration   61
    3.2.1   Restoring grassland plants – a pastoral duty   62
    3.2.2   Restoring tropical forest – abandoned farmland reclaimed for nature   62
3.3   Species traits as predictors of invasion success   65
    3.3.1   Species traits predict invasive conifers   66
    3.3.2   Invasion success – the importance of flexibility   66
    3.3.3   Separating invasions into sequential stages – different traits for each?   68
    3.3.4   What we know and don't know about invader traits   71
3.4   Species traits as predictors of extinction risk   71
    3.4.1   Niche breadth and flexibility – freshwater and forest at risk   72
    3.4.2   When big isn't best – r/K theory, harvesting, grazing and pollution   73
    3.4.3   When competitiveness matters – CSR theory, grazing and habitat fragmentation   77

4   Dispersal, migration and management   81
4.1   Introduction – why species mobility matters   82
4.2   Migration and dispersal – lessons for conservation   84
    4.2.1   For whom the bell tolls – the surprising story of a South American bird   84
    4.2.2   The ups and downs of panda conservation   85
    4.2.3   Dispersal of a vulnerable aquatic insect – a damsel in distress   86
    4.2.4   Designing marine reserves   88
4.3   Restoration and species mobility   89
    4.3.1   Behavior management   89
    4.3.2   Bog restoration – is assisted migration needed for peat's sake?   89
    4.3.3   Wetland forest restoration   91
4.4   Predicting the arrival and spread of invaders   92
    4.4.1   The Great Lakes – a great place for invaders   92
    4.4.2   Lakes as infectious agents   94

	4.4.3	Invasion hubs or diffusive spread?	95
	4.4.4	How to manage invasions under globalization	96
4.5	Species mobility and management of production landscapes		97
	4.5.1	Squirrels – axeman spare that tree	97
	4.5.2	Bats – axeman cut that track	97
	4.5.3	Farming the wind – the spatial risk of pulverizing birds	100
	4.5.4	Bee business – pollination services of native bees depend on dispersal distance	103

Part 2: Applications at the level of populations

5	Conservation of endangered species		108
	5.1	Dealing with endangered species – a crisis discipline	109
	5.2	Assessing extinction risk from correlational data	113
	5.3	Simple algebraic models of population viability analysis	117
		5.3.1 The case of Fender's blue butterfly	117
		5.3.2 A primate in Kenya – how good are the data?	118
	5.4	Simulation modeling for population viability analysis	119
		5.4.1 An Australian icon at risk	120
		5.4.2 The royal catchfly – a burning issue	122
		5.4.3 Ethiopian wolves – dogged by disease	123
		5.4.4 How good is your population viability analysis?	126
	5.5	Conservation genetics	127
		5.5.1 Genetic rescue of the Florida panther	128
		5.5.2 The pink pigeon – providing a solid foundation	128
		5.5.3 Reintroduction of a 'red list' plant – the value of crossing	129
		5.5.4 Outfoxing the foxes of the Californian Channel Islands	130
	5.6	A broader perspective of conservation – ecology, economics and sociopolitics all matter	130
		5.6.1 Genetically modified crops – larking about with farmland biodiversity	131
		5.6.2 Diclofenac – good for sick cattle, bad for vultures	133
6	Pest management		139
	6.1	Introduction	140
		6.1.1 One person's pest, another person's pet	140
		6.1.2 Eradication or control?	141
	6.2	Chemical pesticides	146
		6.2.1 Natural arms factories	146
		6.2.2 Take no prisoners	147
		6.2.3 From blunderbuss to surgical strike	147
		6.2.4 Cut off the enemy's reinforcements	150
		6.2.5 Changing pest behavior – a propaganda war	150
		6.2.6 When pesticides go wrong – target pest resurgence and secondary pests	151
		6.2.7 Widespread effects of pesticides on nontarget organisms, including people	153

	6.3	Biological control	154
		6.3.1 Importation biological control – a question of scale	155
		6.3.2 Conservation biological control – get natural enemies to do the work	156
		6.3.3 Inoculation biological control – effective in glasshouses but rarely in field crops	158
		6.3.4 Inundation biological control – using fungi, viruses, bacteria and nematodes	159
		6.3.5 When biological control goes wrong	160
	6.4	Evolution of resistance and its management	162
	6.5	Integrated pest management (IPM)	164
		6.5.1 IPM against potato tuber moths in New Zealand	165
		6.5.2 IPM against an invasive weed in Australia	166
7		Harvest management	172
	7.1	Introduction	173
		7.1.1 Avoiding the tragedy of the commons	173
		7.1.2 Killing just enough – not too few, not too many	174
	7.2	Harvest management in practice – maximum sustainable yield (MSY) approaches	178
		7.2.1 Management by fixed quota – of fish and moose	178
		7.2.2 Management by fixed effort – of fish and antelopes	181
		7.2.3 Management by constant escapement – in time	182
		7.2.4 Management by constant escapement – in space	183
		7.2.5 Evaluation of the MSY approach – the role of climate	184
		7.2.6 Species that are especially vulnerable when rare	185
		7.2.7 Ecologist's role in the assessment of MSY	186
	7.3	Harvest models that recognize population structure	186
		7.3.1 'Dynamic pool models' in fisheries management – looking after the big mothers	187
		7.3.2 Forestry – axeman, spare which tree?	190
		7.3.3 A forest bird of cultural importance	191
	7.4	Evolution of harvested populations – of fish and bighorn rams	191
	7.5	A broader view of harvest management – adding economics to ecology	193
	7.6	Adding a sociopolitical dimension to ecology and economics	195
		7.6.1 Factoring in human behavior	195
		7.6.2 Confronting political realities	197

Part 3: Applications at the level of communities and ecosystems

8		Succession and management	202
	8.1	Introduction	203
	8.2	Managing succession for restoration	206
		8.2.1 Restoration timetables for plants	206
		8.2.2 Restoration timetable for animals	208
		8.2.3 Invoking the theory of competition–colonization trade-offs	209

		8.2.4	Invoking successional-niche theory	209
		8.2.5	Invoking facilitation theory	210
		8.2.6	Invoking enemy-interaction theory	215
	8.3	Managing succession for harvesting		216
		8.3.1	Benzoin 'gardening' in Sumatra	216
		8.3.2	Aboriginal burning enhances harvests	217
	8.4	Using succession to control invasions		219
		8.4.1	Grassland	219
		8.4.2	Forest	220
	8.5	Managing succession for species conservation		221
		8.5.1	When early succession matters most – a hare-restoring formula for lynx	221
		8.5.2	Enforcing a successional mosaic – first aid for butterflies	222
		8.5.3	When late succession matters most – range finding for tropical birds	223
		8.5.4	Controlling succession in an invader-dominated community	223
		8.5.5	Nursing a valued plant back to cultural health	224
9		Applications from food web and ecosystem theory		229
	9.1	Introduction		230
	9.2	Food web theory and human disease risk		234
	9.3	Food webs and harvest management		236
		9.3.1	Who gets top spot in the abalone food web – otters or humans?	236
		9.3.2	Food web consequences of harvesting fish – from tuna to tiddlers	238
	9.4	Food webs and conservation management		239
	9.5	Ecosystem consequences of invasions		240
		9.5.1	Ecosystem consequences of freshwater invaders	240
		9.5.2	Ecosystem effects of invasive plants – fixing the problem	241
	9.6	Ecosystem approaches to restoration – first aid by parasites and sawdust		242
	9.7	Sustainable agroecosystems		245
		9.7.1	Stopping caterpillars eating the broccoli – so that people can	245
		9.7.2	Managing agriculture to minimize fertilizer input and nutrient loss	245
		9.7.3	Constructing wetlands to manage water quality	247
		9.7.4	Managing lake eutrophication	248
	9.8	Ecosystem services and ecosystem health		249
		9.8.1	The value of ecosystem services	249
		9.8.2	Ecosystem health of forests – with all their mites	252
		9.8.3	Ecosystem health in an agricultural landscape – bats have a ball	253
		9.8.4	Ecosystem health of rivers – it's what we make it	254
		9.8.5	Ecosystem health of a marine environment	255

Part 4: Applications at the regional and global scales

10   Landscape management   261
     10.1   Introduction   262
     10.2   Conservation of metapopulations   267
            10.2.1   The emu-wren – making the most of the conservation
                     dollar   267
            10.2.2   The wood thrush – going down the sink   268
            10.2.3   The problem with large carnivores – connecting with
                     grizzly bears   269
     10.3   Landscape harvest management   270
            10.3.1   Marine protected areas   270
            10.3.2   A Peruvian forest successional mosaic – patching
                     a living together   271
     10.4   A landscape perspective on pest control   272
            10.4.1   Plantation forestry in the landscape   272
            10.4.2   Horticulture in the landscape   273
            10.4.3   Arable farming in the landscape   274
     10.5   Restoration landscapes   274
            10.5.1   Reintroduction of vultures – what a carrion   275
            10.5.2   Restoring farmed habitat – styled for hares   276
            10.5.3   Old is good – willingness to pay for forest
                     improvement   276
            10.5.4   Cityscape ecology – biodiversity in Berlin   277
     10.6   Designing reserve networks for biodiversity conservation   277
            10.6.1   Complementarity – selecting reserves for fish
                     biodiversity   279
            10.6.2   Irreplaceability – selecting reserves in the Cape
                     Floristic Region   279
     10.7   Multipurpose reserve design   280
            10.7.1   Marine zoning – an Italian job   280
            10.7.2   A marine zoning plan for New Zealand – gifts, gains
                     and china shops   283
            10.7.3   Managing an agricultural landscape –
                     a multidisciplinary endeavor   283

11   Dealing with global climate change   290
     11.1   Introduction   291
     11.2   Climate change predictions based on the ecology of individual
            organisms   297
            11.2.1   Niche theory and conservation – what a shame
                     mountains are conical   297
            11.2.2   Niche theory and invasion risk – nuisance on
                     the move   298
            11.2.3   Life-history traits and the fate of species – for better
                     or for worse   300

11.3 Climate change predictions based on the theory of population
       dynamics   303
       11.3.1   Species conservation – the bear essentials   303
       11.3.2   Pest control – more or less of a problem?   303
       11.3.3   Harvesting fish in future – cod willing   304
       11.3.4   Forestry – a boost for developing countries?   305
11.4 Climate change predictions based on community and
       ecosystem interactions   306
       11.4.1   Succession – new trajectories and end points   306
       11.4.2   Food-web interactions – Dengue downunder   307
       11.4.3   Ecosystem services – you win some, you lose some   307
11.5 A landscape perspective – nature reserves under climate
       change   308
       11.5.1   Mexican cacti – reserves in the wrong place   309
       11.5.2   Fairy shrimps – a temporary setback   310

Index   315

Plates

Plates between pages 176 and 177

Plate 1.5   Extinctions in Singapore since the early 1800s.

Plate 10.6   Predicted population growth rate of wood thrush populations in relation to forest cover in the eastern USA.

Plate 10.7   Simulated spread on ocean currents along the north Pacific coast of Canada of particles introduced at a depth of 2 m to represent fish larvae, from each of ten marine protected areas over a 90-day period.

Plate 10.12   Distribution of biodiversity hotspots, showing numbers of species of globally threatened birds plus amphibians mapped on an equal area basis.

Plate 10.14   Map of South Africa's Cape Floristic Region showing site irreplaceability values for achieving a range of conservation targets in the 20-year conservation plan for the region.

Plate 10.16   Present landscape and alternative future scenarios for the Walnut Creek catchment area in Iowa, USA.

Plate 11.1   The northerly shift by about 1000 km of zooplankton species typical of warm temperate conditions in the northeast Atlantic over the last 40 years.

Plate 11.2   Predicted annual changes in average surface air temperature, average precipitation and average sea level rise from 1960–1990 to 2070–2100, when carbon dioxide concentration in the atmosphere is expected to have doubled.

Plate 11.3   Simulated distribution of biomes throughout the world using contrasting climate change models.

Plate 11.4   Results of a bioclimatic model for the plant *Protea lacticolor* in the Cape Floristic Region of South Africa.

Plate 11.6   Predicted changes to distribution of the Argentine ant between now and 2050.

Plate 11.8   Current distribution of sweetgum trees across the eastern USA, predicted distribution of sweetgum in 2100 (assuming that rate of dispersal can keep up with change in habitable area), and actual occupied area in 2100 (which is predicted to be very much less than the potential area, because of dispersal restrictions).

Plate 11.12   Dengue fever risk maps for *Aedes albopictus* and *Aedes aegypti*.

Plate 11.13   Stress status of river catchment areas throughout Europe by 2080 according to four different climate change models.

Plate 11.14   Distribution of biological reserves in the Central Valley region of California, USA. Current distribution of F_{max}, and predicted changes to F_{max} under a cooler and drier scenario and under a warmer and wetter scenario.

Boxes

1.1   Classification of extinction risk   5

1.2   Ecological tidbits   23

2.1   Essential niche theory   37

3.1   Essential life-history theory   60

4.1   Dispersal and migration – the conceptual framework   83

5.1   Population dynamics theory 1   110

5.2   Genetics of small populations   114

6.1   Population dynamics theory 2   143

7.1   Population dynamics theory 3   175

8.1   The theory of ecological succession   204

9.1   Food webs and pathways of energy and nutrients   231

10.1 Landscape theory   263

11.1 Predicting the ecological effects of global climate change   292

Preface

The links between pure and applied ecology – a new approach

Most of the widely used ecological textbooks, including those I have had a hand in, focus on the fundamental concepts and theory of ecology and deal with applications to human problems as a secondary matter. This is not inappropriate, given the danger of attempting to apply ecological theory without understanding it. In this book, though, I have decided to take an entirely different approach. My primary focus is now on ecological applications, with the bare essentials of theory provided as a backdrop and confined to a 'box' in each chapter.

I have two reasons for this new approach. First, everyone must now be aware of all the human actions that damage or destroy natural ecosystems: careful steward-ship and the innovative use of ecological applications are needed to ensure future generations have the same opportunities as past ones to enjoy and use nature. Thus ecological literacy has never been more important and so this book will appeal to the nonspecialist, and should be particularly appropriate for those who will take only a single course in introductory ecology.

My second reason is based on 30 years' experience of teaching. I have found it is much easier to grab someone's interest when real-life problems are to the fore. And, somewhat surprisingly, many students of ecology only gain a proper grasp of the fundamentals of the science after they have appreciated its importance for the welfare of the planet and its inhabitants. For this reason, I fully expect my approach to strike a useful chord even for those who will proceed further to learn about the subject from both pure and applied points of view.

The scope of the book – ecology meets economics and social science

The book encompasses a broad spectrum of management for sustainability including restoration, conservation, biosecurity, pest control, harvest management and the design of reserves. These topics are woven together in a novel analysis that considers, in turn, four levels of ecological organization. In the first section of the book I deal with how ecological theory at the level of individual organisms can be turned to advantage by resource managers. In the second and third sections I address, in a similar manner, the application of ecological knowledge at the level first of populations (all individuals of a species combined) and then of communities and ecosystems (multispecies assemblages in their physicochemical setting). Finally, I turn to the macroecological scale and deal with landscape, regional and global issues.

Another key feature of the book is frequent recourse to economic and sociopoliti-cal dimensions of the sustainable use of natural resources. Understanding the ecological basis of our interactions with nature is essential, but sustainability has three strands – ecological, economic and social – and a one-eyed approach is often doomed to failure.

Pedagogical features

A number of pedagogical features are included to aid study:
• Each chapter begins with a set of key concepts that you should understand before proceeding to the next chapter.
• The essential theory relevant to the topic at hand is presented early in each chapter as a box.
• Each chapter begins with a 'conversation' highlighting a particular perspective on a resource management issue. These exchanges with focal individuals, which didn't actually take place, reflect my experience of the diversity of views that are held in society. They are there to help readers appreciate a broad view of the issues involved.
• Each chapter concludes with my summary and then a conundrum raised by the focal individual.

Ecologist as advocate or adviser?

Readers expecting me to take a high moral stand on the ecological problems facing us may be disappointed. It is true that I accept, as a basic tenet, the idea of sustainable behavior. But my real aim is to present the ecological underpinning of a resource management issue, highlighting also any economic or sociopolitical dimensions, but without necessarily advocating a particular course of action. Like everyone else I have my view on how to deal with an issue and I could choose to advocate that approach. As a scientist, however, I believe my role is to help people understand the ecological ramifications of possible courses of action. As a teacher, I think it is even more inappropriate to give the impression that I know all the answers. It is for you to weigh up the ecological and other dimensions of each issue.

The science of ecology is difficult and the conclusions are often imprecise. Sustainability debates almost invariably involve a range of disciplines, and it would be presumptuous for someone with a detailed grasp of just one of those to pontificate to the rest. Rather, ecological scientists, economists, social, political and management scientists all have their knowledge to contribute, as do those in local communities – who often have more useful knowledge than the rest put together. This is why I have chosen to let others (the focal individuals) have the final word in each chapter.

Colin Townsend

Acknowledgments

Many people played a role in the gestation and birth of this book. At Blackwell, and in the production stage, I was greatly assisted by Nancy Whilton, Elizabeth Frank, Rosie Hayden, Pat Croucher, Rachel Moore and Delia Sandford. Many people have read and made helpful comments on particular sections or even the whole book: I am very grateful to Lloyd Davis, Bethan Wood, Graham Russell, Anita Diaz, James Houpis, Elizabeth Hane, Eric Dibble, Mhairi Harvey, Richard Thacker and Clark Gantzer. My brothers and sisters – Julie and Ralph, Kevin and Anne, and Jeff and Joan – and my son and daughter – Dominic and Jenny – provided vital support and a place to write, as did Breck Bowden and Eugenia Marti. Thanks to Carlos Villanueva who showed me what a biodiversity hotspot is really like. And thanks to Alan Hildrew who helped me learn the ecologist's craft, and showed how much fun it can be, and to John Harper and Mike Begon for honing my writing skills. I am indebted to my wife Laurel Teirney who opened my eyes to the importance of moving beyond scientists and managers to involve local people in decisions about ecological management.

In essence, environmental behavior that is sustainable meets current needs without compromising the ability of future generations to provide for themselves. The crunch question is then whether our descendants will be able to enjoy the same opportunities as us. If not, perhaps they will judge their ancestors to have been far from wise. With this in mind, I dedicate this book to my grandson Brennan.

> *Led, finger clutched in hand, from dusk to dawn.*
> *Grandfather sustained by grandson.*
> *Now my wish for you.*
> *Wild nature to astonish and refresh*
> *and, closer to home,*
> *flower-filled meadows, contented cows (moos) and teeming waters.*
> *Yours the hand of a kaitiaki – nature's caretaker?*
> *Sustain, for the sake of your grandchildren,*
> *and theirs.*

1 Introduction – humans, nature and human nature

The history of the human species as global caretaker has not been good. As *Homo sapiens* subspecies *exploitabilis* we have polluted air, land and water, destroyed large areas of almost all kinds of natural habitat, overexploited living resources, transported organisms around the world with negative consequences for native ecosystems, and driven a multitude of species close to extinction. Our 'evolution' to subspecies *sustainabilis* needs to involve some significant behavior changes underpinned by ecological knowledge.

Chapter contents

1.1 *Homo* not-so-*sapiens*? 2
    1.1.1 *Homo sapiens* – just another species? 3
    1.1.2 Human population density and technology underlie environmental impact 3
1.2 A biodiversity crisis 4
    1.2.1 The scale of the biodiversity problem 6
    1.2.2 Biodiversity, ecosystem function and ecosystem services 7
    1.2.3 Drivers of biodiversity loss – the extinction vortex 11
    1.2.4 Habitat loss – driven from house and home 12
    1.2.5 Invaders – unwanted biodiversity 13
    1.2.6 Overexploitation – too much of a good thing 14
    1.2.7 Habitat degradation – laying waste 17
    1.2.8 Global climate change – life in the greenhouse 18
1.3 Toward a sustainable future? 20
    1.3.1 Ecological applications – to conserve, restore and sustain biodiversity 22
    1.3.2 From an economic perspective – putting a value on nature 28
    1.3.3 The sociopolitical dimension 29

Key concepts

In this chapter you will

note that *Homo sapiens* is not the only species to destroy habitat, overexploit resources or pollute the environment

recognize that human population density coupled with technology underlie our unique impact on nature

understand the scale of current and future impacts on biodiversity from habitat loss, introduced invader species, overexploitation, habitat degradation and global climate change

see the link between biodiversity loss and the provision of ecosystem services of importance to human well-being

grasp that a sustainable society is one able to meet current needs without compromising the ability of future generations to provide for themselves

appreciate that a sustainable future has three dimensions – ecological, economic and sociopolitical

---

1.1 *Homo* not-so-*sapiens*?

Homo sapiens, the name of the most recent in a line of hominids, might well be considered a misnomer. Just how sapient (wise) has *Homo sapiens* been? We have certainly been clever – inventing an amazing array of tools and technologies from the wheel to the nuclear power station. But how much of the natural world has been disrupted or destroyed during this technological 'progress'? And is our way of life actually sustainable? A crunch question is whether your descendants will be able to enjoy the same opportunities as you. If not, perhaps they will judge their ancestors to have been far from wise.

Humans destroy natural ecosystems to make way for urban and industrial development and to establish production ecosystems such as forestry and agriculture. We also exploit the natural world for nonrenewable resources (mining) as well as renewable ones (fisheries and forests). Mining destroys habitat directly, and fishery techniques such as bottom trawling can physically disrupt habitat. The natural ecosystems that remain are also affected by human activities. Our harvesting of species from the wild (whether trees, antelopes or fish) has often led to their decline through overexploitation. Our transport systems allow species from one part of the world to hitch a ride to another where, as 'invaders', their impacts on native biota can be profound. And every human activity, including defecation, transport, industry and agriculture, produces 'pollutants' that can adversely affect the biota locally or globally.

You might imagine there would be consensus about what constitutes reasonable behavior in our interactions with the natural world. But people take a variety of standpoints and there are a host of contradictions. Farmers usually consider weeds that reduce the productivity of their crops to be a very bad thing. But conservationists bemoan the farmers' attack on weeds because these species often help fuel the activities of butterflies and birds. The Nile perch (*Lates nilotica*) was introduced to Africa's Lake Victoria to provide a fishery in an economically depressed region, but it has driven most of the lake's 350 endemic fish species towards extinction (Kaufman, 1992). So gains at our dinner tables can equate to a loss of biological diversity.

Then again, our knowledge of plant physiology allows agricultural ecosystems to be managed intensively for maximum food production. But heavy use of fertilizers means that excess plant nutrients, particularly nitrate and phosphate, end up in rivers and lakes. Here ecosystem processes can be severely disrupted, with blooms of microscopic algae shading out waterweeds and, when the algae die and decompose, reducing oxygen and killing animals. And even in the oceans, large areas around river mouths can be so badly impacted that fisheries are lost. The farmers' gain is the fishers' loss.

Pesticides, too, are applied to land but find their way to places they were not intended to be. Some pass up food chains and adversely affect local birds of prey. Others move via ocean currents and through marine food chains, damaging predators at the ends of the earth (such as polar bears and the Inuit people of the Arctic). And hundreds of kilometers downwind of large population centers, acid rain (caused by emission of oxides of nitrogen and sulfur from power generation) kills trees and drives lake fish to extinction. Ironically, in other parts of the world a new ecology is imposed in previously fishless lakes because of the introduction of fish favored by anglers.

So *Homo sapiens* has a diversity of views and a wide variety of impacts. But are we really so different from other species?

1.1.1 Homo sapiens – *just another species?*

Feces, urine and dead bodies of animals are sometimes sources of pollution in their environments. Thus, cattle avoid grass near their waste for several weeks, burrow-dwelling animals defecate outside their burrows, sometimes in special latrine sites, and many birds carry away the fecal sacs of their nestlings. Humans are not unique, either, in regarding corpses as pollutants to be removed. The 'undertaker' caste of honeybee, for example, recognizes dead bodies and removes them from the hive.

And just like humans, many species make profound physical changes to their habitats. These 'ecological engineers' include beavers that build dams (changing a stream into a pond), prairie dogs that build underground towns and freshwater crayfish that clear sediment from the bed. In each case other species in the community are affected. The impact may be positive (for pond dwellers in the beaver ponds, for species that share the prairie dog town, for insects whose gills are sensitive to clogging by sediment) or negative (stream species, plants displaced by burrowing, insects that feed on sediment).

Overexploitation, where individuals of a population are consumed faster than they can replenish themselves, is also a common feature in natural ecosystems. Sometimes overexploitation is subtle, with preferred prey species less common in the presence of their consumers – as compared to their less tasty or harder-to-catch counterparts. But overexploitation may be more dramatically demonstrated when the disappearance of top predators (such as wolves) allows herbivores (such as moose) to multiply to such an extent that the vegetation is virtually destroyed. And the appalling loss of fish species in Lake Victoria after the arrival of the 'invader' Nile perch provides a graphic example of overexploitation by one fish of others.

Invaders have always been a fact of nature, when by chance some individuals breach a dispersal barrier such as a mountain range or a stretch of ocean. But some species that migrate or disperse over large distances can carry their own invaders with them – just as humans do along transport routes. Examples include diseases carried by dispersing fruits and seeds and migrating mammals and birds. The animals may also have parasites and small hitchhikers in their fur and feathers.

Finally, there are species that, like farmers, increase plant nutrient concentrations in their habitats, and even some that produce 'pesticides'. Leguminous plants have root nodules containing symbiotic bacteria that fix atmospheric nitrogen into a form readily available to plants. The soil in their vicinity, and the water draining into neighboring streams, are both likely to contain higher concentrations of nitrate. And certain plants produce chemicals (allelochemicals) whose function appears to be the inhibition of growth of neighboring plants, giving the producer a competitive advantage.

So humans are hardly unique in their ecological impacts. When population density was low, and before the advent of our ability to harness nonfood energy, human populations probably had no greater impact than many other species that shared our habitats. But now the scale of human effects is proportional to our huge numbers and the advanced technologies we employ.

1.1.2 *Human population density and technology underlie environmental impact*

The expanding human population (Figure 1.1) is the primary cause of a wide variety of environmental problems. Someone has calculated that the total mass of humans is now about 100 million tonnes, in comparison to a paltry 10 million tonnes for all

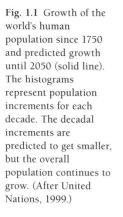

Fig. 1.1 Growth of the world's human population since 1750 and predicted growth until 2050 (solid line). The histograms represent population increments for each decade. The decadal increments are predicted to get smaller, but the overall population continues to grow. (After United Nations, 1999.)

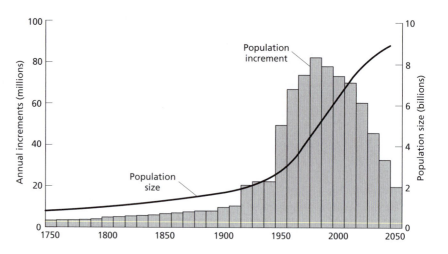

wild mammals combined. We are not unique in destroying habitat and contaminating the environment. But we are distinctive in using fossil fuels, water and wind power, and nuclear fission to provide energy for our activities. These technologies have provided the power to transform much of the face of the planet through urbanization, industrial development, mining, and highly intensive agriculture, forestry and fishing. The loss of habitats and the degradation of what remains are responsible for driving a multitude of species to the verge of extinction. Beavers, prairie dogs and crayfish may fundamentally alter the habitats in which they live, but the burgeoning population of *Homo sapiens*, with attendant technologies, has spread to every continent. The consequences are both intense and widespread, leaving few hiding places for pristine nature to thrive.

Many environmental effects are caused locally, although the same patterns are repeated across the globe (pollution by fertilizers and pesticides, the spread of invaders, and so on). In one very important case, however, the scale of the problem is itself global – climate change resulting from an increase in atmospheric carbon dioxide (produced by burning fossil fuels) together with other 'greenhouse' gases. You will discover that this global pollution problem has implications for every other environmental management issue.

The remainder of this chapter focuses on the scale of human impacts on biological diversity (and the consequences for human welfare – Section 1.2), as well as the knowledge that needs to be harnessed for a sustainable future (Section 1.3). This will form the backdrop to the remainder of the book where, chapter by chapter and topic by topic, I explore how ecological knowledge can be applied to remedy the problems we have caused.

1.2 A biodiversity crisis

It is important to be clear about the meaning of *biodiversity*, and its relationship to *species richness*. Species richness is the total number of species present in a defined area. At its simplest, biodiversity is synonymous with species richness – and this is generally how I will use it. Biodiversity, though, can also be viewed at scales smaller

Box 1.1 Classification
of extinction risk

The *IUCN Red List of Threatened Species* (produced by the World Conservation Union – previously the International Union for the Conservation of Nature, IUCN) highlights species at greatest risk of extinction in each taxonomic group in every part of the world. Overseen by expert specialist teams, plants and animals are defined according to criteria related mainly to population size, distributional range, and whether the population is currently declining. Figure 1.2 illustrates the 'decision tree' used to categorize species status. Of course, not all have been evaluated – most species remain to be identified and named! In some cases an evaluation has been attempted but the data are currently insufficient to classify threat (*Data Deficient*). Some are already considered to be *Extinct* or *Extinct in the Wild* (where individuals remain only in cultivation or captivity).

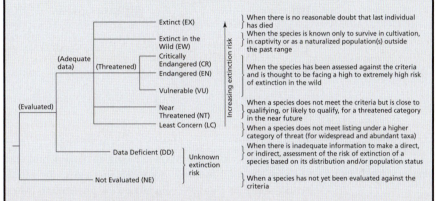

Fig. 1.2 A 'decision tree' showing the nine IUCN Red List categories in order of increasing extinction risk. (From Rodrigues et al., 2006.)

Species considered to be under threat of extinction are listed as *Critically Endangered* (more than a 50% probability of extinction in 10 years or three generations, whichever is longer), *Endangered* (more than a 20% chance of extinction in 20 years or five generations) or *Vulnerable* (greater than a 10% chance of extinction in 100 years). A further category is *Near Threatened* – species close to qualifying for a threat category or judged likely to qualify in the near future. Species that do not meet any of the threat categories are assessed as of *Least Concern*.

An estimated 12% of bird species, 25% of mammals and 30% or more of amphibians are threatened with extinction. The threat classifications help conservation managers prioritize their actions. In addition, conservation plans are supported by the wealth of data collected in the IUCN assessments (Rodrigues et al., 2006).

(genetic diversity within species) and larger than the species (the variety of ecosystem types present – e.g. streams, lakes, grassy glades, mature forest patches).

Human impacts are responsible for driving a multitude of species to such low numbers that much of the world's biodiversity is under threat. In this section I consider just how big this problem is (Section 1.2.1) before discussing the consequences of reduced biodiversity for the way that whole ecosystems function – and for the free 'services' that natural ecosystems provide us (1.2.2). A variety of processes are responsible for species extinctions (1.2.3) and the scale of each of these will be considered in turn: habitat loss (1.2.4), invaders (1.2.5), overexploitation (1.2.6), habitat degradation (1.2.7) and global climate change (1.2.8).

Fig. 1.3 Numbers of species identified and named (dark histograms) and estimates of unnamed species that exist (light histograms). (Modified from Millennium Ecosystem Assessment, 2005a.)

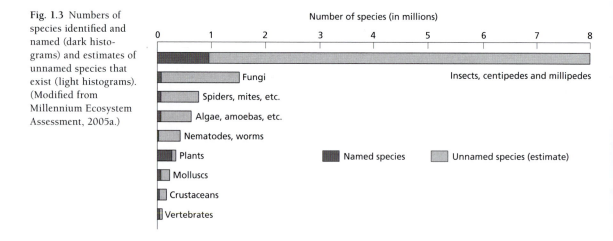

1.2.1 *The scale of the biodiversity problem*

To judge the scale of the problem facing environmental managers it would be useful to know the total number of species that exist, the rate at which these are going extinct and how this rate compares with pre-human times. Not surprisingly, there are considerable uncertainties in our estimates of all these things. For example, only about 1.8 million species have so far been named, but the real number lies between 3 and 30 million. Most biodiversity specialists think it is around 10 million (Figure 1.3).

Palaeontologists estimate that species exist, on average, for between 1 and 10 million years. If we accept this assumption, and taking the total number of species on earth to be 10 million, we can predict that each century between 100 (if species last 10 million years) and 1000 species will go extinct (if species last 1 million years). This represents a 'natural' extinction rate of between 0.001% and 0.01% of species per century. The current estimate of extinction of birds and mammals, the groups for which we have the best information, is about 1% per century. In other words, the current rate may be as much as 100 to 1000 times the 'natural' background rate. And when we bear in mind the number of species believed to be under threat (Box 1.1), the future rate of extinction may be more than ten times higher again (Millennium Ecosystem Assessment, 2005a).

Estimates of extinction rates are beset with difficulties and most extinctions pass unnoticed. Another way to gauge the problem is to focus on long-term assessments of the population sizes of species that have not yet gone extinct. In the case of British birds it is clear that woodland species and, more particularly, farmland species have been in decline for many years (Figure 1.4a). Worldwide, amphibians (Figure 1.4b) and marine and freshwater vertebrates (Figure 1.4c) also show clear signs of widespread population declines.

Consider how instructive it would be to carry out a massive experiment in which a region is allowed to completely fulfill its economic potential while simultaneously documenting the consequences for biodiversity. This decidedly 'unethical' experiment would give us a glimpse of what the world could be like if unlimited population growth and development continue indefinitely everywhere. In fact the 'experiment'

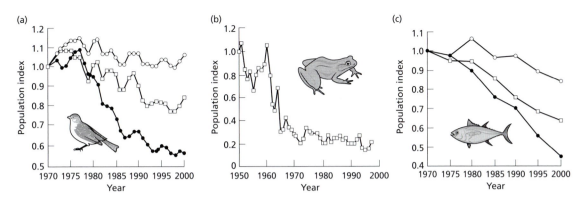

Fig. 1.4 Combined indexes of change in population size for various animal groups for which long-term data are available. In each case the index is standardized at 1.0 for the first year of the dataset. (a) Mean population sizes of British bird species from 1970 to 2000: open circles – all species (105 species), squares – woodland species (33 species), closed circles – farmland species (19 species). (b) Index of change in amphibian populations worldwide from 1950 to 1997, based on accumulated annual changes in 936 populations of 157 species. (c) Index of change in vertebrate populations worldwide: open circles – forest vertebrates (282 populations), closed circles – freshwater vertebrates (195 populations), squares – marine vertebrates (217 populations). (After Balmford et al., 2003, where original references can be found.)

has been done and, moreover, in one of the world's biodiversity hotspots in the East Asian tropics.

The island of Singapore has experienced exponential population growth, from 150 villagers in the early 1800s to more than four million people as it developed into a prosperous metropolis. During this period 95% of Singapore's forest was lost, initially to make way for crops and more recently for urbanization and industrialization. Many extinctions have been documented since 1800 (Figure 1.5 – green histograms). In addition, species lists from nearby Malaysia can be used to infer the likely pristine biodiversity in Singapore and provide an estimate of the number of extinctions that have gone unrecorded (Figure 1.5 – blue histograms). It seems that the majority of the island's species from a wide range of animal and plant groups are now extinct, an unfortunate consequence of the economic 'success story' of modern Singapore. Of course, Singapore is not unique and a similar exercise would produce an equally uncomfortable result for most of the world's cities and nations.

No matter how uncertain the data may be and however imprecise our knowledge of the history of Singapore, or anywhere else in the world, there is no room for complacency – population declines and increased extinction risks need to be confronted.

1.2.2 *Biodiversity, ecosystem function and ecosystem services*

Most people regret any extinction and value species in their own right. But a reduction in biodiversity can also have consequences at a higher level of ecological organization – that of the ecosystem. The *ecosystem* consists of all the species that coexist in an area, together with their physicochemical environment. Ecosystem ecologists pay particular attention to the way solar energy and chemical elements are harnessed by plants in photosynthesis and subsequently pass between living ecosystem compartments (herbivores, carnivores, decomposer organisms) and non-living compartments (dead organic matter in soil or water). Ecosystem processes that might respond to changes in biodiversity include the rate at which plants

Fig. 1.5 Extinctions in Singapore since the early 1800s – green (light) and blue (dark) bars represent recorded and inferred extinctions, respectively. (After Sodhi et al., 2004.) (This figure also reproduced as color plate 1.5.)

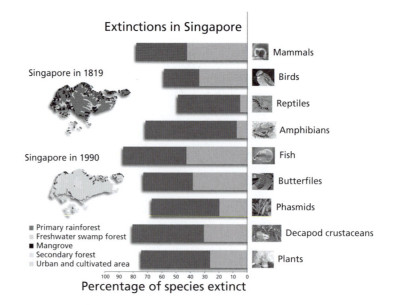

produce new biomass (primary productivity), the rate at which dead organic matter decomposes, and the extent to which nutrients are recycled from dead organic matter back to living organisms. These processes are so fundamental that a substantial change in any one will ramify throughout the food web.

Note, first, that ecosystem properties are not invariably sensitive to a reduction in biodiversity. It may be, for example, that different species carry out similar functional roles and can 'cover for each other' should some be lost. In addition, some species only contribute a little to productivity (or decomposition or nutrient cycling) so their loss would barely register. Other species, however, contribute more than their fair share – the extinction of one of these would be strongly felt (Hooper et al., 2005).

Of most significance is the question of whether species are 'complementary' in the way they operate. If they are, then higher biodiversity will generally equate to higher productivity (or decomposition rate, or nutrient recycling). Take, for example, a set of grassland experiments carried out in Europe (Figure 1.6a). Plant biomass at the end of the growing season was higher when each of three different functional groups was represented (grasses, forbs (nongrass herbs) and nitrogen-fixing legumes). Similarly, the rate of breakdown of tree leaves that fall into streams is higher when the richness of detritivorous insect species is higher (because they 'shred' and feed on the leaves in different ways) (Figure 1.6b). In these cases, then, loss of species is likely to have a detectable impact on the way an ecosystem functions. Managers need to beware loss of biodiversity, both for the sake of the species concerned but also because of consequent changes to ecosystem processes.

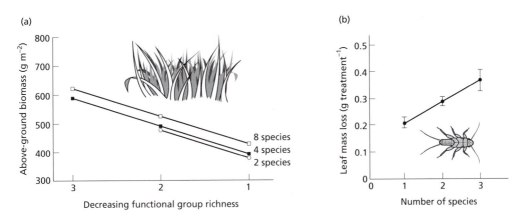

Fig. 1.6 (a) Primary productivity (expressed as above-ground biomass at the end of the growing season) of European grasslands composed of two, four or eight species – note that productivity is somewhat higher when there are more species. However, in cases where all three functional plant groups are represented (grasses, forbs and nitrogen-fixing legumes) productivity is substantially higher than where only two or one of the functional groups are represented. The functional groups differ in the way they garner and convert radiant energy, water and plant nutrients into biomass. (After Hector et al., 1999.) (b) Rate of decomposition of tree leaves that fall into a stream is greater when larvae of three species of stream-shredding stoneflies are present, in comparison to just two or one species. The same total number of stonefly individuals is present in all cases. (After Jonsson & Malmqvist, 2000.)

Beyond the academic quest to understand biodiversity and its role in ecosystems, a utilitarian view of nature focuses on the services that ecosystems provide for people to use and enjoy. *Provisioning services* include wild foods such as fish from the ocean and bushmeat and berries from the forest, medicinal herbs, wood and fiber products, fuel and drinking water. Then there are *cultural services* that nature contributes to human well-being by providing spiritual or aesthetic fulfillment and educational and recreational opportunities. *Regulating services* include the ecosystem's ability to deal with pollutants, the moderation by forest and wetland of disturbances such as floods, the ecosystem's ability to reduce pests and disease risk, and even the regulation of climate (via the capture or 'sequestration' by plants of the greenhouse gas carbon dioxide). Finally, there are *supporting services* that underlie all the other services, such as primary production (by plants), the nutrient cycling upon which productivity is based, and soil formation.

Different ecosystems, both relatively pristine and human-engineered, provide their particular blends of ecosystem services (Figure 1.7). In the case of three 'provisioning' services – production of crops, livestock and aquaculture – human activities have had a positive effect. And in recent times, because of increased tree planting in some parts of the world, there has been an improvement in the sequestration of carbon by trees (a 'regulating' ecosystem service).

But humans have degraded most of the other services. There have been adverse effects on 'provisioning' services in capture fisheries, timber production and water supply (because forest ecosystems moderate river flow, so forest loss increases flow during flooding and decreases it during dry times). We have also seen reductions in many 'regulating' services, including the soil's capacity to detoxify manmade

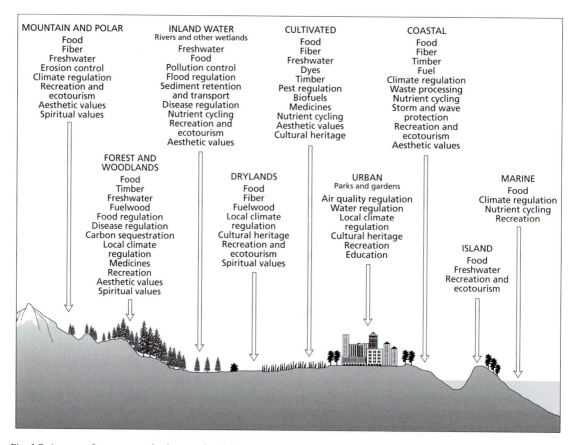

MOUNTAIN AND POLAR
Food
Fiber
Freshwater
Erosion control
Climate regulation
Recreation and
    ecotourism
Aesthetic values
Spiritual values

INLAND WATER
Rivers and other wetlands
Freshwater
Food
Pollution control
Flood regulation
Sediment retention
    and transport
Disease regulation
Nutrient cycling
Recreation and
    ecotourism
Aesthetic values

CULTIVATED
Food
Fiber
Freshwater
Dyes
Timber
Pest regulation
Biofuels
Medicines
Nutrient cycling
Aesthetic values
Cultural heritage

COASTAL
Food
Fiber
Timber
Fuel
Climate regulation
Waste processing
Nutrient cycling
Storm and wave
    protection
Recreation and
    ecotourism
Aesthetic values

**FOREST AND
WOODLANDS**
Food
Timber
Freshwater
Fuelwood
Food regulation
Disease regulation
Carbon sequestration
Local climate
    regulation
Medicines
Recreation
Aesthetic values
Spiritual values

DRYLANDS
Food
Fiber
Fuelwood
Local climate
    regulation
Cultural heritage
Recreation and
    ecotourism
Spiritual values

URBAN
Parks and gardens
Air quality regulation
Water regulation
Local climate
    regulation
Cultural heritage
Recreation
Education

MARINE
Food
Climate regulation
Nutrient cycling
Recreation

ISLAND
Food
Freshwater
Recreation and
    ecotourism

Fig. 1.7 A range of ecosystems, both natural and human engineered, extending from the mountains to the sea. Each ecosystem type provides its own particular set of ecosystem services. (From Millennium Ecosystem Assessment, 2005b.)

chemicals and the ecosystem's capacity to decompose organic waste. Similarly, the loss of riverside vegetation (which can filter nutrient loads) has allowed pollutant levels to increase in aquatic ecosystems. Declines have also occurred in natural hazard protection (loss of natural flood regulation), regulation of air quality and climate, regulation of soil erosion and in many 'cultural' services (Millennium Ecosystem Assessment, 2005b). It is worth noting that ecosystem modification to enhance one service (e.g. intensification of agriculture to produce more crop per hectare – 'provisioning') generally comes at a cost to other services that the ecosystem previously provided (loss of 'regulating' services such as nutrient uptake so pollutant runoff to streams is increased; loss of 'cultural' services such as sites sacred to particular people, streamside walks and valued biodiversity).

All ecosystem services depend directly on elements of biodiversity or on the ecosystem processes supported by biodiversity. Loss of biodiversity, therefore, will often reduce the range of services available to people. There are, in other words, strong economic reasons to manage and conserve nature. This is a point I return to in Section 1.3.2.

1.2.3 *Drivers of biodiversity loss – the extinction vortex*

Extinction may be caused by one of a number of 'drivers', including habitat loss (Section 1.2.4), invasive species (Section 1.2.5), overexploitation (Section 1.2.6) and habitat degradation (including pollution and intensification of agriculture – Section 1.2.7). The relative importance of different drivers for global bird biodiversity is illustrated in Figure 1.8. During the past 500 years bird extinctions can be attributed, in roughly equal measure, to the effects of invasive species, overexploitation by hunters and habitat loss. But now habitat loss is the biggest problem facing threatened species (whether they are classed as critically endangered, endangered or vulnerable). And in the case of 'near threatened' bird species, the ones that managers will increasingly need to attend to in future, habitat loss to agriculture is overwhelmingly the most important driver.

In reality, it seems likely that more than one driver will have played a role in the extinction of any given animal or plant. Thus, a species may be driven to a very small population size by habitat loss/degradation and/or the effect of an invader and/or overexploitation. Then, when numbers become very small there is an increased chance of matings among relatives that produce deleterious effects due to inbreeding depression, causing the population to become smaller still – the so-called extinction vortex (for more detail see Chapter 5, Box 5.1). And a further driver now needs to be added to the list – the global climate change that is predicted to occur over the next century (Section 1.2.8).

Changes to the relative importance of different drivers for all species in various ecosystem types are illustrated in Figure 1.9. Climate change and pollution are predicted to become progressively more important causes of biodiversity loss across all ecosystem types. Habitat change is also set to increase in importance, except in temperate forest, desert, island and mountain ecosystems. Invaders are expected to

Fig. 1.8 Relative importance of different 'drivers' responsible for the loss or endangerment of bird biodiversity. Patterns are shown for five categories of extinction threat (refer to Box 1.1). The values above each histogram are the numbers of species in each threat category globally. In comparison to the past, habitat loss/degradation poses a much bigger risk now (compare histogram for *extinct* birds with histograms for *endangered* and *vulnerable* categories) and this is set to increase in future, in particular via agricultural expansion (histogram for *near threatened* species). (Modified from Balmford & Bond, 2005, where further references can be found.)

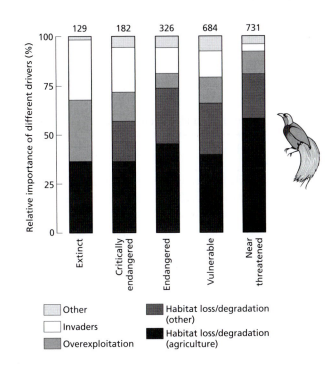

Fig. 1.9 Principal 'drivers' of biodiversity change in various terrestrial and aquatic ecosystems. The cell shade expresses expert panel opinion on the impact of each driver on biodiversity over the last century. The arrows indicate the predicted future trend in each driver's impact. (From Millennium Ecosystem Assessment, 2005b.)

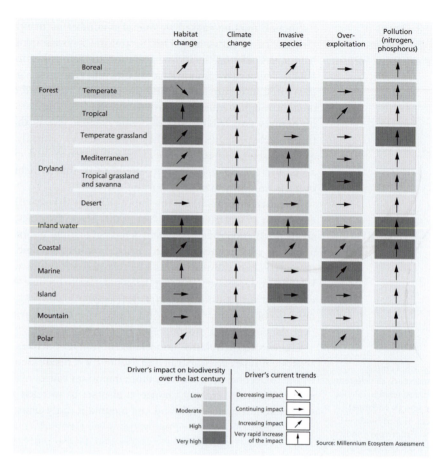

pose a particular risk in forest and dryland habitats (other than desert) and in inland and coastal waters, whereas overexploitation may not figure quite so dramatically in future (except in aquatic environments, polar regions and tropical forests). These scenarios must underpin the plans and priorities of ecosystem managers around the globe.

1.2.4 Habitat loss – driven from house and home

Trends in extinction threat reflect, in large measure, the continuing escalation of the most powerful of human influences – the loss of between 0.5 and 1.5% of wild habitat each year (Balmford et al., 2003). To date, approximately one quarter of the earth's surface has been transformed for agriculture. And in total, well over half of temperate broadleaf and Mediterranean forests have been lost together with 40–50% of tropical and subtropical forests and grasslands (Millennium Ecosystem Assessment, 2005b). Moreover, since 1980 about 35% of mangroves and 20% of tropical coral reefs have gone. In addition to the actual loss of habitat, what remains is almost invariably highly fragmented in its distribution and supports fewer species as a result. Because they are less hospitable to humans, the world's deserts, mountains, boreal forests and tundra have fared less badly.

One obvious management response to habitat loss is to protect as much as possible of what remains, and to include in a network of reserves examples of the variety of natural habitats that exist. In fact, protected areas of various kinds (national parks, nature reserves, sites of special scientific interest, etc.) grew both in number and area during the twentieth century. But only about 7.9% of the world's land area is protected (and 0.5% of sea area – Balmford et al., 2002) and, moreover, there is the disturbing fact that most large reserves are on land that no one else wanted (Figure 1.10).

Protection of wilderness is important and, in one sense, 'relatively' easy to achieve. This is because wilderness is inhospitable to humans and therefore difficult to exploit. (But threats emerge if valuable minerals are discovered in such pristine settings.) However, distributions of endangered plant and animal species sometimes overlap with human population centers. To conserve maximum diversity, it follows that greater focus must be placed in future on areas of higher human value. A global trend toward reduced government subsidies for agriculture and the lowering of international trade barriers may have fortunate consequences for the protection of biodiversity. Thus, in Europe, North America and elsewhere, 'marginal' agricultural land is becoming increasingly uneconomic to farm. Mass-membership organizations, such as the Wilderness Foundation and the Royal Society for the Protection of Birds, have been responding to the opportunity by purchasing some of this land for 're-wilding'. The restoration of biodiverse grasslands and woodlands will add somewhat to the total area of the world that is protected for biodiversity.

1.2.5 Invaders – unwanted biodiversity

Travel has boomed, the world has shrunk and, just like people, plant and animal species have become globetrotters, sometimes transported to a new region on purpose but often as accidental tourists. Only about 10% of invaders become established and perhaps 10% of these spread and have significant consequences but, when they do, the effects can be dramatic. Take, for example, the huge loss of native fish biodiversity in Lake Victoria after the introduction of Nile perch. A more 'subtle' example concerns the arrival in South Africa of the *Varroa* mite, a species that parasitizes the larvae of honeybees in hives and wild nests. Commercial operators can use pesticides to keep the mite in check but 'natural' bee colonies are likely to be wiped out. This will put plant biodiversity at risk because 50–80% of South Africa's native flowers are pollinated by bees (Enserink, 1999).

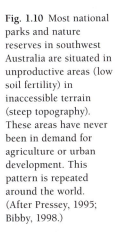

Fig. 1.10 Most national parks and nature reserves in southwest Australia are situated in unproductive areas (low soil fertility) in inaccessible terrain (steep topography). These areas have never been in demand for agriculture or urban development. This pattern is repeated around the world. (After Pressey, 1995; Bibby, 1998.)

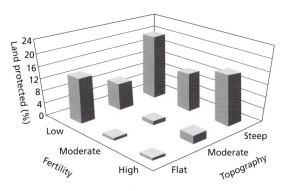

A far-reaching consequence of global transport and the spread of human colonists around the world has been 'homogenization' of the biota. The same set of human camp followers now occur in widely separate regions – sparrows, cockroaches, rats and mice, salmonid fishes and game animals, domestic animals and crop plants (with their associated pests and diseases). Native species often do poorly in the face of this set of invaders so that many parts of North America and the Southern Hemisphere now reflect a European legacy more closely than their native heritage. A graphic example of biotic homogenization (involving fish, molluscs and crustaceans) is provided at either end of a trade link between the Great Lakes of North America and the Baltic Sea. Often spread in the ballast water of the ships that ply their trade along this route, a third of the 170 invaders in the Great Lakes come from the Baltic Sea and a third of the 100 invaders in the Baltic Sea come from the Great Lakes (Millennium Ecosystem Assessment, 2005b).

If native species, endemic to a region, are lost at the expense of a common set of invaders, local biodiversity can remain high but global biodiversity is diminished. And invaders can have far-reaching economic as well as ecological consequences. Table 1.1 breaks down the tens of thousands of exotic invaders in the USA into a variety of taxonomic groups. Among these, the red fire ant (*Solenopsis invicta*) from South America kills lizards, snakes, ground-nesting birds and poultry; in Texas alone, its estimated damage to wildlife, livestock and public health is $300 million per year with a further $200 million spent on control. Large populations of zebra mussel (*Dreissena polymorpha*) from the Caspian Sea threaten native mussels and other animals by reducing food and oxygen availability and by physically smothering them. The mussels also invade and clog water intake pipes, and millions of dollars need to be spent clearing them from water filtration and hydroelectric generating plants. The yellow star thistle (*Centaurea solstitalis*) from the Mediterranean area is a crop weed that now dominates more than 4 million hectares in California, resulting in the total loss of once productive grassland. Rats destroy $19 billion of stored grains nationwide per year, cause fires (by gnawing electric wires), pollute foodstuffs, spread diseases and prey on native species. Overall, pests of crop plants, including weeds, insects and pathogens, are the most costly. Imported human disease organisms, particularly HIV and influenza viruses, are also very expensive to treat and result in 40,000 deaths per year (see Pimentel et al., 2000, for further details). Ecological knowledge is needed to enable us to predict future invasions that are likely to have damaging consequences, so that we can confront the 'invaders', preferably before they arrive (via biosecurity precautions at national borders).

1.2.6
Overexploitation – too much of a good thing

The world once had many more large animals (*megafauna*). Toward the end of the last ice age, for example, Australia was home to giant marsupials, North America had its mammoths and giant ground sloths, and New Zealand and Madagascar were home to giant flightless birds – the moas (Dinornithidae) and elephant birds (Aepyornithidae), respectively. Much of this megafaunal biodiversity disappeared during recent millennia (Figure 1.11a), but at different times in different places (Figure 1.11b). The extinctions seem to mirror patterns of human migration – the arrival in Australia of ancestral aborigines some 50,000 years ago, the appearance of abundant stone spear points in North America about 12,000 years ago, and the arrival of humans around 1000 years ago in New Zealand and Madagascar. The demise of the megafauna may have involved the effects of habitat transformation, particularly by

Table 1.1 Estimated annual costs (billions of US dollars) associated with invaders in the USA. In each case, the cost is made up of loss and damage caused plus dollars spent to control the pests. Taxonomic groups are ordered in terms of the total costs associated with them. (After Pimentel et al., 2000.)

Type of organism	Number of invaders	Major culprits	Loss and damage	Control costs	Total costs
Microbes (pathogens)	>20,000	Crop pathogens	32.1	9.1	41.2
Mammals	20	Rats and cats	37.2	NA	37.2
Plants	5,000	Crop weeds	24.4	9.7	34.1
Arthropods	4,500	Crop pests	17.6	2.4	20.0
Birds	97	Pigeons	1.9	NA	1.9
Molluscs	88	Asian clams, zebra mussels	1.2	0.1	1.3
Fishes	138	Grass carp, etc.	1.0	NA	1.0
Reptiles, amphibians	53	Brown tree snake	0.001	0.005	0.006

NA, data not available.

(a)

(b)

(c)

(a) Yellow star thistle. (© Greg Hudson, Visuals Unlimited.) (b) Red fire ants. (© Visuals Unlimited/ARS.) (c) Zebra mussels. (© Visuals Unlimited/OMNR.)

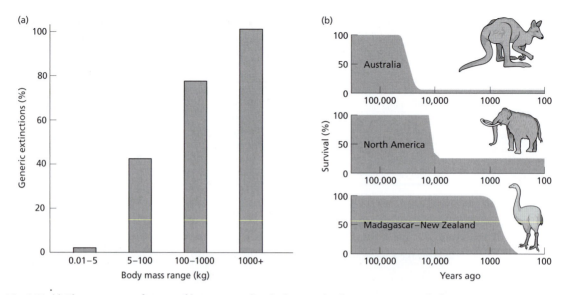

Fig. 1.11 (a) The percentage of genera of large mammalian herbivores that became extinct in the last 130,000 years is strongly related to size (data for North and South America, Europe and Australia combined. (After Owen-Smith, 1987.) (b) Percentage survival of large animals on two continents and two islands (New Zealand and Madagascar) during the past 100,000 years or so. Significant declines in numbers of large animals (mammals, reptiles, birds) occurred at different times in different places, mirroring historical evidence about the arrival of efficient human hunters. (After Martin, 1984.)

fire, and of diseases introduced by humans and their camp followers. But it seems likely that the prime cause was the arrival of efficient human hunters who targeted the largest and most highly profitable prey. Big animals tend to reproduce at a slow rate, making them particularly vulnerable to overexploitation and a downward spiral to extinction.

Prehistoric megafaunal extinctions were particularly dramatic, but in modern times a host of less conspicuous species have been driven by harvesters to the brink of extinction or, at least, to such low numbers that it is no longer profitable to hunt them. The most commonly overexploited species are marine fish and invertebrates (e.g. lobsters and shellfish) as well as trees and terrestrial animals hunted for meat.

Three quarters of the world's industrial fisheries are considered to be fully (50%) or overexploited (25%) (Millennium Ecosystem Assessment, 1995b). And there is a parallel here with the prehistoric extinctions – because species with lower reproductive rates are the most susceptible. Thus, overexploitation of large tuna species is a recognized problem, whereas smaller fish continue to thrive. One consequence of the size–vulnerability relationship, repeatedly observed around the world, is that the mean size of fish taken for human consumption has been declining. Note that it is not just harvesting for food that causes problems – overexploitation may involve plants that provide timber or medicinal products, or live animals and plants collected for the pet and garden trades. The effective regulation of harvesting effort is a difficult business, depending both on a thorough ecological understanding of the dynamics of exploited populations and an ability to regulate the behavior of harvesters.

The impacts of overexploitation are sometimes coupled to harvesting techniques that destroy habitat. A stark example is provided by the dynamiting of coral reef to stun and collect fish. The effects of bottom trawling are less visible but may sometimes be equally destructive. Take the cold-water coral reefs that occur down to depths of 3 km in the offshore waters of at least 41 nations. The technology to study these in close-up recently became available, only to find, for example, that heavy trawling gear has already destroyed up to 40% of the reef off the west coast of Ireland. Managers face the double task of developing harvesting policies that respond to the risk of overexploitation and the threat of physical damage to habitat.

1.2.7 *Habitat degradation – laying waste*

Like the other drivers of biodiversity loss, degradation of habitat by human pollutants continues to show an alarming increase. The chemicals that we release into the atmosphere return to Earth as gases, particles or dissolved in rain, snow and fog. But in the process the pollutants may be carried in the wind for hundreds or thousands of kilometers. Sulfur dioxide (SO_2) and oxides of nitrogen (NO_X) (associated with the burning of fossil fuels) interact with water and oxygen in the atmosphere to produce sulfuric and nitric acids, creating atmospheric pollution which falls as 'acid rain'. Rainwater has a pH of about 5.6, but pollutants lower it to below 5.0 and values as low as 2.1 have been recorded in various industrial areas of the world. Acid rain acidifies the water in lakes and streams, and many species of algae, invertebrates and fish cannot tolerate the extreme conditions. Forest trees can be affected just as badly. Other atmospheric pollutants, including the carbon dioxide produced by the burning of fossil fuels, are now known to cause disturbingly far-reaching climatic effects, with expected changes to global patterns of temperature and precipitation. This will be dealt with in Section 1.2.8.

Our dependence on fossil fuels has other consequences too. More than 4 million tonnes of oil find their way into waterways every year, some seeping naturally from the ocean floor, some from industry, and a large proportion from oil wells and oil tankers. Oil prevents light from reaching aquatic plants and reduces aeration of the water, with adverse effects for seaweeds and invertebrates such as molluscs and crustaceans. Feathers of seabirds become choked with oil and fish gills cease to function. The infamous incident in 1989, when the oil tanker *Exxon Valdez* ran aground in Alaska, spread oil along the coast for a thousand kilometers, contaminating the shores of state parks and other protected areas, and killing an estimated 300 harbor seals, 2800 sea otters and 250,000 birds.

Among the various categories of habitat degradation, agricultural development is set to pose the greatest problems in future. Between 3 and 6% of natural ecosystems around the world have been converted to agriculture since 1950, and this has consequences both for natural habitat loss and for the pollution and degradation of what remains. The scale of the problem is not uniform. As our use of habitat becomes ever more intensive (from protected land, through light grazing of natural grasslands, to cultivation and urban development), biodiversity loss increases for all animal and plant groups (Figure 1.12).

Increasing agricultural intensity is associated with increases in soil erosion, salinization (loss of productive capacity because of salt intrusion) and desertification, and with increased removal of surface and ground water for irrigation. River flow has been reduced so dramatically that, for example, the Nile in Africa, the Yellow River in China and the Colorado River in North America, for parts of the year dry

Fig. 1.12 Estimated average remaining population sizes of various animal and plant taxa in relation to intensity of land use. The index commences at 100 for all taxa in the most natural 'protected' situation, and represents the situation 300 years ago. There is a more-or-less progressive decline in all cases. The surprising increase in amphibian densities associated with urban land use occurs because urban areas in semiarid landscapes of southern Africa provide artificial water-filled habitats needed by frogs and their relatives. (From the Millennium Ecosystem Assessment, 2005b.)

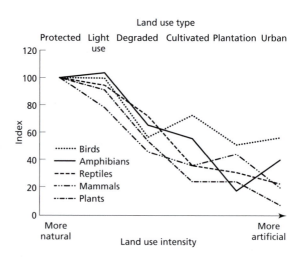

up before they reach the ocean. In addition, excess plant nutrients find their way into waterways, and chemical pesticides affect nontarget species. All these agricultural problems look set to increase over the next 50 years as more land is converted (Figure 1.13). And because greater human population growth is expected in species-rich tropical areas, increased agricultural activity will place biodiversity at high risk. The challenge for managers is to keep land conversion to a minimum (needed to support the human population) and to promote agricultural 'best practices' that minimize ecological fallout.

1.2.8 *Global climate change – life in the greenhouse*

The most far-reaching consequence of our use of fossil fuels has been an increase in the concentration of carbon dioxide in the atmosphere. The level in 1750 (i.e. before the Industrial Revolution), measured in gas trapped in ice cores, was about 280 ppm (parts per million), but this rose to 320 ppm by 1965 and stands at about 380 ppm today (Figure 1.14). It is projected to increase to 700 ppm by the year 2100.

The Earth's atmosphere behaves rather like a greenhouse. Solar radiation warms up the Earth's surface, which reradiates energy outward, principally as infrared radiation. Carbon dioxide – together with other gases whose concentrations have increased as a result of human activity (nitrous oxide, methane, ozone, chlorofluorocarbons) – absorbs infrared radiation. Like the glass of a greenhouse, these gases (and water vapor) prevent some of the radiation from escaping and keep the temperature high. The air temperature at the land surface is now $0.6 \pm 0.2°C$ warmer than in pre-industrial times. Note, however, that temperature change has not been uniform over the surface of the Earth. Up to 1997, for example, Alaska and parts of Asia experienced rises of 1.5–2°C, while the New York area experienced little change, and temperatures actually fell in Greenland and the northern Pacific Ocean. Given the expected further rises in greenhouse gases, temperatures are predicted to continue to rise by a global average of between 1.8°C and 4.0°C by 2100 (IPCC, 2007; Millennium Ecosystem Assessment, 2005b), but to different extents in different places. Such changes will lead to a melting of glaciers and icecaps, a consequent rise

Fig. 1.13 Predicted increases in nitrogen (N) and phosphorus (P) fertilizers, irrigated land, pesticide use and global areas under crops and pasture by 2020 (dark bars) and 2050 (light bars). (From Laurance, 2001, based on data in Tilman et al., 2001a.)

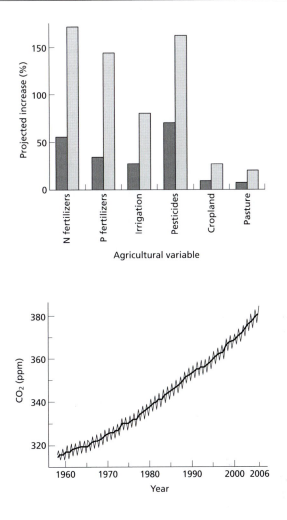

Fig. 1.14 The concentration of atmospheric CO_2 measured at the Mauna Loa Observatory, Hawaii showing the seasonal cycle (peaking each northern summer when photosynthetic rates are maximal in the Northern Hemisphere) and, more significantly, the long-term increase that is due largely to the burning of fossil fuels. (Courtesy of the Climate Monitoring and Diagnostics Laboratory (CMDL) of the National Oceanic and Atmospheric Administration (NOAA).)

in sea level, and large changes to global patterns of precipitation, winds, ocean currents and the timing and scale of storm events.

The principal cause of increased greenhouse gases has been the combustion of fossil fuels, but other factors also come into play. Adding the carbon dioxide released when limestone is kilned to produce cement (about 0.1 Pg of carbon per year) to fossil fuel use (5.6 Pg per year), a net increase of 5.7 (±0.5) PgC per year was added to the atmosphere during the period 1980–1995 (1 petagram = 10^{15} g) (Houghton, 2000). Landuse change is believed to have pumped a further 1.9 (±0.2) PgC into the atmosphere each year. In particular, the exploitation of tropical forest causes a significant release of carbon dioxide, particularly if the forest is cleared and burnt to make way for agriculture. Much of the carbon goes up in smoke, followed by further carbon dioxide release as vast stores of soil organic matter decompose.

Where does the extra 7.6 PgC per year of carbon end up? The observed increase in atmospheric carbon dioxide accounts for 3.2 (±1.0) PgC (i.e. 42% of the human inputs), while much of the rest, 2.1 (±0.6) PgC, dissolves in the oceans. This leaves 2.3 PgC per year, which is generally attributed to a terrestrial 'sink' – probably

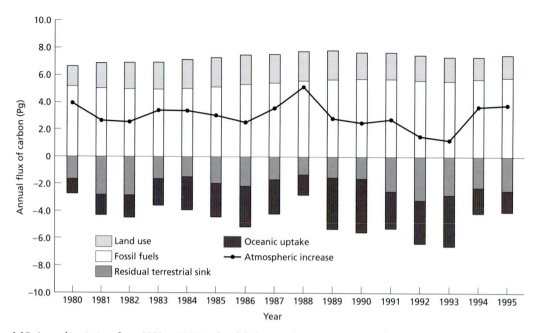

Fig. 1.15 Annual variations from 1980 to 1995 in the global atmospheric increase in carbon dioxide (circles and black line) and in carbon released (histograms above the midline) or accumulated (histograms below the midline) by changes to the burning of fossil fuels, land use, oceanic uptake and an uncertain terrestrial sink (probably related to increased plant productivity). (From Begon et al., 2006, after Houghton, 2000.)

involving carbon dioxide uptake associated with increased terrestrial productivity in northern mid-latitude regions (Houghton, 2000).

There is considerable year-to-year variation in the estimates of carbon sources and sinks, and of increases in the atmosphere (Figure 1.15). Declines in the rate of increase between 1981 and 1982 followed sharp rises in oil prices, while declines in 1992 and 1993 followed the collapse of the Soviet Union. In 1997–8 (not shown in Figure 1.15 but evident, if you look carefully, in Figure 1.14), massive forest fires in a small part of the globe (Indonesia) doubled the growth rate of CO_2 in the atmosphere. The accurate prediction of future changes in carbon emissions is difficult because so many variables play a role (climatic, political and sociological). And predictions of consequences for patterns in global temperature and precipitation are no less straightforward. However, the matter is pressing because we can be sure that climate change will further complicate all the other environmental issues so far discussed.

1.3 Toward a sustainable future?

Can there really be people (exploiters) interested only in short-term financial gain and with absolutely no thought for adverse environmental consequences? And can others (preservationists) be so naive as to argue that nature should be protected always and everywhere? Human nature is such that we tend to identify more with one pole than the other, and then to assume that those at the other end are extremists. Of course there are some fundamentalists, but the vast majority are not so polarized. Those who take the middle ground appreciate the necessity to produce food and industrial goods, to harvest natural resources and to control pests. But they also value nature and recognize the need to protect biodiversity. These nonextremists, of course, also have a variety of views – some are closer to the preservationist end of the spectrum (happy to have a lower standard of living for the sake of the natural world) and others are more exploitationist in their view (requiring that only a small portion of the natural world be protected – to be enjoyed as a recreational walker, an ecotourist or in a natural history film). Reconciling these views remains a challenge. But given the accumulating evidence of adverse effects of our activities, where many impacts only become apparent in the long term, we need to rise to the challenge.

Is it possible to take a completely objective approach to determine just how far we can push the drive to exploit – or, conversely, to decide how much of the natural world should be maintained in a completely pristine state, or be protected at some level? This is not easy, to say the least, but we can get close by pursuing a particular aim – that of sustainability. To call an activity 'sustainable' means that it can be continued or repeated in future. If people wish to have tuna to eat in future, they cannot continue to harvest them from the sea faster than the population can replace those that are lost. Nor can farmers continue to use fertilizers indiscriminately if people want to retain the ecosystem services provided by rivers, lakes and oceans that are impacted by the agricultural excess. And on the largest scale of all, our present reliance on fossil fuels puts at risk, through global climate change, the sustainability of all our activities, whether exploitative or protective. In essence, a sustainable society is one able to meet current needs without compromising the ability of future generations to provide for themselves. Sustainable behavior, in other words, provides the best outcomes for both human and natural systems – now and in future.

One of the problems with the idea of sustainability is that it can only be defined on the basis of what is known now. But what about the many factors that are unknown or unpredictable? Things might take a turn for the worse – when locally adverse oceanographic conditions damage a fishery already suffering from overexploitation, or global climate change increases flood risk in a region already prone to flooding because of deforestation. On the other hand, some people tend to downplay the risk – because *Homo sapiens* is so smart. Thus, they believe that technological advances will allow activities to be sustained that previously seemed unsustainable – the invention of processes to remove pollutants from the outflows of power stations and industry, or of a pesticide more precisely targeted on the pest and without harm to innocent bystanders. But it would be risky indeed to have faith that there will always be a technological 'fix' to solve current environmental problems. *Homo sapiens* needs to become truly wise, factoring in all conceivable risks to sustainability scenarios.

The recognition of the importance of sustainability as a unifying idea can be said to have come of age in 1991. This was the year that the Ecological Society of America published *The sustainable biosphere initiative: an ecological research agenda* (Lubchenco et al., 1991), and the World Conservation Union, the United Nations Environment Programme and the World Wide Fund for Nature jointly published *Caring for the Earth: a strategy for sustainable living* (IUCN/UNEP/WWF, 1991).

The emphasis shifted more recently from a purely ecological perspective to one that incorporates economic and social conditions that influence sustainability

(Milner-Gulland & Mace, 1998), a theme that has gathered pace in the new millennium. In 2002, for example, 190 countries committed themselves 'to achieve by 2010 a significant reduction of the current rate of biodiversity loss at the global, regional and national levels as a contribution to poverty alleviation and to the benefit of all life on Earth' (UNEP, 2002). The Millennium Ecosystem Assessment, launched in 2001, is also based on contributions from a large number of natural and social scientists (Millennium Ecosystem Assessment, 2005b). Its aim is to provide both the general public and decision makers with 'a scientific evaluation of the consequences of current and projected changes in ecosystems for ecosystem services and human well-being' (Balmford & Bond 2005). In the remaining sections of this chapter, I introduce the ecological (Section 1.3.1), economic (1.3.2) and sociopolitical (1.3.3) dimensions of sustainability.

1.3.1 *Ecological applications – to conserve, restore and sustain biodiversity*

The body of ecological theory is organized in a hierarchical way. My focus in this book will be on ecological applications, but these will be presented in a sequence that mirrors the underlying theory.

At the lowest level is the ecology of *individual organisms* – their niche requirements (resource needs and tolerance of physicochemical conditions – Chapter 2), their life-history characteristics (Chapter 3) and their dispersal/migratory behavior (Chapter 4). Knowledge at this level is crucial when reintroducing species that have become locally extinct, restoring natural grassland and forest, or predicting invaders likely to pose a major problem. See Box 1.2 for ecological tidbits from each chapter.

Next comes the *population* level of ecological organization. The population comprises all the individuals of a single species in a particular place, and the focus is on factors that determine the density and genetic diversity of populations. Population theory is central to the management of endangered species (Chapter 5), pests (Chapter 6) and harvests (Chapter 7).

Moving up another step in the ecological layer cake, *community* ecology concerns itself with all the species that coexist. Two areas of community ecology of particular importance for environmental management are succession (the predictable temporal pattern in species composition after a disturbance such as a storm or fire – Chapter 8) and patterns in the feeding interactions of food webs. When the spotlight is turned on the community in relation to its physicochemical environment, and specifically the flux of energy and matter through the food web, we talk of the *ecosystem* level of organization (Chapter 9). Ecological theory associated with communities and ecosystems helps managers devise plans to conserve and restore natural communities, counteract invasions, increase the range of harvestable products and make agroecosystems sustainable. And of course, ecosystem theory underpins the whole idea of ecosystem services and their contribution to human well-being.

The last part of the book combines examples from all levels in the ecological hierarchy but shifts emphasis to a larger spatial scale, dealing with landscape ecology (the patchwork of habitats in the landscape as a whole – Chapter 10) and finally with the global environment, where global climate change becomes the focus (Chapter 11). Landscape ecology is crucial when designing networks of nature reserves, but often also when planning conservation, restoration, harvests and pest control. And global climate change has implications for everything else – whether at the level of individuals, populations, communities, ecosystems or landscape

Box 1.2 Ecological
tidbits

This book, about how ecological theory can be harnessed to protect biodiversity and ecosystem services, is divided into sections that relate to four areas of ecological theory (see chapter grouping in the main contents list). Here are some morsels from the four chapter groupings to tempt your appetite – one per chapter.

1 Ecological applications at the level of individual organisms

Niche theory and the translocation of New Zealand's takahe

Source: Ross Armstrong/Alamy.

One of the few remaining representatives of a guild of large, flightless birds, the takahe declined almost to extinction because of human hunting and the effects of invaders that compete with them (deer) or prey upon them (stoats). The surviving individuals were restricted to a remote and climatically extreme mountain area. A conservation plan called for captive breeding and release of birds in suitable habitat elsewhere. Selecting the correct release sites, a crucial step, depends on understanding the bird's niche requirements – but was their mountain distribution a reflection of ideal conditions or simply their last outpost? Managers used evidence of fossil remains to map the takahe's historical distribution and throw light on its optimal niche requirements. Translocation of individuals to lowland areas on offshore islands has proved successful (Section 2.3.2).

Species traits can predict invasive trees and threatened natives

Photo: M.P. Frankis.

Of a hundred pine species that have been introduced to the USA, a small proportion have caused problems by spreading into native habitats. Their 'success' turns out to depend on certain traits, including small seed size, a short interval between successive large seed crops and rapid maturity to reproductive adult. Conversely, native pine species that are particularly prone to extinction, such as *Pinus maximartinezii*, have precisely the opposite traits. Such patterns give managers something to work on when it comes to predicting problematic invaders or identifying natives that are most likely to need protection (Section 3.3.1).

Migratory behavior and the design of nature reserves for giant pandas

Source: blickwinkel/Alamy.

Giant pandas depend for food on just a few species of bamboo. From June to September in China's Qinling Province, home to 20% of the world's remaining animals, the pandas eat *Fargesia spathacea*, which grows from 1900 to 3000 m. But as colder weather sets in, they move to lower elevations and between October and May feed primarily on *Bashania fargesii*, a bamboo that grows from 1000 to 2100 m. Existing reserves did not cater for the needs of pandas at both ends of this seasonal migration, putting the population at risk. But now, with a fuller knowledge of migratory behavior and the distribution of bamboo species, a network of connected reserves has been established (Section 4.2.2).

2 Applications at the level of populations

Countering the threat of extinction – genetic rescue of the Florida panther

Source: Mark J. Barrett/Alamy.

The last remaining population of the Florida panther (a subspecies of *Puma concolor*) became so small that genetic variation was remarkably low and deleterious forms of genes occurred at high frequency, causing features such as undescended testes, kinked tails and the poorest semen quality of any cat species. Managers decided to translocate individuals from another subspecies, the Texas cougar, in an attempt to eliminate deleterious variants and restore more normal levels of genetic variation. Now panthers with some cougar ancestry show dramatic reversals in the frequency of undesirable features and the signs are good that the probability of extinction has been reduced (Section 5.5.1).

Pest control on the island of St Helena – a ladybird beetle saves the day

Reproduced with permission of the copyright holder, CAB International.

An invasive scale insect had put at risk the national tree of a small South Atlantic island. Only 2500 St Helena gumwoods were left when, in 1991, the South American scale insect *Orthezia insignis* was found to be mounting an attack on the trees, killing more than 100 of them by 1993 and expected to wipe them out by 1995. A known predator of the scale insect, the ladybird *Hyperaspis pantherina*, was cultured and released on the island in 1993 and as its numbers increased the scale insects quickly declined. Since 1995 no scale insect outbreak has been reported and culturing of the ladybirds has been discontinued because the population is maintaining itself at low density in the wild – as good biological control agents should (Section 6.3.1).

Harvest management – counteracting evolution towards small size

© D.P. Wilson/FLPA.

Harvesting is often size-selective, whether for bighorn sheep with the largest horns or for the biggest fish to sell at market. In an unharvested population, a few small individuals may mature earlier than the rest. If the population is then harvested in a way that takes mainly large individuals, the few small but mature animals are likely to provide a disproportionate number of offspring to the next generation and future generations become dominated by smaller animals. A graphic example is provided in the north Atlantic where the size at which cod mature has suffered a dramatic decline as a result of heavy fishing pressure. The consequences for harvest yields can be profound and managers need to devise harvesting rules that counteract this evolutionary trend (Section 7.4).

3 Applications at the level of communities and ecosystems

Succession theory – nursing a community back to health

Source: Kalpana Kartik/Alamy.

After a disturbance, such as a hurricane, volcanic eruption or forest clearance for agriculture, the community proceeds through a predictable successional species sequence until mature forest is regained. A dilemma for managers wishing to restore Mediterranean forest was whether to remove *or* encourage the pioneer shrubs in the succession. By-passing the early species might speed up the transition to forest, but not if the shrubs *facilitate* the success of later species in this hot, dry region. In fact, experiments showed that the pioneer shrubs act as 'nurse' plants for the vulnerable tree seedlings, shading and protecting them from the scorching Mediterranean sun. When pioneer species are facilitators of successional change the management prescription must be to leave them in place (Section 8.2.5).

Food web theory – minimizing human disease risk

Source: PHOTOTAKE Inc./Alamy.

If left untreated, Lyme disease can damage heart and nervous system and lead to a type of arthritis. The illness is caused by the bacterium *Borrelia burgdorferi*, which is carried by ticks in the genus *Ixodes* and transmitted when the tick feeds, first on small animals such as the white-footed mouse, and later on large mammals, including people wandering through the forest. In unusual years of massive acorn production by oak trees, the mouse population thrives and the parasite–host dynamics are such that two years later the risk of Lyme disease is considerably increased. This knowledge helps forest managers to provide a timely warning to hikers. Many small animal species are hosts to the ticks, but most are much less efficient than the mice at passing on the bacterium. This means that disease risk to humans is lower where the biodiversity of squirrels, birds and lizards is high – providing a compelling reason to conserve biodiversity (Section 9.2).

4 Applications at the regional and global scales

A marine zoning plan for sustainability

Source: Chad Ehlers/Alamy.

Maori, recreational and commercial fishers, tourism operators, marine scientists and environmentalists got over their differences, learnt from each other and produced a comprehensive ecosystem management plan for the large Fiordland region of New Zealand. A significant feature of the proposal was the concept of *gifts* and *gains* by the various groups. Thus the plan called for new fishing behavior: a reduction in bag limits for recreational fishers, the withdrawal of commercial fishers from the inner fiords and a voluntary suspension of certain customary fishing rights by Maori. In addition, a number of marine reserves and protected areas were identified to protect representative ecosystems and *china shops* – areas with outstanding but vulnerable natural values. These gains in sustainability and conservation were balanced by the gift from environmentalists to refrain from pursuing their original goal of a much more extensive marine reserve program. The New Zealand government agreed to implement the plan in its entirety and has passed the new legislation necessary (Section 10.7.2).

Global climate change – predicting future invasions

© MARK MOFFETT/Minden Pictures/FLPA.

The Argentine ant is now established on every continent except Antarctica. It can achieve extremely high densities, with adverse effects on biodiversity and unpleasant consequences for domestic life – the ants tend to swarm over human foodstuffs and even sleeping babies. A niche model was developed for the ant, according to current distributions in its native and invaded ranges. Then,

> based on expected global changes to temperature and precipitation, the likely future distribution of the ant was predicted. The species is set to contract its range in tropical areas but expand into temperate areas. Ironically, therefore, the Argentine ant looks set to do less well in Argentina and South America but become a major pest in North America and Europe. Efforts to eradicate Argentine ants have rarely been successful. The management response is therefore to increase biosecurity precautions in regions expected to become progressively more invadable as climate change takes hold (Section 11.2.2).

ecology, and whether concerned with conservation, restoration, harvest management, pest control, biosecurity or sustainable agroecosystems.

1.3.2 From an economic perspective – putting a value on nature

There is an economic side to every resource management argument. Sometimes this is obvious to everyone and 'relatively' straightforward to quantify. A decision to become a player in a fishery will take into account the costs of buying and maintaining fishing boats, crew, gear and shore facilities in relation to the value of the sustainable catch. The decision of a farmer to invest in pest control also depends on weighing up the dollars spent killing pests in comparison to the gains to be made in extra product at the farm gate. And the economic value of what is put at risk by the arrival of invaders can be set against the costs of biosecurity border operations to keep them out.

It can also be relatively easy to work out the cost of saving a species from extinction – in terms of purchasing a reserve, predator control, and so on. But how can managers determine the value of the species so they can decide whether the cost is justified? Then again, you have seen that many economic activities put ecosystem services at risk (Section 1.2.2). How do we determine the value of lost services so these can be set against the economic gains associated with the activity? Even the 'straightforward' economics of fishing and farming turn out to be fraught with difficulty. This is because traditional economics have not taken into account the associated environmental costs that are borne by society in general. Take, for example, the destruction of cold-water coral reefs while trawling, or the reduction in river water quality resulting from a farmer's indiscriminate application of fertilizer.

Economic valuation of nature is inherently a human preoccupation, being based on the contributions that biodiversity makes to our well-being. There are also 'deeper' reasons for conserving biodiversity – species may be considered of value in their own right, a value that would be the same if people were not around to exploit or appreciate it. But, to be effective, it seems inevitable that conservation arguments must ultimately be framed in cost–benefit terms. This is because governments decide policies against a background of the priorities accepted by their electorates, and the money they have to spend. The importance of the concept of ecosystem services, which is relatively new, lies in its focus on how biodiversity provides for human well-being. Now, economic value can be ascribed to biodiversity in a way that can be understood by everyone.

A range of techniques are available to put a value on nature. Goods and services for which there is a market are the most straightforward – values can quite readily be ascribed to clean water for drinking or irrigation, to fish from the sea and medicinal herbs from the forest. In other cases, a more imaginative approach is required.

For example, *travel cost* paid by people to access a natural area provides a minimum value of this recreational service. *Contingent valuation* may be determined in a survey of people's *willingness to pay* for each of a set of hypothetical landuse scenarios (perhaps in terms of a hypothetical 'nature tax'). *Replacement cost* is an estimate of how much would need to be spent to replace an ecosystem service with a manmade alternative – such as substituting the natural waste disposal capacity of a wetland by building a treatment works. *Avoided cost* is an estimate of the cost that would have occurred had a service not been available – such as flood damage if a protective off-shore reef were not present.

And when an ecosystem service has already been lost, the real costs – in loss of property, livelihoods, health and so on – can be determined. Take, for example, the collapse through overexploitation of the Newfoundland cod fishery in the early 1990s – this cost at least $2 billion in income support and retraining for the thousands of people who lost their jobs. Another graphic example is provided by the largely deliberate burning of 50,000 km^2 of Indonesian vegetation in 1997 – the economic cost comprised $4.5 billion in lost forest products and agriculture, increased greenhouse gas emissions, reductions in income from tourism, and healthcare expenditure on 12 million people affected by the smoke (Balmford & Bond, 2005).

Viewed from the broadest perspective of all, the total value of the world's ecosystem services has been roughly estimated at $38 trillion ($10^{12}$) – more than the gross domestic product of all nations combined (Costanza et al., 1997). The 'new economics' provides persuasive reasons for taking great care of biodiversity. You can dip into this book's smorgasbord of examples where economic arguments are prominent in Sections 2.4.3, 4.4, 4.5.3, 4.5.4, 5.6, 7.5, 7.6, 8.3 and 10.5.3.

1.3.3 *The sociopolitical dimension*

Many ecologists feel outside their comfort zone when asked to confront economic realities. But the situation is more complex still, because environmental issues almost always have a sociopolitical angle too. Sociologists can help managers identify the best approaches to reconcile the desires of all interested parties, from farmers and harvesters to tourism operators and conservationists. And political scientists help address the twin problems of whether sustainable management should be fostered by penalties or inducements, and be set in law or encouraged by education. Moreover, there are both sociological and political dimensions to the question of how the needs and perspectives of indigenous people can be taken into account. Sustainable environmental management clearly has a *triple bottom line* – ecological, economic and sociopolitical.

At the local level, the knowledge and ideals of the community can be of great value in improving sustainable behavior. So-called *social capital* is a measure of connectedness in a community, reflecting relationships of trust, willingness to share information, and to develop common rules about biodiversity protection and the sustainable use of nature (Pretty & Smith, 2004). By getting together, rural people, for example, can improve their understanding of the relationship between agriculture and nature and find ways to deal with adverse effects – by fencing waterways, replanting riparian (stream-side) vegetation buffer strips, and implementing more careful use of plough, fertilizer and pesticide. This process of social learning increases 'social capital' and helps new ideas to spread more rapidly through the community and to other places. Community groups reach a zenith of achievement when diverse interest sectors, which previously failed to engage with each other, come together to confront a sustainability issue. Once the barriers are down (aided by a skilled facilitator if necessary), people as diverse as commercial fishers and environmentalists can learn from each other and identify the real sustainability problems (see, for example, the marine zoning plan in Box 1.2). When people are well connected in groups and networks, and when their knowledge is sought and incorporated during environmental management planning, it seems they are more likely to retain a care-taking role in the long term (Pretty & Smith, 2004). At one end of the scale of community participation, government agencies merely keep people informed of plans, or consult by asking questions, but fail to concede to the community a share in decision-making. At the other end of the scale, and better by far, is full participation by the community in analysis, planning, implementing and policing a management strategy for which they take ownership.

If an environmental problem occurs at too large a scale for local communities and governments to solve, the sociopolitical machinations need to occur globally. Estimates of future greenhouse gas emissions, the concentrations to be expected in the atmosphere, and the resulting changes to global temperature vary considerably. Some of this variation reflects uncertainties in climate science. But the predicted patterns of increase, and in some cases eventual decreases, depend on how fast the human population continues to grow, where the population will peak, changing attitudes to the use of energy sources, the technological advances that come to pass and attitudes to the importance of ecosystem services. There is a profound sociopolitical dimension to all these things.

An analysis of four quite detailed sociopolitical scenarios in Table 1.2 explores likely trends in climate change, pollution problems and the state of ecosystem services. If there is little change in our sociopolitical outlook, the *order from strength* scenario may be our fate, with poor economic growth, degradation of all ecosystem services and a large increase in global temperature. A more globally connected society (*global orchestration*) could produce the highest economic growth and strongest improvement for the poorest people, but at the cost of many ecosystem services and with the largest predicted temperature increase. The global outcome of a world driven by local communities focusing on sound environmental management (*adapting mosaic*) will lead to the smallest economic growth, improvements to all ecosystem services and an intermediate rise in global temperature. Finally, the *technogarden* scenario, with its environmentally sound but highly managed ecosystems, and crucially with a climate change policy (stabilizing CO_2 at 550 ppm), leads to the smallest rise in temperature, reduces nutrient pollution of waterways and improves ecosystem services – except cultural ones, because so many ecosystems are managed and relatively unnatural. Which of these, or other, scenarios comes to pass depends on a wide range of sociopolitical factors.

Anyone wishing to make a difference to the fate of biodiversity will need to take on board the diversity of perspectives in their community and internationally. To encourage this broad perspective, and foster an approach that values the environmental knowledge existing in all sectors of society, I use a particular device at the beginning of each remaining chapter. Here you will encounter a viewpoint on an environmental issue that may be alien to your own or, at least, that engenders a more circumspect approach to the issue at hand. You may not agree with what the 'focal person' says, but what can you learn from them and how could you engage

Table 1.2 Four scenarios that explore plausible futures for ecosystems and human well-being based on different assumptions about sociopolitical forces of change and their interactions. Greenhouse gas emissions (carbon dioxide, methane, nitrous oxide and 'other') are expressed as gigatons of carbon-equivalents (a gigaton is one thousand million tons). (Based on Millennium Ecosystem Assessment, 2005b.)

	Greenhouse gas emissions to 2050 and predicted temperature rise	Land use and nitrogen transport in rivers	Ecosystem services
Global Orchestration – a globally connected society focused on global trade and economic liberalization. Assumes a reactive approach to ecosystem problems. Takes strong steps to reduce poverty and inequality and to invest in public goods such as infrastructure and education. Economic growth is the highest of the four scenarios, while population in 2050 is lowest (8.1 billion)	CO_2: 20.1 GtC-eq CH_4: 3.7 GtC-eq N_2O: 1.1 GtC-eq Other: 0.7 GtC-eq 2050 +2.0°C 2100 +3.5°C	Slow forest decline to 2025, 10% more arable land Increased nitrogen in rivers	Provisioning services improved, regulating and cultural services degraded
Order from Strength – a regionalized and fragmented world, concerned with security and protection, emphasizing primarily regional markets, paying little attention to public goods, and taking a reactive approach to ecosystem problems. Economic growth rate is the lowest (particularly in developing countries) while population growth is the highest of the scenarios (9.6 billion in 2050)	CO_2: 15.4 GtC-eq CH_4: 3.3 GtC-eq N_2O: 1.1 GtC-eq Other: 0.5 GtC-eq 2050 +1.7°C 2100 +3.3°C	Rapid forest decline to 2025, 20% more arable land Increased nitrogen in rivers	All ecosystem services heavily degraded
Adapting Mosaic – river catchment-scale ecosystems are the focus of political and economic activity. Local institutions are strengthened and local ecosystem management strategies are common, with a strongly proactive (and learning) approach. Economic growth is low initially but increases with time. Population in 2050 is high (9.5 billion)	CO_2: 13.3 GtC-eq CH_4: 3.2 GtC-eq N_2O: 0.9 GtC-eq Other: 0.6 GtC-eq 2050 +1.9°C 2100 +2.8°C	Slow forest decline to 2025, 10% more arable land Increased nitrogen in rivers	All ecosystem services improved
TechnoGarden – a globally connected world relying on environmentally sound technology, using highly managed, often engineered, ecosystems to deliver ecosystem services, and taking a proactive approach to ecosystem management. Economic growth is relatively high and accelerating, while the 2050 population is midrange (8.8 billion). This is the only scenario to assume a climate policy (stabililizing CO_2 at 550 ppm)	CO_2: 4.7 GtC-eq CH_4: 1.6 GtC-eq N_2O: 0.6 GtC-eq Other: 0.2 GtC-eq 2050 +1.5°C 2100 +1.9°C	Forest increase to 2025, 9% more arable land Decreased nitrogen in rivers	Provisioning and regulating services improved, cultural services degraded

them in an effective dialogue? These are points to bear in mind as you move from chapter to chapter. You can get a taste of the sociopolitical dimension of sustainability by dipping into the first section of each chapter and also in Sections 2.3.1, 4.4, 5.6, 6.4, 7.3, 7.6, 8.3, 9.8, 10.5, 10.7 and 11.4.

Summary

Homo sapiens – *not just another species*

Humans destroy natural ecosystems to make way for urban and industrial development and to establish production ecosystems such as forestry and agriculture. Moreover, the natural ecosystems that remain are also affected by our activities – via overexploitation of harvested species, the spread of invaders, local pollution and global climate change. In one sense, we are not so different from many other species in our effects on other animals and plants. But human impacts are very much more profound because of the size of our population and the technologies we use.

The biodiversity crisis

To judge the scale of the human threat to biodiversity we need to know the total number of species that exist, the rate at which these are going extinct and how this compares with pre-human times. Roughly speaking, the current rate may be as much as 100–1000 times the historical rate. Bearing in mind the number of species believed to be under threat, the future rate of extinction may be more than ten times higher again.

A reduction in biodiversity can have consequences for the ecosystem as a whole. Species vary in the contribution they make to overall productivity, nutrient cycling or decomposition rates in an ecosystem – the loss of some will barely register. Of particular significance are situations where species are 'complementary' in the way they contribute to ecosystem function. Where this is the case, lower biodiversity will generally equate to impaired ecosystem functioning and losses to *ecosystem services* – whether *provisioning* (e.g. fish from the sea), *cultural* (e.g. recreational opportunities), *regulating* (e.g. flood control) or *supporting* (e.g. soil formation).

Causes of biodiversity loss

Extinction may be caused by one or a combination of *drivers* that include habitat loss, invasive species, overexploitation and habitat degradation (pollution and agricultural intensification). Historically, habitat loss, habitat degradation and overexploitation have been of most significance. In future, climate change and the pollution associated with agricultural intensification are predicted to become progressively more important causes of biodiversity loss across all ecosystem types.

Increasing agricultural intensity is associated with increases to soil erosion, desertification and removal of water for irrigation (so that some major rivers no longer reach the sea). In addition, excess plant nutrients find their way into waterways, and chemical pesticides affect nontarget species, often long after they are first applied. Because greater human population growth is expected in species-rich tropical areas, increased agricultural activity will place biodiversity at high risk.

The most far-reaching consequence of our use of fossil fuels has been an increase in the atmospheric concentration of carbon dioxide, a greenhouse gas. As a result, air temperature at the land surface is now $0.6 \pm 0.2°C$ warmer than in pre-industrial times, and is predicted to continue to rise by a global average of between $2.0°C$ and $5.5°C$ by 2100. Such changes will lead to a melting of glaciers and icecaps, sea-level rise, and large changes to global patterns of precipitation, winds, ocean currents and the timing and scale of storm events. The ecological consequences for biodiversity and ecosystem services will be profound.

Toward a sustainable future

An activity is 'sustainable' if it can be continued into the future. If we want to eat tuna in future, we cannot continue to harvest them faster than the population can replace those that are lost. Nor can farmers continue to use fertilizers indiscriminately if people want to retain the ecosystem services provided by rivers, lakes and oceans that are impacted by the agricultural excess. The recognition of the importance of sustainability as a unifying idea came of age in the early 1990s. Since then the focus has shifted from a purely ecological perspective to one that incorporates the economic and social conditions that influence sustainability. Thus, sustainability has ecological, economic and sociopolitical dimensions.

The ecological dimension

From the ecological point of view, sustainability topics can be organized according to the underlying structure of ecology theory. At the lowest level is the ecology of individuals – niche requirements, life-history traits and dispersal/migratory behavior. Knowledge at this level is crucial when reintroducing species that have gone locally extinct, restoring natural grassland and forest, or predicting the arrival of damaging invaders. Next comes the population level – all individuals of a single species in a particular place. Population theory is central to the management of endangered species, pests and harvests. Then there is community (species composition) and ecosystem (energy and nutrient flux) ecology. Theory at this level helps managers devise plans to restore natural communities, counteract invasions, increase the range of harvestable products and make agroecosystems sustainable. Finally, at the largest scales, landscape ecology is crucial when designing networks of nature reserves, and global climate change has implications for just about everything else.

The economic dimension

There is an economic side to every resource management argument. Sometimes the costs and benefits are relatively straightforward to compute. But imaginative approaches are needed to determine the value of a species or an ecosystem service (e.g. *travel cost* paid by people to access a natural area provides a minimum value of this recreational service). Viewed from the broadest perspective of all, the total value of the world's ecosystem services has been roughly estimated at $38 trillion – more than the gross domestic product of all nations combined. The 'new economics' provides persuasive reasons for taking great care of biodiversity.

The sociopolitical dimension

Environmental issues almost always have a sociopolitical angle too. Sociologists can help managers reconcile the desires of all interested parties. And political scientists help determine whether sustainable management should be fostered by penalties or inducements, or be set in law or encouraged by education. At the local level, when people are well connected in groups and networks, and when their knowledge is sought and incorporated during environmental management planning, they are more likely to retain a care-taking role in the long term. If an environmental problem occurs at too large a scale for local solutions, the sociopolitical machinations need to occur globally. Estimates of future greenhouse gas emissions and the resulting changes to global temperature vary according to sociopolitical factors – our predictions need to be based on models that take these things into account.

References

Balmford, A. & Bond, W. (2005) Trends in the state of nature and their implications for human well-being. *Ecology Letters* 8, 1218–1234.

Balmford, A., Bruner, A., Cooper, P., and 16 others (2002) Economic reasons for conserving wild nature. *Science* 297, 950–953.

Balmford, A., Green, R.E. & Jenkins, M. (2003) Measuring the changing state of nature. *Trends in Ecology and Evolution* 18, 326–330.

Begon, M., Townsend, C.R. & Harper, J.L. (2006) *Ecology: from individuals to ecosystems*, 4th edn. Blackwell Publishing, Oxford.

Bibby, C.J. (1998) Selecting areas for conservation. In: *Conservation Science and Action* (W.J. Sutherland, ed.), pp. 176–201. Blackwell Science, Oxford.

Costanza, R., D'Arge, R., de Groot, R. et al. (1997) The value of the world's ecosystem services and natural capital. *Nature* 387, 253–260.

Enserink, M. (1999) Biological invaders sweep in. *Science* 285, 1834–1836.

Hector, A., Shmid, B., Beierkuhnlein, C. et al. (1999) Plant diversity and productivity experiments in European grasslands. *Science* 286, 1123–1127.

Hooper, D.U., Chapin, F.S., Ewel, J.J. and 11 others (2005). Effects of biodiversity on ecosystem functioning: a consensus of current knowledge. *Ecological Monographs* 75, 3–35.

Houghton, R.A. (2000) Interannual variability in the global carbon cycle. *Journal of Geophysical Research* 105, 20121–20130.

IPCC (2007) *Fourth Assessment Report of the Intergovernmental Panel on Climate Change.* Working Group 1, Intergovernmental Panel on Climate Change, Geneva.

IUCN/UNEP/WWF (1991) *Caring for the Earth. A strategy for sustainable living.* Gland, Switzerland.

Jonsson, M. & Malmqvist, B. (2000) Ecosystem process rate increases with animal species richness: evidence from leaf-eating, aquatic insects. *Oikos* 89, 519–523.

Kaufman, L. (1992) Catastrophic change in a species-rich freshwater ecosystem: lessons from Lake Victoria. *Bioscience* 42, 846–858.

Laurance, W.F. (2001) Future shock: forecasting a grim fate for the Earth. *Trends in Ecology and Evolution* 16, 531–533.

Lubchenco, J., Olson, A.M., Brubaker, L.B. et al. (1991) The sustainable biosphere initiative: an ecological research agenda. *Ecology* 72, 371–412.

Martin, P.S. (1984) Prehistoric overkill: the global model. In: *Quaternary Extinctions: a prehistoric revolution* (P.S. Martin & R.G. Klein, eds), pp. 354–403. University of Arizona Press, Tuscon, AZ.

Millennium Ecosystem Assessment (2005a) *Ecosystems and Human Well-being: Biodiversity synthesis.* World Resources Institute, Washington, DC.

Millennium Ecosystem Assessment (2005b) *Living Beyond our Means: natural assets and human well-being. Statement of the Board.* World Resources Institute, Washington, DC.

Milner-Gulland, E.J. & Mace, R. (1998) *Conservation of Biological Resources.* Blackwell Science, Oxford.

Owen-Smith, N. (1987) Pleistocene extinctions: the pivotal role of megaherbivores. *Paleobiology* 13, 351–362.

Pimentel, D., Lach, L., Zuniga, R. & Morrison, D. (2000) Environmental and economic costs of nonindigenous species in the United States. *BioScience* 50, 53–65.

Pressey, R.L. (1995) Conservation reserves in New South Wales: crown jewels or leftovers. *Search* 26, 47–51.

Pretty, J. & Smith, D. (2004) Social capital in biodiversity conservation and management. *Conservation Biology* 18, 631–638.

Rodrigues, A.S.L., Pilgrim, J.D., Lamoreux, J.F., Hoffman, M. & Brooks, T.M. (2006) The value of the IUCN Red List for conservation. *Trends in Ecology and Evolution* 21, 71–76.

Sodhi, N.S., Koh, L.P., Brook, B.W. & Ng, P.K.L. (2004) Southeast Asian biodiversity: an impending disaster. *Trends in Ecology and Evolution* 19, 654–660.

Tilman, D., Fargione, J., Wolff, B. et al. (2001a) Forecasting agriculturally driven global environmental change. *Science* 292, 281–284.

Tilman, D., Reich, P.B., Knops, J., Wedin, D., Meilke, T. & Lehman, C. (2001b) Diversity and productivity in a long-term grassland experiment. *Science* 294, 843–845.

UNEP (2002) Report on the Sixth Meeting of the Conference of the Parties to the Convention on Biological Diversity (UNEP/ CBD/COP/6/20/Part 2) Strategic Plan Decision VI/26. Available at: http://www.biodiv.org/doc/meetings/cop/ cop-06/official/cop-06–20-part2-en.pdf.

United Nations (1999) *The World at Six Billion*. United Nations Population Division, Department of Economic and Social Affairs, United Nations Secretariat, New York.

2 Ecological applications of niche theory

The ecological niche is a summary of an organism's abiotic tolerances and its relationships with resources and enemies – knowledge that allows managers to predict where potential invaders might do well, to choose locations for reintroductions and design reserves for endangered species, or restore degraded habitats.

Chapter contents

2.1 Introduction	37
2.2 Unwanted aliens – lessons from niche theory	41
2.2.1 Ecological niche modeling – predicting where invaders will succeed	42
2.2.2 Are we modeling fundamental or realized niches?	44
2.2.3 When humans disrupt ecosystems and make it easy for invaders	44
2.3 Conservation of endangered species – each to its own niche	46
2.3.1 Monarch's winter palace under siege	46
2.3.2 A species off the rails – translocation of the takahe	48
2.4 Restoration of habitats impacted by human activities	49
2.4.1 Land reclamation – prospecting for species to restore mined sites	49
2.4.2 Agricultural intensification – risks to biodiversity	51
2.4.3 How much does it cost to restore a species?	52
2.4.4 River restoration – going with the flow	53

Key concepts

In this chapter you will

recognize that to establish and maintain a population, individuals must tolerate abiotic conditions, find sufficient resources and persist in the face of enemies

see the value of ecological niche modeling for predicting where invaders might establish and where desirable natives can be translocated

realize the risky nature of predicting niches of rare species when data are scarce

understand how knowledge of the niche relations of plants can assist when planning land reclamation and habitat restoration

appreciate that environmental management usually involves confronting economic and sociopolitical as well as ecological perspectives

---

2.1 Introduction

Kelly is a really keen angler. His friends have to smile at the sign on the door of his room: 'Fish worship – can it be wrong?'. Kelly knows as much about the habits and habitats of river fish as many experts, and enjoys both the solitude of being in a quiet backwater and the challenge of landing the big one. But he sometimes finds himself in situations where others denigrate his sport. *They tell me that angling is cruel – and I say so is killing stock at the slaughterhouse.* Then they switch to a conservation argument, pointing out how exploitation can put species at increased risk of extinction. *I say that I fish for introduced fish – brown trout in my case – and they should be pleased because this takes the pressure off native species that have been dwindling in numbers. (I know this is a bit dishonest – I would fish for trout even if they were native to my part of the USA!) How can people hold such contrasting views? Isn't there some middle ground? I can't find it, and it bugs me.*

Members of society are certainly not unanimous in the way they value the natural world – the spectrum of opinions is broad, with zealots at both extremes. On the one hand, there are some industrialists, fishers, farmers and foresters who accept none of the conservationist case and are not prepared to look objectively at the scientific evidence; while on the other side are the environmental zealots – preservationists who seem unwilling to accept any exploitation of the natural world. The middle ground is occupied by both exploiters and conservationists whose basic philosophy holds that natural resources can be used, but this should be in a sustainable and balanced manner. A good understanding of ecology can provide the scientific basis for what, in its broader context, is a question of ethics.

A successful species invasion (of Kelly's introduced brown trout, for example) depends on the ability to establish and maintain a population in a new location. To do this, individuals must tolerate the abiotic conditions they encounter, find sufficient resources (nutrients, water, food, nest sites and so on), and persist in the face of enemies (competitors, predators, parasites). This is the domain of niche theory (Box 2.1). Niche information is crucial not only when predicting the spread of invaders (Section 2.2), but also to design reserves for endangered species (Section 2.3), as well as for species recovery in habitat restoration projects (Section 2.4). In fact, whenever there is a focal species – an invader, an endangered species or one crucial to habitat recovery – the better our understanding of its ecological niche, the more likely the success of a management strategy.

Box 2.1 Essential niche theory

Conditions and resources – what's the difference?
Before explaining the term 'niche', it is important to distinguish between two kinds of ecological factor – conditions and resources.

Conditions are physicochemical features of the environment such as temperature, humidity and pH – these may be altered by the presence of organisms but cannot be consumed. Extreme conditions (e.g. high and low temperatures) may be lethal, with a continuum of more favorable values between the extremes. In some cases, the timing and duration of extremes may be more important to an organism's success than absolute values. Species differ in the shapes of their response curves – some, for example, because of particular metabolic effectiveness or enzyme structures, can tolerate higher or lower temperatures (or greater temperature ranges) than others.

Resources, in contrast to conditions, are consumed by organisms so that availability to others may be affected – plants consume photons, nutrients, water and carbon dioxide; animals consume water, food (whether plant or animal, living or dead) and may also be said to consume nest sites and hiding places (making them unavailable while occupied). The fact that resources can be consumed has a critical consequence for the way that organisms interact with each other. By consuming resources, individuals may reduce resource availability to others of their own or different species. Intraspecific competition for a limited resource occurs among individuals of a single species; its counterpart, involving competition among individuals of different species, is interspecific competition. The ultimate effect of competition is on the survival, growth and reproduction of the competitors. In extreme cases, interspecific competition can lead to the extinction of one of the competing species or its restriction to a more limited set of circumstances.

The fundamental niche – an idea with many dimensions

Where an organism lives is its habitat. Its fundamental niche, on the other hand, is not a place but an idea – a summary of the organism's tolerances and requirements, within which the species can maintain a population in the long term (and without the need for inward migration of individuals). The modern concept has its roots in Hutchinson's (1957) proposal that the niche defines the conditions and resources needed by a species to practice its way of life. Temperature, for instance, limits the growth and reproduction of all organisms, but organisms of different species tolerate different ranges of temperature – this range is one *dimension* of an organism's ecological niche. Figure 2.1a shows how some species of plants vary in this niche dimension – specifically, how they vary in the range of temperatures at which they can successfully photosynthesize. Of course, there are many

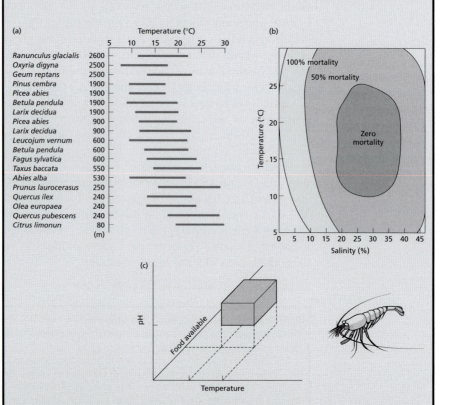

Fig. 2.1 (a) A niche in one dimension. The range of temperatures at which a variety of plant species from the European Alps can achieve net photosynthesis at low intensities of radiation (70 W m⁻²). (After Pisek et al., 1973.) (b) A niche in two dimensions for the sand shrimp (*Crangon septemspinosa*) showing the fate of egg-bearing females in aerated water at a range of temperatures and salinities. (After Haefner, 1970.) (c) A diagrammatic niche in three dimensions for an aquatic organism showing a volume defined by temperature, pH and availability of food. (After Townsend, et al., 2003.)

other dimensions of a species' niche: its tolerance of various other conditions (relative humidity, pH, wind speed, water flow and so on), and its need for various resources. In reality, the niche of a species must be multidimensional.

It is easy to visualize the first steps of building such a multidimensional niche. Figure 2.1b illustrates how two niche dimensions (temperature and salinity) together define a two-dimensional area that is part of the niche of a sand shrimp. Temperature, pH and the availability of a particular food may define a three-dimensional niche volume (Figure 2.1c). In fact, we consider a niche to be an *n-dimensional hypervolume*, where *n* is the number of dimensions that make up the niche. It is hard to imagine (and impossible to draw) this more realistic picture of a volume defined by the boundaries that limit where a species can live, grow and reproduce. But the concept has become a cornerstone of ecological thought.

The niche characteristics of a species, which depend above all on its physiology, are reasonably constant over extended evolutionary time periods so that the details of a species' niche in one place are broadly transferable to another. This is important to managers, who increasingly seek to define the niche characteristics of key species. Many recent applications have followed a process of ecological niche modeling (e.g. Peterson, 2003), where occurrence patterns in a species' native range are used to build a model that can be projected to identify other areas that are potentially habitable (Figure 2.2).

At first sight, there may seem to be a paradox between the view that the niche is not a habitat but an idea (summarizing the tolerances and requirements of a species), as opposed to niche modeling, which explicitly identifies geographical areas that are habitable. The niche concept deals

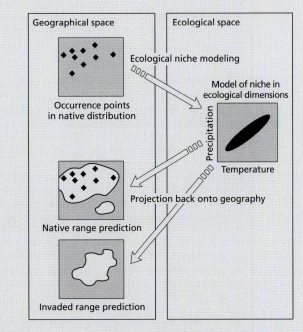

Fig. 2.2 The process of ecological niche modeling. This involves, first, characterizing a species' distribution in two-dimensional geographical space. Next the niche is modeled in ecological space (in terms of a restricted number of dimensions of the *n*-dimensional hypervolume – temperature, precipitation, elevation, and so on). Finally, the occupation of ecological space is projected back into geographical space. This step can be used to predict distributions in the native range (checking how good the model is) and in new locations of interest (e.g. where an invader may be successful). Areas where occurrence is predicted to be likely are shown in white. (Modified from Peterson, 2003.)

in the basic ecological factors that determine a species' success, and therefore account for its distribution in time and space. Niche modeling, on the other hand, uses native geographical distribution to determine some important dimensions of the niche (usually abiotic conditions to do with altitude, temperature, humidity, etc.; sometimes called 'climate matching' – Peterson, 2003) that can then be extrapolated to determine two-dimensional geographical areas elsewhere that lay within the abiotic requirements of the species. Such climate envelope models perform best when they are based on a large number of data points for the focal species (50–75 or more), spread over the whole climatic range that it occupies (Kadmon et al., 2003). The models are inevitably based on incomplete pictures, but they often seem to be adequate for successful prediction.

In fact, the complete *n*-dimensional niche has not been determined for any species – it remains very much an idea! However, the niche dimensions of some species are more completely known than others, with information coming from laboratory studies, where conditions are experimentally varied and biological performance is measured, as well as field-based descriptive approaches such as climate matching.

The realized niche – enemy action and mutual aid

Provided that a location has conditions within acceptable limits for a given species, and provided also that it contains all necessary resources, then the species can, potentially, occur and persist there. Whether or not it does so depends on a further factor – its occurrence may be precluded by the action of individuals of other species that compete with, prey upon, or parasitize it. Usually, a species has a larger ecological niche in the absence of enemies than it has in their presence. In other words, there are certain combinations of conditions and resources that can allow a species to maintain a viable population, but only if it is not being adversely affected by other species. This led Hutchinson to distinguish between the *fundamental* and the *realized* niche. The former describes the overall potentialities of a species; the latter describes the more limited spectrum of conditions and resources that allow it to persist, even in the presence of competitors, predators and parasites.

As an example of the role of competition in reducing the size of the fundamental niche, consider Connell's (1961) classic study in Scotland of two species of barnacle, *Chthamalus stellatus* and *Balanus balanoides*. These coexist on the same shore but their distributions, when looked at on a finer scale, overlap very little – *Balanus* outcompetes (for space from which to filter food) and excludes *Chthamalus* from the lower zones of rocky shores where, if *Balanus* is absent (experimentally removed by Connell), *Chthamalus* can maintain a population. In similar vein, the native fish *Galaxias depressiceps* has a fundamental niche that includes small streams and larger rivers in southern New Zealand; but predation by exotic brown trout (*Salmo trutta*) now restricts its range to headwaters above the waterfalls that prevent upstream trout migration (Townsend, 2003). Finally, the extinction of nearly 50% of the endemic birds of Hawaii, which were broadly spread across the islands, and the restricted distributions of those that remain, have been attributed in part to introduced bird parasites (malaria and bird pox; van Riper et al., 1986).

Just as negative species' interactions can play a role in determining distributions, so can the positive effects of mutualists. Take, for example, the tropical anemone fish *Amphiprion percula*, which retreats amongst the tentacles of the sea anemone *Heteractis magnifica* when danger threatens, but also provides protection to the anemone from grazers (Elliott & Mariscal, 2001). Either species may tolerate the abiotic conditions at a particular location, but the success of each also depends on the presence of its mutualist. Similarly, most higher plants have intimate mutualisms between their roots and fungi (mycorrhiza) that capture nutrients from the soil, which they transport to the plants in exchange for plant photosynthetic products. Many plant species can live without their mycorrhizal fungi in soils where nutrients and water are in good supply, but in the harsh competitive world of natural plant communities the presence of its fungus is often necessary if a plant is to prosper (Buscot et al., 2000).

The interspecific complications of distribution patterns (expressed in realized as opposed to fundamental niches) are not incorporated in the climate matching models described in the previous section, but in some cases we ignore them at our peril.

2.2 **Unwanted aliens – lessons from niche theory**

The arrival on the scene of new species, dispersing from elsewhere, has always been a fact of ecological life. However, human transport has greatly enhanced the opportunities for species to travel long distances and to bridge ancient barriers, such as deserts, mountain ranges and oceans. The trickle of naturally invading species has become a flood of human-assisted aliens. Many newcomers fail to launch self-sustaining populations and others establish but without negative effect. But a large number have adverse ecological or economic impacts, as pointed out in Chapter 1.

The alien plants of the British Isles illustrate a number of points about invaders and the niches they fill (Godfray & Crawley, 1998). Species whose niches encompass areas where people live and work are more likely to be transported to new regions, where they will tend to be deposited in habitats like those where they originated; thus, more invaders are found in disturbed habitats close to transport centers and fewer in remote mountain areas (Figure 2.3a). Moreover, more invaders arrive from

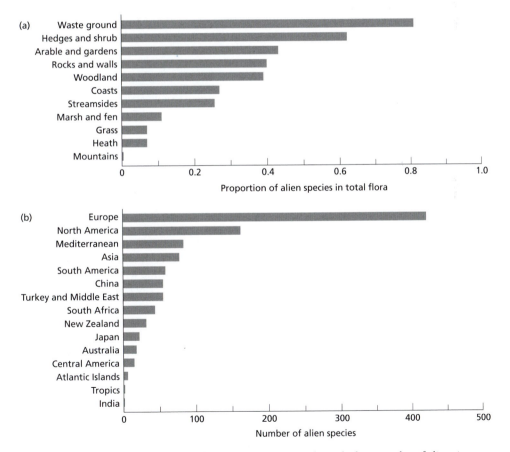

Fig. 2.3 Alien plants of the British Isles: (a) according to community type (note the large number of aliens in open, disturbed habitats close to where people live); (b) in terms of geographical origin (reflecting proximity, trade and climatic similarity). (From Begon et al., 2006, after Godfray & Crawley, 1998.)

nearby locations (e.g. Europe) or from remote locations whose climate (and therefore the invader's niche) matches what is found in Britain (Figure 2.3b). Note the small number of alien plants from tropical environments; these species usually lack the frost-hardiness required to survive the British winter.

Not all invaders cause obvious harm; indeed some ecologists distinguish exotic species that establish without significant consequences from those they consider 'truly invasive' – whose populations expand 'explosively' in their new environment, with significant impacts for indigenous species. Managers need to differentiate among potential new invaders both according to their likelihood of establishing should they arrive in a new region and in relation to the probability of having dramatic consequences in the receiving community. The likelihood of establishment is largely dependent on niche requirements, relevant at the ecological scale of individual organisms, and the subject of the present chapter. Community consequences will be dealt with in Chapter 9 (Section 9.5) because these are relevant at the larger scale of communities and ecosystems. Management strategies to control invaders that have achieved pest status usually require an understanding at the intermediate ecological scale of population dynamics and will be covered in Chapter 6.

In this section I describe a key process for predicting where invaders will do well – ecological niche modeling (Section 2.2.1). Next I address the question whether it is fundamental or realized niches that are being modeled (Section 2.2.2). Finally, I consider how human disturbance may sometimes facilitate invasions (Section 2.2.3)

2.2.1 *Ecological niche modeling – predicting where invaders will succeed*

Ecological niche modeling has been applied to a diverse range of invading organisms from aquatic and terrestrial plants, to fish, wood-boring beetles, butterflies and vultures. As an example of the approach, let's turn to studies of four invasive plants in North America. *Hydrilla verticillata*, an aquatic weed from South America, affects lakes by forming dense canopies that often shade out native vegetation, interfere with swimming and boating, and can clog power-generation intake pipes. *Sericea lespedeza*, or Chinese bush clover, is a perennial legume native to eastern Asia that can reduce or eliminate competing native plants, as can *Elaeagnus angustifolia*, or Russian olive, a small thorny shrub native to southeastern Europe and western Asia. Finally, *Alliaria petiolata* is a cool-season herb from Europe, known as garlic mustard because the leaves give off a garlic odor when crushed. Once introduced to an area, garlic mustard outcompetes many native springtime wild flowers, and can adversely impact rare native insects, such as the white butterfly *Pieris virginiensis*, through the loss of their food plants.

The first step in modeling the ecological niches of these species is to obtain precise locations ('geo-referenced') at which individuals are known to occur (from published accounts and herbarium records). Next the niches are modeled using a computer program that relates physicochemical characteristics at the known occurrence points to those of points sampled randomly from the rest of the study region. This plant study invoked a large number of 'niche dimensions' including elevation, slope and aspect, annual averages of temperature (diurnal range, mean, minimum and maximum), number of frost days and wet days, precipitation and solar radiation. The best model for each species in its native range is then projected onto a map of North America, using geo-referenced data on the physicochemical variables across the continent, to provide predictions of potential distributions (Figure 2.4). These

Fig. 2.4 Predictions of native distributional areas and potential invaded distributional areas for *Hydrilla verticillata*, *Sericea lespedeza*, *Elaeagnus angustifolia* and *Alliaria petiolata*. The native areas analyzed were southern and eastern Asia for *Hydrilla*, eastern Asia for *Sericea* and Europe and western Asia for *Elaeagnus* and *Alliaria*. White symbols on native distribution maps indicate occurrence data used to build ecological niche models. Black polygons on the introduced distribution maps indicate known occurrences in counties or river catchment areas. Increasingly dark shading indicates greater confidence in prediction of presence (model agreement). (From Peterson et al., 2003.)

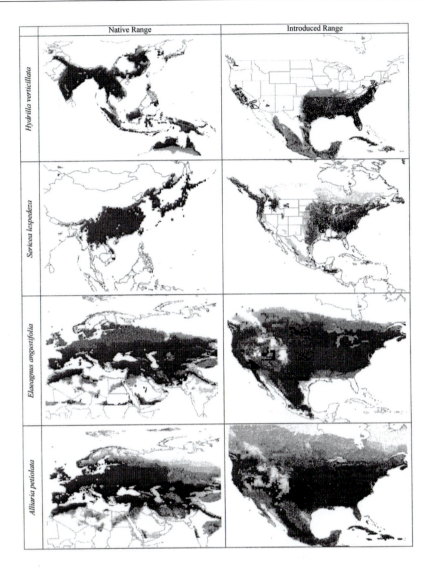

particular plants have already invaded North America so we can test the accuracy of predictions, something it is critical to know if we want to use the same approach for potential invaders that have not yet arrived.

The projected distribution areas in North America were relatively small for *Hydrilla* and Chinese bush clover but much more extensive for Russian olive and garlic mustard. In each case, however, there was a high degree of coincidence between the projected and observed invaded ranges. In the case of *Hydrilla*, for example, the model successfully predicted occurrence in 201 of the 206 counties where it is known to have established. Just as important, such models indicate areas where a species has not established but where there is a high probability it will be able to do so. Managers can use this kind of information to devise plans to reduce the likelihood that invaders will arrive in the susceptible locations.

2.2.2 *Are we modeling fundamental or realized niches?*

The models of Peterson et al. (2003) make acceptable predictions about invasion success despite being based on a restricted number of niche dimensions: they only include an incomplete set of conditions and have no dimensions at all explicitly related to resource use (nutrient availability, etc.) or the presence of enemies. Whether these are actually models of fundamental or realized niches is debatable. For example, if the same critical enemies (competitors, predators, parasites) are present in both native and invaded ranges it could be claimed that the models represent realized rather than fundamental niches. However, when the same set of species is not present in both locations (often the case for invaders, for example, and one of the reasons some of them do so well), heavy reliance on climate matching models may be misplaced.

Large parts of the globe now experience much higher nutrient concentrations, particularly nitrogen (because of fertilizer application and atmospheric deposition resulting from the burning of fossil fuels). While some plant species are able to use the additional nitrogen for increased biomass production (such as the crops sown by farmers), other meadow species, including herbs, sedges and bryophytes, cannot efficiently use higher nitrogen inputs (Zechmeister et al., 2003). The fundamental niches of these meadow species are not usually limited by high nutrient concentrations, but their realized niches are. This is because at higher nutrient levels they become subject to interspecific competition and are excluded by more competitive species.

Furthermore, the availability of nutrients, particularly phosphorus, can be enhanced by mutualistic relationships with mycorrhizal fungi (Box 2.1). In a controlled laboratory experiment involving the shrub *Ardisia crenata*, an invader of the southeastern United States, inoculation with mycorrhizal fungi from the roots of an established shrub increased the growth rate of seedlings. In addition, the outcome of competition between seedlings of the invader and those of a native tree, *Prunus caroliniana*, depends on mycorrhizal fungi (Bray et al., 2003). The presence of mycorrhizae allowed *Prunus* seedlings to invest more to leaf area, enabling them to compete more effectively with the invading *Ardisia*.

At least in some cases then, climate matching may not provide an adequate description of likely patterns of invader establishment. Managers may need to take other characteristics into account, such as relative competitive ability, which can itself depend on the details of nutrient availability (sometimes modified by fungal mutualism). Grazing intensity, too, can modify the likelihood of invader success because plant species vary in their vulnerability to grazing (this is dealt with in detail in Chapter 3, Section 3.4.3).

2.2.3 *When humans disrupt ecosystems and make it easy for invaders*

Shea and Chesson (2002) use the phrase *niche opportunity* to describe the potential provided in a given region for invaders to succeed – in terms of a high availability of resources and appropriate physicochemical conditions, coupled with a lack or scarcity of natural enemies. They note that human activities often disrupt conditions in a way that provides niche opportunities for invaders – river regulation, dealt with next, is a case in point.

I noted in Section 2.1 the difficulty of visualizing the multidimensional niche of a species when more than three dimensions are involved. However, a mathematical technique called *ordination* allows us to simultaneously display species and environmental variables on the same graph, the two dimensions of which combine the

Fig. 2.5 Plot of results of canonical correspondence analysis (CCA) showing native species of fish (black circles), introduced invader species (open triangles) and five influential environmental variables (arrows represent the correlation of the physical variables within the two dimensions of the graph. Species with similar niches appear as points close together on the graph. Note how the native and exotic species occupy different regions of niche space. (From Begon et al., 2006, after Marchetti & Moyle, 2001.)

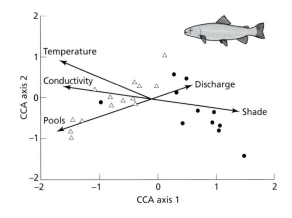

most important of the niche dimensions (see Ter Braak & Smilauer, 1998, for details). Species with similar niches appear as points close together on the graph, while influential environmental factors appear as arrows indicating their direction of increase within the two dimensions of the graph. Marchetti and Moyle (2001) used an ordination method called canonical correspondence analysis to describe how a suite of fish species, 11 native and 14 invaders, are related to environmental factors at sites in a regulated stream in California (Figure 2.5).

The native and invasive species clearly occupy different parts of multidimensional space: most of the native species occur in places associated with higher mean discharge ($m^3 s^{-1}$), good canopy cover (higher levels of % shade), lower concentrations of plant nutrients (lower conductivity, μS), cooler temperatures (°C) and less pool habitat in the stream (i.e. greater % of fast-flowing, shallow riffle habitat). This combination of variables reflects the natural, undisturbed state of the stream and encompasses the niches of the native species.

The pattern for introduced species is generally the opposite; invaders are favored by the present combination of conditions where water regulation has reduced discharge and increased the representation of slower-flowing pool habitat, riparian vegetation has been removed leading to higher stream temperatures, and nutrient concentrations have been increased through agricultural and domestic runoff. We do not know what proportion of the native species would be successful in areas occupied by the exotic species if the latter were absent (or vice versa). In other words we cannot be sure whether the researchers are dealing with realized or fundamental niches; experimental removal studies would be needed to establish this. However, the general conclusion of Marchetti and Moyle (2001) holds good – restoration of more natural flow regimes (providing for the niche requirements of native fish) is needed to limit the advance of invaders and halt the continued downward decline of native fish in this part of the western USA.

It should not be imagined, however, that invaders inevitably do less well in 'natural' flow regimes. Invasive brown trout (*Salmo trutta*) in New Zealand streams do better in the face of floods than native fish such as *Galaxias anomalus*; indeed river reaches whose discharge has been substantially reduced because of water abstraction for land irrigation seem to provide refuges for the natives in the face of predation by brown trout (Leprieur et al., 2006). The low discharge sites would not be optimal habitat in the absence of trout. However, the fundamental niche of *G. anomalus*, in contrast to that of trout, encompasses the higher temperatures and lower oxygen concentrations associated with reduced stream flow.

2.3 Conservation of endangered species – each to its own niche

Conservation of species at risk may involve establishing protected areas (Section 2.3.1) or translocating individuals to new locations (Section 2.3.2). Both approaches involve identifying geographical areas that provide for the niche requirements of the species concerned; they therefore have much in common with the invader stories in the previous section.

Before proceeding further, however, I should highlight an uncomfortable fact of life faced by conservation managers – not everyone thinks like them. Opposing views are frequently firmly held because conservation often involves setting aside areas where exploitation is restricted or prohibited. The hunters, fishers, loggers and farmers can then reasonably demand indisputable evidence for the need to curtail their activities, something that can be difficult to prove to the satisfaction of all. Section 2.3.1 presents a case study where this reality is confronted by an expert panel approach to defining a conservation reserve network that, as a simultaneous aim, seeks to minimize the exclusion of loggers from forest they value.

2.3.1 *Monarch's winter palace under siege*

Overwintering habitat in Mexico is absolutely critical for the monarch butterfly (*Danaus plexippus*), which breeds in southern Canada and the eastern USA. The butterflies from a huge 2.6 million km^2 area east of the Rocky Mountains form dense winter colonies in forests of oyamel trees (*Abies religiosa*) on just 11 mountains in central Mexico. A group of experts was assembled to define objectives, assess and analyze the available data, and produce alternative feasible solutions to the problem of maximizing the protection of overwintering habitat while minimizing the use of land of value for logging (Bojorquez-Tapia et al., 2003).

The critical dimensions of the butterfly's overwintering niche include relatively warm and humid conditions (permitting survival and conservation of energy for their return north) and availability of streams (resource) from which the butterflies drink on clear, hot days. The majority of known colony sites are in forest on moderately steep slopes, at high elevation (>2890 m), facing towards the south or southwest, and within 400 m of streams (Figure 2.6a–d). According to the degree to which locations in central Mexico matched the optimal habitat features, and taking into account the desire to minimize inclusion of prime logging habitat, a modeling approach similar to that outlined in Section 2.2.1 was then used to delineate three scenarios. These differed according to the area the government might be prepared to set aside for monarch butterfly conservation (4500 ha, 16,000 ha, or no constraint; Figure 2.7a–c). The experts preferred the no-constraint scenario, which called for 21,727 ha of reserves (Figure 2.7c), and despite the fact that their recommendation was the most expensive it was accepted by the authorities.

A similar approach has been used for the rare rufous bristlebird (*Dasyornis broadbenti*), threatened by habitat loss in southwestern Australia (Gibson et al., 2004). Bristlebird locations were recorded using a global positioning system on early morning visits in the breeding season. Then the relationships between bird presence and absence and a variety of niche dimensions were determined. Negative associations between bristlebird presence and increasing elevation, distance-to-creek, distance-to-coast and solar radiation indicated preferences for areas of relatively low

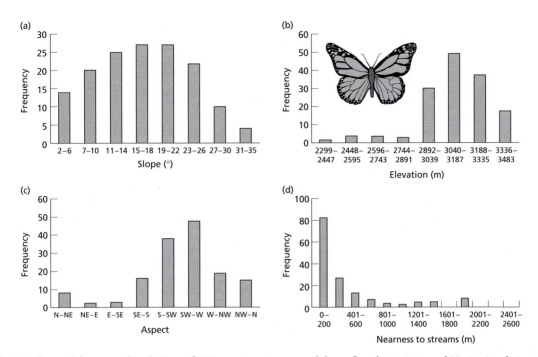

Fig. 2.6 Observed frequency distributions of 149 overwintering monarch butterfly colonies in central Mexico in relation to (a) slope, (b) elevation, (c) aspect and (d) proximity to a stream. (From Begon et al., 2006, after Bojorquez-Tapia et al., 2003.)

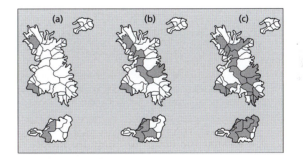

Fig. 2.7 Optimal distribution in the mountains of central Mexico of overwintering monarch butterfly reserves (shaded areas) according to three scenarios: (a) area constraint of 4500 ha; (b) area constraint of 16,000 ha; (c) no area constraint (area included is 21,727 ha). Black lines are the boundaries between river catchment areas. Scenario (c) was accepted by the authorities for the design of Mexico's 'Monarch Butterfly Biosphere Reserve'. (From Begon et al., 2006, after Bojorquez-Tapia et al., 2003.)

altitude, close to the coastal fringe or waterways, which do not receive direct sunlight. A positive association with increasing habitat complexity also suggested a preference for areas containing tree and shrub canopy as well as herbaceous ground cover. Of the total study area, 16% was predicted as suitable habitat that needs to be protected.

2.3.2 *A species off
the rails –
translocation of the
takahe*

Unraveling the fundamental niche of species that have been driven to extreme rarity
may not be straightforward. The takahe (*Porphyrio hochstetteri*), a giant rail, is one
of only two remaining species of the guild of large, flightless herbivorous birds that
dominated the pre-human New Zealand landscape. Indeed, it was also believed
extinct until the discovery in 1948 of a small population in the remote and
climatically extreme Murchison Mountains in the southeast of the South Island
(Figure 2.8).

Since then intense conservation efforts have involved habitat management, captive
breeding, wild releases into the Murchison Mountains and nearby ranges, and
translocation to offshore islands that lack the mammals introduced by people and
which are now widespread on the mainland (Lee & Jamieson, 2001). Some ecologists
argued that because the takahe is a grassland specialist (tall tussocks in the genus
Chionochloa are its most important food) and adapted to the alpine zone it would
not fare well outside this niche (Mills et al., 1984). Others pointed to fossil evidence
that the species was once widespread and occurred mainly at altitudes below 300 m
(often in coastal areas; Figure 2.8) where they were associated with a mosaic of
forest, shrublands and grasslands. Thus, takahe might be well suited for life on offshore islands that are free of mammalian invaders.

It turned out that the sceptics were wrong in thinking that translocated island
populations would not become self-sustaining (takahe have been successfully introduced to four islands). But they seem to have been right in thinking that islands
would not provide optimal habitat – island birds have poorer hatching and fledging
success than mountain birds (Jamieson & Ryan, 2001). The fundamental niche of
takahe probably encompasses a large part of the landscape of the South Island, but

Fig. 2.8 Location of
fossil bones of takahe
in the South Island of
New Zealand. The
population had become
restricted to just one
site in the Murchison
Mountains, giving a
misleading picture of its
niche requirements.
(After Trewick &
Worthy, 2001.)

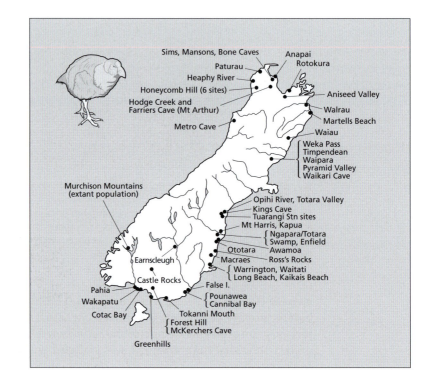

the species became confined to a much narrower realized niche by people who hunted them, mammalian invaders such as red deer (*Cervus elaphus scoticus*) that compete with them for food, and stoats (*Mustella erminea*) that prey upon them. The current distributions of species like takahe, which have been driven very close to extinction, may provide misleading information about niche requirements. It is probable that neither the Murchison Mountains nor offshore islands (with pasture rather than tussock grasses) coincide with the optimal set of conditions and resources of the takahe's fundamental niche. Historical reconstructions of the ranges of endangered species can be expected to help managers identify the best sites for reserves.

2.4 Restoration of habitats impacted by human activities

The term 'restoration ecology' can be used, rather unhelpfully, to encompass almost every aspect of applied ecology (recovery of overexploited fisheries, removal of invaders, revegetation of habitat corridors to assist endangered species, etc.) (Ormerod, 2003). I restrict my discussion to restoration of landscapes and waterscapes whose physical nature has been affected by human activities, dealing specifically with mining (Section 2.4.1), intensive agriculture (Sections 2.4.2, 2.4.3) and water abstraction from rivers (Section 2.4.4).

2.4.1 *Land reclamation – prospecting for species to restore mined sites*

Land that has been damaged by mining is usually unstable, liable to erosion and devoid of vegetation. The simplest solution to land reclamation is the re-establishment of vegetation cover, because this will stabilize the surface, be visually attractive and self-sustaining, and provide the basis for natural or assisted succession to a more complex community (Bradshaw, 2002). Candidate plants for reclamation are those that are tolerant of the toxic heavy metals present; such species are characteristic of naturally metal-rich soils (e.g. *Alyssum bertolonii* in serpentine geological regions in Italy) and have fundamental niches that incorporate the extreme conditions.

Of particular value are ecotypes – different genotypes, within a species, that have different fundamental niches – which have evolved resistance in mined areas. Antonovics and Bradshaw (1970) were the first to note how the intensity of selection for tolerant genotypes changes abruptly at the edge of contaminated areas – populations on contaminated areas may differ sharply in their tolerance of heavy metals over distances as small as 1.5 m (e.g. sweet vernal grass, *Anthoxanthum odoratum*). Against this background, metal-tolerant grass genotypes (or cultivars) have been selected for commercial production in the UK for use on neutral-to-alkaline soils contaminated by lead or zinc (*Festuca rubra* cultivar 'Merlin'), acidic lead and zinc wastes (*Agrostis capillaris* cultivar 'Goginan') and acidic copper wastes (*A. capillaris* cultivar 'Parys') (Baker, 2002).

Since plants lack the ability to move, many species characteristic of metal-rich soils have evolved biochemical systems for nutrient acquisition, detoxification and the control of local geochemical conditions – in effect, they help create the conditions appropriate to their fundamental niche. *Phytoremediation* involves placing such plants in contaminated soil with the aim of reducing the concentrations of heavy metals and other toxic chemicals.

Phytoremediation can take a variety of forms (Susarla et al., 2002). *Phytoaccumulation* occurs when the contaminant is taken up by the plants but is not degraded rapidly or completely; these plants, such as the zinc-accumulating herb *Thlaspi caerulescens*, are harvested to remove the contaminant and then replaced. Similarly,

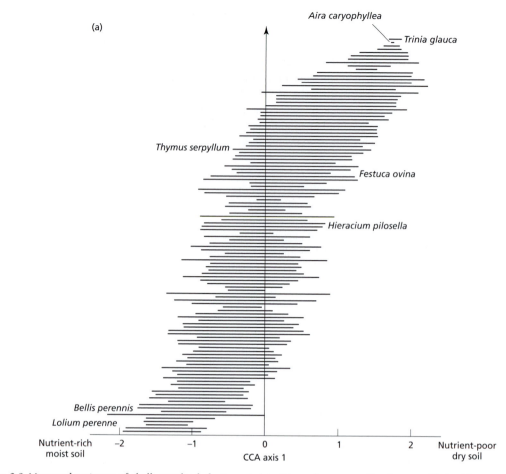

Fig. 2.9 Mean and variances of chalk grassland plant species' positions along two ordination axes (derived from canonical correspondence analysis, CCA). (a) Axis 1 is negatively correlated with soil nutrient richness and water content (more nutrient-poor and drier to the right). (b) Axis 2 is positively correlated with high intensities of mowing and grazing (more intensive mowing/grazing to the right). The midpoint of each species line is its mean position along the axis, while the length of the line expresses its niche width (variance in site scores where the species occurs). (After Barbaro et al., 2004.)

so-called halophytic plants such as *Atriplex prostrata*, which live in naturally salt-rich areas and accumulate salt in their tissues, are prime candidates for restoring land contaminated by saltwater spills from broken pipelines in oil-production areas (Keiffer & Ungar, 2002). *Phytostabilization*, on the other hand, takes advantage of the ability of root exudates to precipitate heavy metals and render them biologically harmless. Finally, *phytotransformation* involves elimination of a contaminant by the action of plant enzymes; for example, hybrid poplar trees *Populus deltoides × nigra* have the remarkable ability to degrade TNT and show promise for restoration of munition dump areas. Note that microorganisms are also used for remediation in polluted situations. A dramatic example is provided by the bacterium *Pseudomonas putida*, a bug that can be said to cut the mustard. The concentration of a nasty compound called 1,4-perhydrothiazine, a toxic remnant of the mustard gas

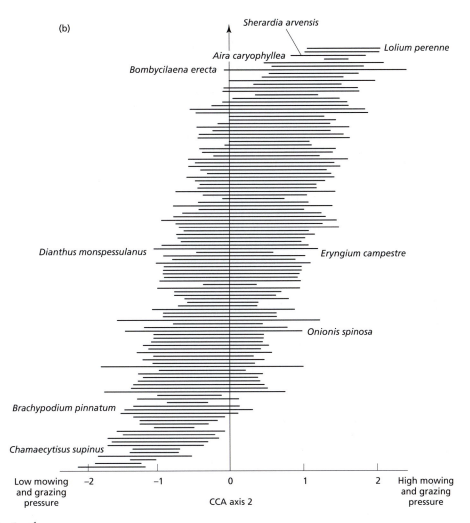

Fig. 2.9 *Continued*

stockpiled in huge quantities during the cold war, is reduced by more than half in cultures containing the bacterium (Ermakova et al., 2005).

2.4.2 *Agricultural intensification – risks to biodiversity*

Agriculture in most places around the world has become progressively more intense, with dramatic increases in fertilizer use, irrigation, mowing frequency and grazing intensity (Chapter 1). To avoid extinctions related to agricultural management (that is, to make agriculture more sustainable), and to identify native plant species that are appropriate for restoration efforts, we need to know as much as possible about the niches of meadow species. You saw in Section 2.2.3 how the mathematical process known as ordination can encapsulate species' niches along axes that combine important niche dimensions. Barbaro et al. (2004) have taken this a stage further by plotting not just the mean locations of a large number of chalk grassland plant species along two ordination axes, but also their variance (from scores for each species at the different sites where they occur). This variance is a measure of how narrow (specialist) or broad (generalist) their niches are.

The first of the two ordination axes corresponds to a gradient of soil conditions from nutrient-enriched and moist (left) to nutrient-poor and dry (right) in 102 locations in southeast France (Figure 2.9a). The second axis corresponds to the mowing/grazing management regime at these sites, from locations that are rarely mown and lightly grazed (left) to sites that are heavily mown and very heavily grazed (right) (Figure 2.9b).

Species restricted to very dry, nutrient-poor soils (to the extreme right in Figure 2.9a) had narrow niches along this axis (e.g. *Aira caryophyllea* and *Trinia glauca*), in contrast to the majority of species associated with intermediate fertility and water availability (e.g. *Thymus serpyllum* and *Festuca ovina*). The enriched end of the spectrum favored species such as *Lolium perenne* and *Bellis perennis*. Species of highly disturbed grassland (to the right in Figure 2.9b) tended to have narrow niches, with high tolerance of trampling (e.g. *Lolium perenne*), grazing (e.g. *Bombycylaena erecta*) or cutting disturbance (e.g. *Aira caryophyllea*). Species associated with intermediate mowing or grazing pressure had narrow or broad niches (e.g. *Dianthus monspessulanus* and *Eryngium campestre*, respectively) and included several forbs with morphological defenses to grazing (e.g. *Eryngium campestre* and *Onionis spinosa*). Finally, in the least disturbed part of the axis (to the left) were found species with narrow (e.g. *Chamaecytisus supinus*) or broad niches (e.g. *Brachypodium pinnatum*) that can only tolerate minimal grazing pressure or total absence of management. This information can be used to define adequate conservation management for the rarest species, and particularly those with narrow niche breadths along both habitat and management gradients. In addition, attempts to restore sustainable agricultural regimes in particular areas will benefit from this detailed knowledge about species appropriate for introducing under the prevailing conditions.

There is a supreme irony in one ecologist working to restore a species in its native range while another seeks its eradication in an invaded location. This is the case for *Hieracium pilosella* (Figure 2.9a), an innocuous species with a broad niche that is characteristic of intermediate fertility and water availability in European grassland. In New Zealand, on the other hand, it is an aggressive invader that poses a heavy economic burden to farmers and an equally heavy ecological cost to grassland biodiversity (Espie, 1994). The reasons for its success partly reflect its broad fundamental niche, but in addition it is resistant to vertebrate grazing pressure and, moreover, has arrived in New Zealand without its normal complement of natural enemies. In its new home, in other words, it is not restricted to the relatively narrow realized niche expressed in Europe.

2.4.3 *How much does it cost to restore a species?*

Another grassland study, with an interesting economic twist, was conducted by Zechmeister et al. (2003) in 31 Austrian meadows. They found that specialist species (with narrow niches in terms of soil moisture and nutrient concentrations) declined in comparison to generalist species as agricultural practices intensified (increased fertilizer input and mowing). Moreover, the farmers' profit margins were found to correlate negatively with overall plant species richness, with meadows that provided low or no profit margins retaining highest species richness (Figure 2.10). This emphasizes how biodiversity can be a matter of economics – the actual cost to the farmer of preserving a single species was estimated to be between 150 and 200 euros.

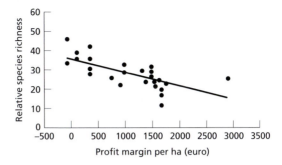

Fig. 2.10 An index of species richness of vascular plants and bryophytes in Austrian meadows in relation to farming profit margin in euros per hectare. Relative species richness expresses for each meadow the number of plants that occur there as a percentage of the total known to occupy meadows in the landscape as a whole. Evidently, moving to a less intensive farming regime would both increase species richness and reduce profit margins. Thus, the average cost of maintaining each species can be estimated. (From Zechmeister et al., 2003.)

If species richness is to be restored, Zechmeister and his colleagues believe that farmers will need increased financial incentives, through agroenvironmental subsidies, to counter the decline in profit margin that managing meadows for species conservation would involve.

2.4.4 *River restoration – going with the flow*

One of the most pervasive of human influences on river ecosystems has been regulation of discharge. Water abstraction for agricultural, industrial and domestic use changes the hydrographs (discharge patterns) of rivers both by reducing discharge (volume per unit time) and altering daily and seasonal patterns of flow. River restoration often involves re-establishing features of the natural flow regime.

The rare Colorado pikeminnow (*Ptychocheilus lucius*), which eats other fish, is now restricted to the upper reaches of the Colorado River. Its present distribution is positively correlated with prey fish biomass, which in turn depends on the biomass of invertebrates upon which the prey fish rely, and this, in its turn, is positively correlated with algal biomass, the energy base of the food web (Figure 2.11a–c). Osmundson et al. (2002) argue that the rarity of pikeminnows can be traced to the accumulation of fine sediment on the riverbed, where it reduces algal productivity in downstream regions of the river. Fine sediment is not part of the fundamental niche of pikeminnows. Historically, spring snowmelt often produced flushing discharges with the power to remove much of the silt and sand that would otherwise accumulate. As a result of river regulation, however, the mean recurrence interval of such discharges has increased from once every 1.3–2.7 years to only once every 2.7–13.5 years (Figure 2.11d), extending the period of silt accumulation.

High discharges can influence fish in other ways too, for example, by maintaining side channels and other elements of habitat heterogeneity, and by improving substrate conditions for spawning. These are all elements of the fundamental niche of particular species. Managers must aim to incorporate ecologically influential aspects of the natural hydrograph of a river into river restoration efforts, but this is easier said than done. Jowett (1997) describes three approaches commonly used to define minimum discharges: historic flow, hydraulic geometry and habitat assessment. The first of these assumes that some percentage of the mean discharge is needed to

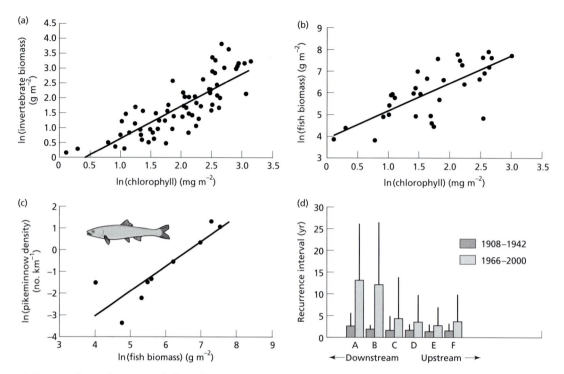

Fig. 2.11 Interrelationships among biological parameters measured in a number of reaches of the Colorado River to determine the ultimate causes of the declining distribution of Colorado pikeminnows. (a) Invertebrate biomass vs algal biomass (chlorophyll *a*). (b) Prey fish biomass vs algal biomass. (c) Pikeminnow density vs prey fish biomass (from catch rate per minute of electrofishing). (d) Mean recurrence intervals in six reaches of the Colorado River (for which historical data were available) of discharges necessary to remove silt and sand that would otherwise accumulate, during recent (1966–2000) and pre-regulation periods (1908–1942). Lines above the histograms show maximum recurrence intervals. (From Begon et al., 2006, after Osmundson et al., 2002.)

maintain a 'healthy' river ecosystem: 30% is often used as a rule of thumb. Hydraulic methods relate discharge to the geometry of stream channels (based on multiple measurements of river cross-sections); river depth and width begin to decline sharply at discharges less than a certain percentage of mean discharge (10% in some rivers) and this inflection point is sometimes used as a basis for setting a minimum discharge. Finally, habitat assessment methods are based on discharges that meet specified ecological criteria, such as a critical amount of food-producing habitat for particular fish species.

Managers need to beware the simplified assumptions inherent in these various approaches because, as you saw with the pikeminnows, the integrity of a river ecosystem may require something other than setting a minimum discharge, like infrequent but high flushing discharges. Simply maintaining a minimum flow may be a poor substitute for what is actually required to be ecologically effective (Palmer et al., 2005) – a change to patterns of water release from dams to move the river back to a more natural hydrograph.

Summary

Niche theory

Conditions are physicochemical features of the environment such as temperature, humidity and pH. Extreme conditions may be lethal, with a continuum of more favorable values between the extremes, but species differ in their tolerance ranges. Resources, in contrast to conditions, are consumed (e.g. photons by plants, plants by herbivores). The *fundamental* niche defines the conditions and resources needed by a species to practice its way of life. Because there are so many factors affecting each species, the fundamental niche is multidimensional. Usually, a species has a larger ecological niche in the absence of enemies (competitors, predators, parasites) than in their presence. The *realized* niche describes the more limited spectrum of conditions and resources that allow a species to persist even in the presence of enemies.

Unwanted aliens

Managers need to differentiate among potential new invaders according to their likelihood of establishing should they arrive in a new region. This is largely dependent on their niche requirements. Managers increasingly use the process of *ecological niche modeling* (generally of the fundamental niche), where occurrence patterns in a species' native range are used to build a model that can be projected to identify other areas that are potentially habitable. Habitat disturbance by human activities may sometimes affect conditions or resources in a way that facilitates the success of invaders.

Conservation of endangered species

Conservation may involve establishing protected areas or translocating individuals to new, safer locations. Both approaches involve identifying geographical areas that provide for the niche requirements of the species concerned. In the case of migratory species, ecological niche modeling may be required at both ends of the migration route, or at the end that is under particular threat. In cases where species' ranges have shrunk to a tiny part of their original distribution, a misleading impression of their fundamental niche may be gained – historical data may then be crucial.

Restoration of habitats

Land that has been damaged by mining is usually unstable, liable to erosion and devoid of vegetation. Candidate plants for restoration are those that are tolerant of the toxic heavy metals present – species, in other words, that have fundamental niches incorporating the extreme conditions. Many plants characteristic of metal-rich soils have evolved biochemical systems for nutrient acquisition and detoxification. Managers can turn to *phytoremediation*, a process that involves introducing such plants to contaminated soil, with the aim of reducing the concentrations of heavy metals and other toxic chemicals.

Grasslands that have lost biodiversity because of intensive agriculture can be restored (after reducing use of fertilizer/pesticide and/or grazing pressure) by reintroducing species with appropriate fundamental niches. The agricultural production lost during restoration can be equated to the price of adding biodiversity (150–200 euros per grassland species in one study).

One of the most pervasive of human influences on river ecosystems has been to change patterns of discharge. River restoration often involves re-establishing features of the natural flow regime to provide for the fundamental niches of native species (and to counter the spread of invaders).

The final word

Kelly, our angler (Section 2.1), is struck by the idea that many ecological problems have a social aspect. *'Aren't different interest groups (exploiters, conservers) more likely to come up with a satisfactory solution to an environmental issue if they all contribute to problem solving? I can imagine that the farmers who took part in the cost-of-biodiversity study would be likely to adopt a more sustainable approach because they were personally involved from the start (Section 2.4.3). I wonder if I should try to start a community group for river habitat improvement – I may be a fanatical angler, but it would be great to work with conservationists and farmers to improve the health of the river generally.'*

Imagine you are Kelly. Devise, in outline, a community action plan to improve the health of a heavily impacted river. This should have the following components: identify all the different interest groups; list the issues that will need to be addressed; identify remedies that the different groups are likely to agree upon; and finally focus on the controversial issues – how do you think the different groups might be brought to successful compromise outcomes?

References

Antonovics, J. & Bradshaw, A.D. (1970) Evolution in closely adjacent plant populations. VIII. Clinical patterns at a mine boundary. *Heredity* 25, 349–362.

Baker, A.J.M. (2002) The use of tolerant plants and hyperaccumulators. In: *The Restoration and Management of Derelict Land: modern approaches* (M.H. Wong & A.D. Bradshaw, eds), pp. 138–148. World Scientific Publishing, Singapore.

Barbaro, L., Dutoit, T., Anthelme, F. & Corcket, E. (2004) Respective influence of habitat conditions and management regimes on prealpine calcareous grassland. *Journal of Environmental Management* 72, 261–275.

Begon, M., Townsend, C.R. & Harper, J.L. (2006) *Ecology: from individuals to ecosystems*, 4th edn. Blackwell Publishing, Oxford.

Bojorquez-Tapia, L.A., Brower, L.P., Castilleja, G. et al. (2003) Mapping expert knowledge: redesigning the monarch butterfly biosphere reserve. *Conservation Biology* 17, 367–379.

Bradshaw, A.D. (2002) Introduction – an ecological perspective. In: *The Restoration and Management of Derelict Land: modern approaches* (M.H. Wong & A.D. Bradshaw, eds), pp. 1–6. World Scientific Publishing, Singapore.

Bray, S.R., Kitajima, K. & Sylvia, D.M. (2003) Mycorrhizae differentially alter growth, physiology, and competitive ability of an invasive shrub. *Ecological Applications* 13, 565–574.

Buscot, F., Munch, J.C., Charcosset, J.Y., Gardes, M., Nehls, U. & Hampp, R. (2000) Recent advances in exploring physiology and biodiversity of ectomycorrhizas highlight the functioning of these symbioses in ecosystems. *FEMS Microbiology Reviews* 24, 601–614.

Connell, J.H. (1961) The influence of interspecific competition and other factors on the distribution of the barnacle *Chthamalus stellatus*. *Ecology* 42, 710–723.

Elliott, J.K. & Mariscal, R.N. (2001) Coexistence of nine anemonefish species: differential host and habitat utilization, size and recruitment. *Marine Biology* 138, 23–36.

Ermakova, I.T., Safrina, N.S., Starovoitov, I.I. et al. (2005) Microbial degradation of the detoxification products of mustard from the Russian chemical weapons stockpile. *Journal of Chemical Technology and Biotechnology* 80, 495–501.

Espie, P.R. (1994) Integrated management strategies for *Hieracium* control. *Proceedings of the New Zealand Grasslands Association* 56, 243–247.

Gibson, L.A., Wilson, B.A., Cahill, D.M. & Hill, J. (2004) Spatial prediction of rufous bris-tlebird habitat in a coastal heathland: a GIS-based approach. *Journal of Applied Ecology* 41, 213–223.

Godfray, H.C.J. & Crawley, M.J. (1998) Introduction. In: *Conservation Science and Action* (W.J. Sutherland, ed.), pp. 39–65. Blackwell Science, Oxford.

Haefner, P.A. (1970) The effect of low dissolved oxygen concentrations on temperature–salinity tolerance of the sand shrimp, *Crangon septemspinosa. Physiological Zoology* 43, 30–37.

Hutchinson, G.E. (1957) Concluding remarks. *Cold Spring Harbour Symposium on Quantitative Biology* 22, 415–427.

Jamieson, I.G. & Ryan, C.J. (2001) Island takahe: closure of the debate over the merits of introducing Fiordland takahe to predator-free islands. In: *The Takahe: fifty years of conser-vation management and research* (W.G. Lee & I.G. Jamieson, eds), pp. 96–113. University of Otago Press, Dunedin, New Zealand.

Jowett, I.G. (1997) Instream flow methods: a comparison of approaches. *Regulated Rivers: Research and Management* 13, 115–127.

Kadmon, R., Farber, O. & Danin, A. (2003) A systematic analysis of factors affecting the performance of climatic envelope models. *Ecological Applications* 13, 853–867.

Keiffer, C.H. & Ungar, I.A. (2002) Germination and establishment of halophytes on brine-affected soils. *Journal of Applied Ecology* 39, 402–415.

Lee, W.G. & Jamieson, I.G. (eds) (2001) *The Takahe: fifty years of conservation management and research*. University of Otago Press, Dunedin, New Zealand.

Leprieur, F., Hickey, M.A., Arbuckle, C.A., Closs, G.P., Brosse, S. & Townsend, C.R. (2006) Hydrological disturbance benefits a native fish at the expense of an exotic fish. *Journal of Applied Ecology* 43, 930–939.

Marchetti, M.P. & Moyle, P.B. (2001) Effects of flow regime on fish assemblages in a regulated California stream. *Ecological Applications* 11, 530–539.

Mills, J.A., Lavers, R.B. & Lee, W.G. (1984) The takahe – a relict of the Pleistocene grassland avifauna of New Zealand. *New Zealand Journal of Ecology* 7, 57–70.

Ormerod, S.J. (2003) Restoration in applied ecology: editor's introduction. *Journal of Applied Ecology* 40, 44–50.

Osmundson, D.B., Ryel, R.J., Lamarra, V.L. & Pitlick, J. (2002) Flow–sediment–biota rela-tions: implications for river regulation effects on native fish abundance. *Ecological Applica-tions* 12, 1719–1739.

Palmer, M.A., Bernhardt, E.S., Allan, J.D. et al. (2005) Standards for ecologically successful river restoration. *Journal of Applied Ecology* 42, 208–217.

Peterson, A.T. (2003) Predicting the geography of species' invasions via ecological niche modeling. *Quarterly Review of Biology* 78, 419–433.

Peterson, A.T., Papes, M. & Kluza, D.A. (2003) Predicting the potential invasive distributions of four alien plant species in North America. *Weed Science* 51, 863–868.

Pisek, A., Larcher, W., Vegis, A. & Napp-Zin, K. (1973) The normal temperature range. In: *Temperature and Life* (H. Precht, J. Christopherson, H. Hense & W. Larcher, eds), pp. 102–194. Springer-Verlag, Berlin.

Shea, K. & Chesson, P. (2002) Community ecology theory as a framework for biological inva-sions. *Trends in Ecology and Evolution* 17, 170–177.

Susarla, S., Medina, V.F. & McCutcheon, S.C. (2002) Phytoremediation: an ecological solution to organic chemical contamination. *Ecological Engineering* 18, 647–658.

Ter Braak, C.J.F. & Smilauer, P. (1998) CANOCO for Windows version 4.02. Wageningen, the Netherlands.

Townsend, C.R. (2003) Individual, population, community and ecosystem consequences of a fish invader in New Zealand streams. *Conservation Biology* 17, 38–47.

Townsend, C.R., Begon, M. & Harper, J.L. (2003) *Essentials of Ecology*, 2nd edn. Blackwell Science, Oxford.

Trewick, S.A. & Worthy, T.H. (2001) Origins and prehistoric ecology of takahe based on morphometric, molecular, and fossil data. In: *The Takahe: fifty years of conservation management and research* (W.G. Lee & I.G. Jamieson, eds), pp. 31–48. University of Otago Press, Dunedin, New Zealand.

van Riper, C., van Riper, S.G., Goff, M.L. & Laird, M. (1986) The epizootiology and ecological significance of malaria in Hawaiian land birds. *Ecological Monographs* 56, 327–344.

Whittingham, M.J., Swetnam, R.D., Wilson, J.D., Chamberlain, D.E. & Freckleton, R.P. (2005) Habitat selection by yellowhammers *Emberiza citrinella* on lowland farmland at two spatial scales: implications for conservation management. *Journal of Applied Ecology* 42, 270–280.

Zechmeister, H.G., Schmitzberger, I., Steurer, B., Peterseil, J. & Wrbka, T. (2003) The influence of land-use practices and economics on plant species richness in meadows. *Biological Conservation* 114, 165–177.

3 Life-history theory and management

To understand why one species is successful in a particular location while another is not, we can focus on the sequence of events that occur in the organism's life cycle and the suite of species traits they possess. This information can be put to good use when restoring degraded habitat or prioritizing invaders and endangered species for special management attention.

Chapter contents

3.1 Introduction – using life-history traits to make management decisions	60
3.2 Species traits as predictors for effective restoration	61
3.2.1 Restoring grassland plants – a pastoral duty	62
3.2.2 Restoring tropical forest – abandoned farmland reclaimed for nature	62
3.3 Species traits as predictors of invasion success	65
3.3.1 Species traits predict invasive conifers	66
3.3.2 Invasion success – the importance of flexibility	66
3.3.3 Separating invasions into sequential stages – different traits for each?	68
3.3.4 What we know and don't know about invader traits	71
3.4 Species traits as predictors of extinction risk	71
3.4.1 Niche breadth and flexibility – freshwater and forest at risk	72
3.4.2 When big isn't best – r/K theory, harvesting, grazing and pollution	73
3.4.3 When competitiveness matters – CSR theory, grazing and habitat fragmentation	77

Key concepts

In this chapter you will

recognize that species can be classified according to features of their life history (e.g. annual/perennial) and other ecological traits (e.g. size, growth rate)

understand that theory has been developed to link suites of traits to particular kinds of environments (r/K selection; CSR)

realize that managers do not need to know the restoration, invasion or extinction potential of every conceivable species – decisions may be based just on knowledge of traits

weigh up the value of databases of gardeners, ornithologists and anglers that record both successful and unsuccessful introductions

appreciate that species that are endangered in one place may be undesirable aliens in another

recognize that habitat generalists are often excellent candidates for successful restoration (a good thing) or invasion (a bad thing) whereas habitat specialists are likely to be particularly vulnerable to extinction

understand that r-selected species figure prominently in restoration and invasions whereas K-selected species are often the most vulnerable to human-caused habitat change

note the useful distinction between competitive (C), ruderal (R) and stress tolerant (S) strategies in predicting species' behavior, particularly in the management of plants but sometimes animals too

---

3.1 Introduction – using life-history traits to make management decisions

Carl runs a garden center. *'I trace my love of things botanical to childhood memories of farm meadows bursting into flower every spring. My very first recollection, full of color and perfume and the chirping of grasshoppers, is toddling through a flower meadow when the grass and the flowers were taller than me! You never see these multicolor farm meadows nowadays.'* When he can, Carl takes an early summer holiday in the Canadian Rockies, searching for the high meadows that, for a few weeks, put on similar breathtaking displays. *'Funnily enough, some of the biggest sellers in my catalog are seed-mixes to recreate in your own garden that lost splendor of meadows in bloom. Ironic, isn't it, that all this was once free?'* Carl can identify with the idea of niche-matching for species success (Chapter 2) – his brochure lists speciality seed mixes such as 'made-for-shade', 'mountain-meadow-glory' and 'seaside-flowerburst'. However, Carl also has a guilty secret. In the early days, his 'marsh-gold' seed-mix included colts-foot (*Tussilago farfara*) and lupin (*Lupinus polyphyllus*), which in our part of the world turned out to be enemies in disguise. Instead of simply contributing to the floral symphony and staying obediently in place, they spread and invaded wild habitats where they ousted delicate native plants. These invaders are now banned from sale, but Carl still worries: *'How many other aliens are we helping, inadvertently or otherwise?'*

Carl can take comfort from the fact that research is going on around the world to work out the best ways to restore natural biodiversity to agricultural landscapes. Much of this depends on understanding the life-history traits of candidate species – whether they are annuals or perennials, specialists or generalists, fast or slow growers, with or without persistent seed-banks, and so on (Box 3.1; Section 3.2). And this kind of life-history information proves to be just as important when predicting which species are likely to be problematic invaders (Section 3.3), and also which need special protection because of their vulnerability to human impacts such as harvesting, habitat fragmentation, logging and pollution (Section 3.4).

Box 3.1 Essential life-history theory

Life cycles

To understand why one species is successful in a particular location while another is not, we often need to focus on the sequence of events that occur in the organism's life cycle. At its simplest, the life cycle of a plant or animal comprises birth, followed by a pre-reproductive juvenile period, a period of reproduction, possibly a post-reproductive period, and then death (although various causes of mortality may intervene at any time). Some species squeeze several or many generations into a single year, some have one generation each year (annuals), but others (perennials) have life cycles that last for several or many years. Many plants and some animals (such as the fairy shrimp *Streptocephalus vitreus*) spend part of the year in a dormant phase (e.g. as seeds or eggs), and their 'seed-banks' can persist for years. All these features help determine the success of a particular species in a particular type of environment.

Species traits

Beyond contrasts in the characteristics of their life cycles, species differ in many other fundamental ways. Individuals may be small or large, invest few or many resources in their offspring (in seed biomass or parental care), have a short or long juvenile period, grow fast or slowly, and be more or less vulnerable to various sources of mortality. Some species are very specialized in their requirements, while at the other extreme are animals that are good at learning to exploit novel resources or plants with plastic tolerance limits that allow them to succeed in a range of circumstances. Species also vary in their competitiveness for limited resources, their tolerance of physically stressful conditions or of environmental disturbances, and their ability to find and exploit new habitats.

Several schemes have been developed that link particular groups of traits to particular kinds of environments (see Chapter 4 in Begon et al., 2006, for a detailed treatment). Two of these schemes will figure here – the r/K and CSR concepts.

The r/K concept

The potential of a species to multiply rapidly – producing large numbers of progeny early in life – is advantageous in environments that are short-lived (created, for example, by a disturbance such as the falling of a forest tree, or the storm-battering of a rocky reef), allowing the organisms to quickly colonize and exploit the new habitat. Such species have been called r-species because they spend most of their lives in a near-exponential phase of population growth where their intrinsic rate of increase (r) is being fully expressed. The habitats where they are favored have been called r-selecting.

At the other end of the scale are organisms with life histories that enable them to survive where there is often intense competition for limited resources. In this case the individuals leaving most descendants are those that capture a larger share of resources. They are called K-species because they spend most of their lives bumping up against the environment's carrying capacity (K). Habitats where they are favored have been called K-selecting.

The r/K concept (MacArthur & Wilson, 1967; Pianka, 1970) can be useful in the interpretation of contrasting patterns in nature. For example, many forest trees are excellent examples of K-species. They compete for light in the canopy and the survivors are those that put their resources into growth so they can overtop their neighbors. They usually delay reproduction until their branches have secured a place in the forest canopy, and hold on to their position and live for a very long time. Overall, they make a relatively low allocation to reproduction but many produce large, well-provisioned seeds. By contrast, in the more disturbed circumstances of r-selecting habitats, plants tend to conform to a contrasting group of r characteristics: a greater reproductive allocation, but smaller seeds, smaller size, earlier reproduction and a shorter life.

The CSR concept

Grime (1974, Grime et al., 1988) produced a different, but not unrelated, classification of habitats and plant life histories. Habitats are seen as varying in two significant ways – in their level of disturbance (brought about by grazing, disease, trampling or adverse weather) and in the extent to which they experience 'stress' (shortages of light, water or nutrients that limit photosynthesis). Grime argued that a stress-tolerant strategy (S) is appropriate when 'stress' is severe but disturbance is uncommon. Conversely, a so-called ruderal strategy (R) is appropriate when disturbance levels are high but conditions are benign and resources abundant (R-species are essentially good colonizers). Finally, a competitive strategy (C) is appropriate when disturbance is rare, resources are abundant and crowded populations develop. Grime then classified a large number of plants on the basis of ecological characteristics he thought suited them for one or other of these strategies (C, S or R), or some intermediate combination (CR, CS, SR or even CSR). Grime suggests that strong competitors have a high relative growth rate, the ability to spread by vegetative means and a tall stature. Stress tolerators, on the other hand, are small in stature with a low relative growth rate. Finally, ruderal species are generally annuals or short-lived herbaceous perennials with a capacity for rapid seedling establishment and growth, and a tendency for a high proportion of photosynthate to be directed into seeds. Intermediate species possess combinations of these traits.

3.2 Species traits as predictors for effective restoration

Native biodiversity is often compromised as a result of human activity, whether by mining, agriculture, forestry or urban development. The desire to restore these places to something approaching their pristine state may lead, through a political process, to legislation or economic incentives that foster recovery. In other cases, economic circumstances change and previous activities are no longer sustainable – mines close, agricultural land is put out of production, forestry is no longer viable. Whatever the background, effective restoration needs to be based on knowledge of which species will do well. Let's consider examples of agricultural restoration where the end-points are very different – species-rich grassland (Section 3.2.1) or forest (Section 3.2.2).

3.2.1 Restoring grassland plants – a pastoral duty

Undisturbed native grasslands are actually quite rare, most having been converted to an unnatural state by pastoral farming or agricultural development. Thus, the restoration of native grasslands has become a priority.

The first study of any ecological phenomenon provides just so much new information that it is risky to derive general conclusions that go beyond the focal organisms and sites. As science progresses, however, studies accumulate in the scientific literature until a more effective search for generalizations can be made using a 'meta-analysis'. Pywell et al. (2003) assembled the results of 25 published experiments dealing with the restoration of species-rich European grasslands from land that had previously been 'improved' for pasture (ploughed, planted with pasture species and fertilized) or used for arable farming. Restoration involved sowing a range of desirable species into areas where agricultural intensity had been reduced. The idea was to relate the performance of different species to their life-history traits. On the basis of the results of the first 4 years of restoration, they calculated a performance index for commonly sown grasses (13 species) and forbs (45 species; forbs are defined as herbaceous plants that are not grass-like). The index, calculated for each of the 4 years, was simply the percentage of quadrats (0.4×0.4 m or larger) originally sown with a species that still retained it. Their life-history analysis included 38 plant traits, including longevity of seeds in the seed-bank, seed viability, seedling growth rate, Grime's life-history strategies (C, S or R – Box 3.1), and the timing of life-cycle events (germination, flowering, seed dispersal).

The best performing grasses include *Festuca rubra* and *Trisetum flavescens* (performance indexes averaged for the 4 years of 77%); and among the forbs, *Leucanthemum vulgare* (50%) and *Achillea mellefolium* (40%) are particularly successful. In fact, grass species in general do better than forbs in restoration attempts, but for grasses only ruderality (R) is positively correlated with performance. With the forbs, on the other hand, good establishment is linked to colonization ability, percentage germination of seeds, autumn germination, vegetative growth, seed-bank longevity and habitat generalism. Interestingly, competitive ability and seedling growth rate become increasingly linked with success as the intensity of competitive interactions increased with time (Table 3.1). Stress tolerators, habitat specialists and species of infertile habitats perform badly, partly reflecting the high residual nutrient availability in many restored grasslands.

Pywell's team notes that restoration efficiency could be increased in future by only sowing species with the identified ecological traits. However, because this would lead to uniformity among restored grasslands, they also suggest that desirable but poorly performing species could be assisted by phased introduction several years after restoration begins, when environmental conditions are more favorable for their establishment.

3.2.2 Restoring tropical forest – abandoned farmland reclaimed for nature

Surprisingly large areas of the tropics now consist of abandoned agricultural land. Close to the Panama Canal, for example, most of the biodiverse tropical forest had been cut and burned to open up land for subsistence agriculture by the mid 1980s. Small farms were quickly abandoned, however, probably because of low productiv-

Table 3.1 Ecological traits of forbs (nongrass herbaceous plants) that showed a significant relationship with successful performance of the plants in years 1–4 after sowing in grassland restoration experiments. (After Pywell et al., 2003.)

Trait	N	Year 1	Year 2	Year 3	Year 4
Ruderality (colonization ability)	39	+*			
Autumn germination	42	+*			
% germination	43	+**	+*	+*	
Seedling growth rate	21		+*	+**	+*
Competitive ability	39	+*	+**	+***	+***
Vegetative growth	36	+**	+*	+*	+*
Seed-bank longevity	44	+*	+*	+*	+*
Stress tolerance	39	−**	−**	−***	−***
Generalist habitat	45	+**	+**	+**	+**

N, number of species in analysis.

+, Positive relationship; −, negative relationship. A positive sign means that the trait in question is associated with good performance in grassland restoration.

Asterisks indicate the level of statistical significance of the results for each trait in each year – more asterisks mean a greater level of confidence in the result: $*p < 0.05$, $**p < 0.01$, $***p < 0.001$, blank not significant. Thus, for example, species with good competitive ability are highly likely to be successful in restoration experiments (particularly in years 3 and 4). Stress tolerators, on the other hand, do very poorly.

ity. Now the areas have been invaded by tall, impenetrable stands of the exotic grass *Saccharum spontaneum*, a serious invader in places as far apart as India, the Philippines and Puerto Rico.

As was the case with native grass seeds in Section 3.2.1, the lack of naturally arriving tree seeds is a major barrier to unassisted regeneration in Panama. However, it would be wasteful simply to throw seeds in an indiscriminate manner into these areas. Hooper et al. (2002) tested the potential of seeds of 20 native trees to successfully germinate, survive and grow within the *Saccharum*-dominated community (Figure 3.1). Two kinds of species trait have particular significance in determining tree seed success in the *Saccharum* grassland – seed size (small: <0.15 g; medium: 0.15–1.0 g; large: >1.0 g) and shade tolerance (low, medium and high; determined from separate germination experiments in a greenhouse where light levels were varied). To demonstrate the major patterns, I present results separately according to these different traits.

Species with the smallest seeds generally perform most poorly. Those considered 'pioneer' species, so named because they are characteristic of open and disturbed locations (and, thus, have low shade tolerance), do a little better than those with intermediate or high shade tolerance, but nowhere near so well as species with medium seeds (in the intermediate shade tolerance class) or large seeds (in both medium and high shade tolerance classes) – just compare their respective performance indexes in Figure 3.1.

You might suppose, from first principles, that the *r*-selected, small-seeded pioneer trees (which include *Trema micrantha* and *Jacaranda copaia*) would be ideal candidates for restoration, because of their good powers of colonization and rapid growth in disturbed situations. However, in Panama at least, they do not tolerate the constraints imposed by the *Saccharum* grass.

At the other end of the scale, species with large or very large seeds (2.9 to a massive 50.4 g each) include the moderately shade tolerant *Dipteryx panamensis* and the very shade tolerant *Calophyllum longifolium*, *Carapa guianensis* and *Virola surinamensis*. The latter three germinate immediately upon planting in the wet

Fig. 3.1 Performance of seeds of 20 tropical trees planted in *Saccharum spontaneum* grassland in Panama. The species have been grouped in three seed classes (small, medium and large) and three shade tolerance classes (low, which require very high light levels to germinate, medium, and high, which are able to germinate in very low light levels). Different classes are represented by different numbers of trees, and none of the trees fell into the medium seed, low shade tolerance grouping. The different histograms relate to percentage germination, percentage survival of germinated seeds, an index of growth rate (cm per day from May to July multiplied by 1000) and an overall performance index (which is the product of the other three: i.e. % germination × % survival × growth rate). Species whose seedlings did not survive have a growth rate and performance index of zero. Note how species with medium or large seeds generally do best. However, small-seeded species that are intolerant of shade also do quite well. (Based on data from Hooper et al., 2002.)

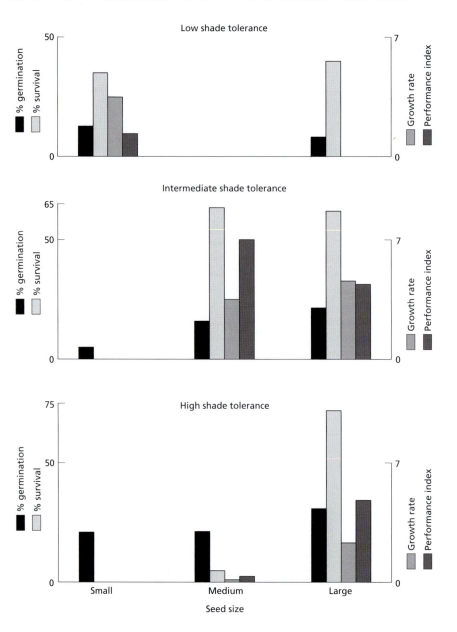

season and, despite slow growth, become relatively tall by the final census (>10 cm), while *Dipteryx* has dormant seeds that germinate the following dry season. All four of these highly competitive species survive well and have good performance indexes.

Hooper and colleagues make several suggestions for restoration management. First and foremost they note that *Saccharum* grassland is prone to fire, which would be a major barrier to tree regeneration. Fire breaks need to be created as an integral part of restoration. Even in the absence of fire, however, natural regeneration will not occur because small seeds – which are generally transported the longest distances and therefore are most likely to arrive at a site – are unlikely to produce viable seedlings in the grassland setting. Thus, the larger seeds of species most likely to do well need to be collected from elsewhere and carried in to the site. But Hooper's team also discovered that shade cover greatly enhances tree seedling success – so they suggest artificial shade should be provided at first. The establishment of the hand-planted species can be expected to catalyze forest regeneration, soon producing its own shade and assisting other species to establish. To increase biodiversity once natural shade has been established, small-seeded species in the shade-tolerant class (including *Genipa americana* and *Heisteria concinna*) can be introduced.

3.3 Species traits as predictors of invasion success

A number of species, plants and animals, have invaded widely separated places on the planet, including the shrub *Lantana camara* (Figure 3.2), the starling *Sturnus vulgaris* and the rat *Rattus rattus*. This prompts the question – do successful invaders share traits that raise the odds of successful invasion (Mack et al., 2000)? If it were possible to produce a list of traits associated with invasion success, managers would be in a good position to assess the probabilities of successful invasions, and thus to prioritize potential invaders and devise appropriate biosecurity procedures (Wittenberg & Cock, 2001).

Here I take advantage of databases available for taxonomic groups where some species are known to be successful invaders, and others not. These can be used to shed light on the species traits associated with a high probability of invasion success. My focus will be on pine trees in Section 3.3.1, New Zealand birds in Section 3.3.2 and parrots, plants and fish in Section 3.3.3.

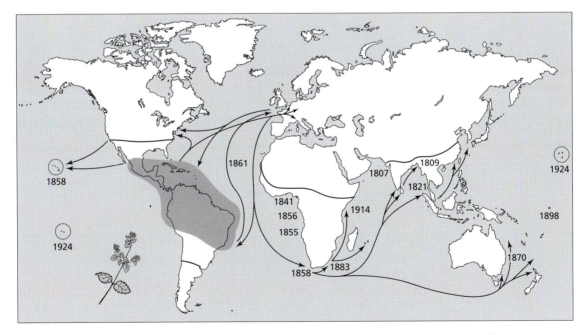

Fig. 3.2 The shrub *Lantana camara*, an example of a very successful invader, was deliberately transported from its native range (shaded area) to widely dispersed subtropical and tropical locations where it spread and increased to pest proportions. Year of arrival is shown where invasions have occurred. (After Cronk & Fuller, 1995.)

3.3.1 *Species traits predict invasive conifers*

The success of some invasive taxa has a strong element of predictability. Of 100 or so introduced pine species in the USA, for example, those that have successfully spread into native habitats are characterized by small seeds, a short interval between successive large seed crops and a short juvenile period (Rejmanek & Richardson, 1996). It is interesting to note in passing that conifer species classified as rare and endangered, such as *Pinus maximartinezii*, are the precise converse of their invader cousins, having large seeds, long juvenile periods and long intervals between large seed crops (Richardson & Rejmanek, 2004). Among pine taxa, in other words, the invaders tend to have *r*-selected traits while the endangered species exhibit *K*-selected traits (Box 3.1).

Grotkopp et al. (2002) delved more deeply into the suite of traits responsible for invasiveness in pine species. They germinated seeds of 29 species under carefully standardized laboratory conditions and harvested seedlings periodically after each species had reached the stage of expanding their first true leaves. Then for 10 weeks they recorded the weights of leaves and other plant parts, and measured the combined area of the seedlings' leaves, as well as their mean width and density (dry weight per unit volume). These data allowed them to calculate a number of plant performance traits (including, among others, relative growth rate (RGR: $mg\,g_{plant}^{-1}$ day^{-1}) and leaf area ratio (LAR: $cm^2_{leaf}\,g_{plant}^{-1}$)). Finally, they performed a Principal Components Analysis (akin to the ordination analyses you encountered first in Section 2.2.3) to explore the relationships between plant traits and invasiveness (Figure 3.3).

The invasive pine species are clearly clumped together to the right on PCA axis 1 and the noninvasive taxa to the left in Figure 3.3. Invasiveness is correlated most strongly, and positively, with relative growth rate (also to the right on axis 1), and negatively with seed mass and juvenile period (to the left on axis 1). In other words, invaders tend to have high growth rates, small seeds and short juvenile periods (i.e. short generation times). The influential high relative growth rate is itself a consequence of high net assimilation rates, leaf mass ratios and, in particular, specific leaf areas. Our example shows how the details of plant physiology and morphology can help explain, and predict, invasion potential.

3.3.2 *Invasion success – the importance of flexibility*

In New Zealand there is a precise record (similar to that of the pines in Section 3.3.1) of successes and failures of attempted bird introductions. These species were shipped out by nineteenth-century European colonists to make their new surroundings more like 'home'.

Sol and Lefebvre (2000) found that invasion success of the introduced birds increases with introduction effort (number of attempts and number of individuals introduced since European colonization). This is hardly surprising because the success of even the best fitted of potential invaders is probabilistic rather than definite. Invasion success is also higher for species whose young do not need to be fed by their parents (such as game birds), species that do not migrate and, in particular, birds with large brains. Animals with relatively large forebrains are thought to cope better with environmental complexity and to respond more rapidly to changes in their environment. In fact, the successful bird invaders have more reports in the international literature of adopting novel food or feeding techniques (mean for 28 species 1.96, SD 3.21) than the unsuccessful species (mean for 48 species 0.58, SD 1.01). For example, among closely related species the successfully invading rook

Fig. 3.3 Results of a Principal Components Analysis (PCA) showing (a) the relationships among species traits of 29 species of pine (*Pinus*) (b) known to be invasive or noninvasive. The positions of the species and the trends in species traits are shown in separate panels for simplicity of presentation. Key for plant traits: relative growth rate (RGR: $mg\,g_{plant}^{-1}\,day^{-1}$), net assimilation rate (NAR: $mg\,cm_{leaf}^{-2}\,day^{-1}$), leaf mass ratio (LMR: $g_{leaf}\,g_{plant}^{-1}$), relative leaf production rate (RLPR: $leaf\,leaf^{-1}\,day^{-1}$), leaf area ratio (LAR: $cm_{leaf}^{2}\,g_{plant}^{-1}$) and specific leaf area (SLA: $cm_{leaf}^{2}\,g_{leaf}^{-1}$). The invasive and noninvasive species cluster in different parts of the diagram. (After Grotkopp et al., 2002.)

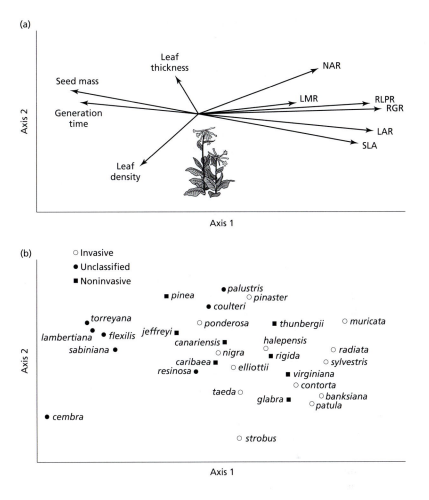

(*Corvus frugilegus*) has five reported novel behaviors while the unsuccessful *Corvus monedula* only has two. Similarly, the successful house sparrow (*Passer domesticus*) and mallard duck (*Anas platyrhynchos*) have ten and five reported novel behaviors, respectively, compared with just two each for their unsuccessful counterparts *Passer montanus* and *Anas penelope*.

In this disparate group of birds, and in contrast to the pine species, *r/K* status does not appear to predict invasive potential. Success seems to be linked with high competitive status (analogous to Grime's C categorization in Box 3.1), an advantage that is conferred by behavioral flexibility.

Something similar may be the case for certain plant invaders too. Japanese grass (*Microstegium vimineum*) is an important annual weed of Asian origin that is now widespread east of the Mississippi River in the USA and in many other countries. Gibson et al. (2002) studied four populations in quite different environments in Illinois, USA. They concluded that the species' invasive potential is fueled by a remarkably flexible response to local microhabitat conditions, being able to tolerate acidic, sandy and silty soils in both open and shaded locations. Such plasticity, akin to the behavioral flexibility of birds, may be a property of many invaders.

Physiological and behavioral flexibility can be equated to possession of a broad (or generalist) niche (Sections 2.4.2 and 3.2.1).

3.3.3 Separating invasions into sequential stages – different traits for each?

Species invasion is a complex process with several identifiable stages: (1) transport from the native range; (2) release into a new location; (3) establishment of a self-sustaining population in the new location; (4) spread beyond the site of first establishment; and (5) impact in the receiving community. Both the invasive pine study (Section 3.3.1) and that of the successfully introduced New Zealand birds (Section 3.3.2) fit into category (4). Generally speaking, managers are most concerned about species like these, regarding as truly invasive those that spread beyond their original establishment site. However, to increase predictability in invasion ecology, Kolar and Lodge (2001) suggest that species' characteristics should be examined in relation to all the stages of invasion.

Cassey et al. (2004) assessed the first three of the five phases – (1) transport, (2) release and (3) establishment – in an analysis of all of the world's 350 parrot species. Transporting and selling parrots is big business and populations that establish in new locations are easily spotted and quickly recorded by ornithologists. For these reasons parrots provide an excellent case study to determine species traits that correlate with transport (species commonly transported outside their natural range score 1, others 0), release (species intentionally released outside their range score 1, those transported but not released score 0) and establishment (the proportion of releases leading to successful population establishment).

Cassey's team found that different sets of variables were related to the probability that a species will enter each stage on the pathway through transport to establishment (Table 3.2). Large species are more likely to be transported. More significantly, species that are transported and released outside their native range tend to be widespread ones that are traded to be kept in captivity or as pets. Presumably this simply reflects the need for a predictable supply of desirable trade species. Threatened species do not figure in the groups that are transported and released (note the negative correlation) – illegal trade in some rare species certainly occurs, but such individuals are not released into new areas.

Of more interest was the finding that species with broader diets and larger altitudinal ranges (i.e. with broader niches) were more likely to establish in a new area, as were those with longer fledging periods, perhaps because of the head start gained from an extended period of parental care. Migratory species, as with the New Zealand birds, were less likely to establish. You might have predicted, on first principles, that migratory species would, like ruderal plants, possess some good colonizer traits and be effective invaders. However, it turns out that the complex requirements of migratory species – innate migration cues, learnt migration routes, appropriate juxtaposition of seasonal habitats – do not suit them for invasions, at least not those of the human-assisted kind.

Pysek et al. (2003) also have something to say about the first three stages of invasion based on their analysis of first records of 668 alien plants in the Czech Republic since the year 1500. Deliberately introduced species with utilitarian value (e.g. culinary or medicinal) arrived before ornamental species, and 'accidental' arrivals have been most recent. Moreover, species of European origin arrived earlier than those from North America, with Asian and Australasian species turning up in more modern times. In their analysis of the relationship between timing of first records

Table 3.2 Tests of the degree to which species variables correlate with the likelihood of parrot species being transported from their native range, released into a new area and establishing in the new location. (Modified from Cassey et al., 2004.)

Variables	Transport	Release	Establishment
Ecological flexibility			
Diet breadth	+*	+*	+**
Altitudinal range			+*
Latitudinal range	+***	+***	
Life history			
Body mass	+*		
Annual fecundity			
Age at maturity			
Fledging period			+**
Migration		−*	−**
Dichromatism		+*	
Movement			
International trade	+***	+***	+**
To be kept in captivity	+***	+***	
To be kept as pets	+***	+***	
Population size/extent			
Threat status	−**	−*	
Population size		+**	+*
Geographical range	+***	+**	

+, Positive relationship; −, negative relationship.
Asterisks indicate the level of statistical significance of the results for each trait in each year − more asterisks mean a greater level of confidence in the result: $*p < 0.05$, $**p < 0.01$, $***p < 0.001$, blank not significant.

Fig. 3.4 Increase with time in the cumulative number of alien plants in the Czech Republic corresponding to Grime's CSR strategy combinations. Refer to Box 3.1 for definitions of C (competitive), S (stress-tolerant) and R (ruderal). (From Pysek et al., 2003.)

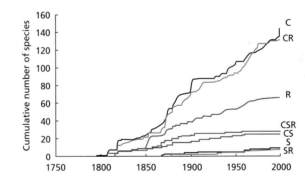

and species traits, Pysek's team considered type of life history (annual, biennial, herbaceous perennial, woody perennial), flowering time, mode of dispersal (water, wind, animal assisted) and Grime's CSR strategy. The strongest relationship was with Grime's strategy: species having the unusual but highly flexible CSR strategy (most features of competitiveness, stress tolerance and ruderality incorporated in a single organism) arriving earlier than CS species, which in turn arrived earlier than C or CR species. The pattern of accumulation of species with each strategy is shown in Figure 3.4. (Note that the early arrival of CSR strategists is not obvious in this figure because of the small number of species of this type.) By far the most commonly represented strategies among species now established are competitive (C) or competitive-ruderal (CR) combinations of traits.

Marchetti et al. (2004) take up where the parrot and plant studies leave off. They compiled information on invasions and failed introductions of fish in catchment areas throughout California and separate the invasion process into the three final stages of our scheme: (3) establishment (of a self-sustaining population); (4) spread (assessed as number of catchment areas occupied); and (5) impact (assessed as average abundance attained). Their analysis includes a variety of species traits, from physiological tolerance and adult size to prior invasion success. The process involves a comparison of all conceivable models to predict species 'success' at the three stages, on the basis of the species traits. The importance, or weight, of particular traits was estimated across all models, and the larger the weight, the more important the variable in predicting the response of interest (Table 3.3).

All eight traits in the analysis are important for predicting species establishment. This is not really surprising as the traits were chosen, according to first principles, because they might be expected to be influential. Establishment is more likely for larger species (more competitive), those showing broad physiological tolerances and diets, parental care, a large native range (broader niches), a nearby native source, a record of successful invasions elsewhere and where large numbers were involved in introduction attempts.

Prediction of the propensity to spread, on the other hand, depends mainly on physiological tolerance and adult size, as well as the effort put in to establishing the species in the first place. Once again, and unlike the pine species, successful fish invaders do not fall consistently at one end of the r/K continuum. Competitive status (large size) seems to play a role, but the overriding ecological variable is broad physiological tolerance, probably increasing the likelihood, in the probabilistic

Table 3.3 Weight of evidence to suggest that a particular response variable is important in predicting 'success' at each stage of invasions of freshwater fish in Californian river catchment areas. The larger the 'weight', the more important is the variable in predicting invasion across all possible models created; maximum weight is 1.0. (After Marchetti et al., 2004.)

Response variables	Establishment (self-sustaining population)	Spread (number of catchment areas)	Impact (average abundance)
Propagule pressure (numbers involved in introduction attempts)	1.00	0.83	0.34
Physiological tolerance	1.00	0.82	0.08
Adult maximum size	1.00	0.38	0.09
Distance from source	1.00	0.23	0.02
Size of native range	1.00	0.10	0.07
Prior invasion success	1.00	0.10	0.76
Trophic status	1.00	0.00	0.00
Parental care	1.00	0.00	0.00

world of invasion success, that these species would encounter a match with appropriate environmental conditions.

Finally, the likely size of invader impact, assessed in terms of the abundance achieved by populations, is not well predicted by species traits, but rather in terms of prior invasion success and establishment effort.

3.3.4 *What we know and don't know about invader traits*

We know rather a lot about the relationship between species traits and invasion success but, regrettably, only for a small proportion of the world's biota. The most informative databases are those concerning taxonomic groups for which information is available about species that have become successful invaders as well as those that have not. This tends to apply to taxa that are important to particular segments of society, and for which particularly good records have been kept – horticulturalists and gardeners (pine trees and Czech plants), those in the pet trade (parrots), colonials who wished to make themselves at home (New Zealand Acclimatization Society – introduced birds) and angling organizations (freshwater fish). These databases are enormously valuable, providing some of the best information we have to assess the causes of invader success.

However, managers need to beware of unwarranted generalization. We do see indications of predictability of invasion success for some taxa, related to high reproductive output (e.g. pine seed production), flexible lifestyles (bird behavior), broad niches (e.g. parrots) and competitive strategies (e.g. Czech plants). But exceptions to the 'rules' are common and there are many cases where no relationships have been found, prompting Williamson (1999) to wonder whether invasions are any more predictable than earthquakes. Commonly, the best predictor of invasion success is previous success as an invader elsewhere. Looking on the bright side, even this provides invasion managers with useful pointers when prioritizing potential invaders to their regions.

3.4 **Species traits as predictors of extinction risk**

Having discovered the life-history traits that are linked to the probability of invasion or of success in restoration projects, I now turn attention to the suites of traits that can be used to predict the vulnerability of native species to human pressures. This question is vital to conservation biologists who must understand what makes certain species prone to extinction.

Successful invaders naturally have much in common with successful restoration. The only distinction is that we have different names for the species we do (restoration) or don't (invasion) want to become established. In Sections 3.2 and 3.3 you saw that traits associated with niche breadth and ecological flexibility could sometimes predict the probability of successful establishment, as could suites of traits associated with the r/K concept (small, rapidly reproducing vs large, slow-growing, competitive organisms) or the CSR concept (particularly R species with good powers of colonization and C species with strong competitive traits).

Vulnerability to extinction is a different kettle of fish because now we wish to ensure that a species is not lost from its native environment in the first place. The list of human pressures that pose risks (and that are responsible for past extinctions, both global and local) is long and diverse. It includes habitat loss and habitat fragmentation as a result of land-use change (e.g. pastoral grazing, forestry, freshwater habitat disruption), habitat degradation as a result of pollution and overharvesting of wild populations. Moreover, many species are confronted by more than just one of these pressures. In this section I ask whether the varied situations have anything in common. Are there, in fact, suites of species traits that seem to 'protect' native species or, alternatively, that make them more likely to succumb to the pressures? And are these traits the same as those that predict successful invasion or restoration?

First, I consider cases where niche breadth and flexibility play an important role (Section 3.4.1) before turning to the many situations where the *r*/*K* concept again seems to come into its own (Section 3.4.2). In a concluding section, the importance of CSR traits, among others, will be considered (Section 3.4.3).

3.4.1 *Niche breadth and flexibility – freshwater and forest at risk*

Managers would be better able to prioritize species for conservation intervention if it were possible to predict, on the basis of species traits, those most at risk of extinction. With this in mind Angermeier (1995) analyzed the traits of 197 native freshwater fish in Virginia, USA, paying special attention to the characteristics of the 17 species now extinct in Virginia and nine others considered at risk because their ranges have shrunk significantly. There has been a mix of human pressures at work, from water abstraction for irrigation and the building of dams, through loss of riverside vegetation, to pollution by agricultural and urban runoff.

Of particular interest is the greater vulnerability of ecological specialists in the face of the various human pressures. A higher probability of local extinction attaches to species whose niches include only one geological type (of several present in Virginia), those restricted to flowing water (as opposed to occurring in both flowing and still water), and those with only one food category in their diet (i.e. wholly piscivorous, insectivorous, herbivorous or detritivorous as opposed to omnivorous on two or more food categories).

When it comes to vulnerability to extinction, do forest birds and freshwater fish have something in common? In his study in a jarrah (*Eucalyptus marginata*) forest in southwestern Australia, Craig (2002) found that passerine species vary considerably in the extent to which their populations are affected by selective logging (where only high quality trees are removed). Western yellow robins (*Eopsaltria griseogularis*) and rufous treecreepers (*Climacteris rufa*) did not decline after logging, unlike golden whistlers (*Pachycephala pectoralis*) and white-naped honeyeaters (*Melithreptus lunatus*). Craig carefully assessed foraging behavior, making observations in three seasons, in different subhabitats, and before and after logging took place. He recorded foraging location (aerial, on the ground, associated with living vegetation, dead logs, etc.), foraging height, and type of maneuver (glean, hover, pounce, etc.).

The species that are least vulnerable to logging show much more flexibility in their foraging behavior, with both seasonal variation and changes in response to logging. Thus, the robins (four changes) forage more on the ground and pounce more in winter (than in other seasons) and also after logging. The treecreepers (four changes) forage more on marri trees (*Corymbia calophylla*, which coexist with the jarrah trees), on horizontal surfaces and lower in the canopy in winter, and forage more in forest gaps after logging. By contrast, each of the species that declined after logging (whistlers and honeyeaters) only display a single variation in foraging behavior. Craig recommends that unlogged buffer areas should be provided to safeguard the whistler and honeyeater populations. Greater behavioral flexibility, which we can equate with a broader niche, seems to buffer the robins and treecreepers from the worst effects of habitat disruption – a pattern analogous to that for Angermeier's (1995) fish.

Habitat fragmentation is another major consequence of human behavior – what remains of previously extensive habitat has become increasingly fragmented into isolated pockets. This has consequences for many but not all species and it is reasonable to ask whether vulnerability to habitat fragmentation can be linked to particular life-history traits. Henle et al. (2004) addressed this question by analyzing accumulated information for diverse plant and animal taxa, both from controlled fragmentation experiments and the many uncontrolled 'natural' experiments documented around the world. Habitat specialists might be expected to have a higher extinction risk because the chance that their niche will be represented in a remaining fragment should be smaller than for generalists. Studies on highly endangered species like the red-cockaded woodpecker (*Picoides borealis*) (Haig et al., 1993) show that the absence of suitable microhabitat is a principle factor determining their sensitivity to forest fragmentation. This kind of habitat specialization also turns out to be a risk factor, in relation to habitat fragmentation, for forest trees, birds, mammals and reptiles.

3.4.2 *When big isn't best* – r/K *theory, harvesting, grazing and pollution*

A pattern that has repeatedly emerged is that extinction risk tends to be highest for species with a large body size. Figure 3.5 illustrates this for Australian marsupials that have gone extinct in the last 200 years or are currently considered endangered. Again the human pressures have been diverse, including hunting, habitat degradation and habitat destruction. Some climatic regions (arid compared to mesic zone) and some taxa (e.g. potoroos, bettongs, bandicoots and bilbies) have experienced higher extinction/endangerment rates than others, but the strongest relationship is between body size and risk of extinction (Cardillo & Bonham, 2001). Recall that body size is part of the life-history syndrome (essentially *r/K*) that associates large size, late maturity and small reproductive allocation (Box 3.1). In fact, you can blame the vulnerability of larger species on their relatively low reproductive rate – they

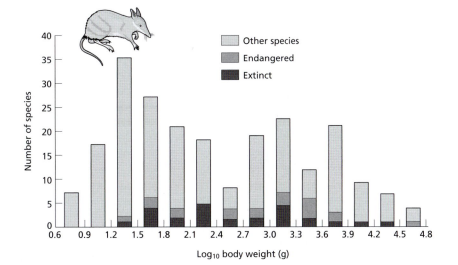

Fig. 3.5 Body-size frequency distribution of the Australian terrestrial marsupial fauna including 25 species that have gone extinct in the last 200 years (black). Sixteen species currently considered endangered are shown in dark grey. (After Cardillo & Bonham, 2001.)

simply cannot compensate for the extra mortality caused by our actions. This is made clearer in the next example.

Cortes (2002) has explored, using a mathematical modeling approach, the relationship for 38 shark species between body size, age at which maturity is reached, generation time and the annual rate of population increase (λ). λ is essentially the multiplier that converts the population size of one generation into the population size of the next generation. Populations increase when $\lambda > 1$ and decrease when $\lambda < 1$.

A three-dimensional plot of λ against generation time and age at maturity shows what Cortes calls a 'fast–slow' continuum, with species characterized by early age at maturity, short generation times and generally high λ at the fast end of the spectrum (bottom right hand corner of Figure 3.6a). Species at the slow end of the spectrum displayed the opposite pattern (left of Figure 3.6a) and also tended to be large bodied (Figure 3.6b). Cortes then assessed the various species' ability to respond to increased mortality due, for example, to pollution or harvesting. 'Fast' sharks, such as *Sphyrna tiburo*, could compensate for a 10% increase in mortality rate by increasing birth rate. On the other hand, particular care must be taken when considering the state of generally large, slow-growing, long-lived species, such as *Carcharhinus leucas*. Here, even moderate increases in mortality require a level of compensation in the form of successful reproduction that the species simply cannot provide.

Skates and rays (Rajidae) provide a graphic illustration of Cortes' warning. Of the world's 230 species, only four are known to have undergone local extinctions and significant range reduction (Figure 3.7a–d). These are among the largest of their group (Figure 3.7e) and Dulvy and Reynolds (2002) propose that seven further species, each as large or larger than the locally extinct taxa, need to be carefully monitored in future.

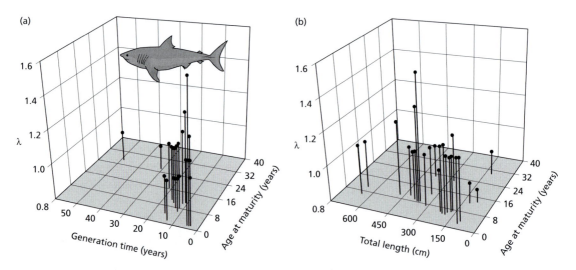

Fig. 3.6 Mean population growth rates λ of 41 populations from 38 species of shark in relation to (a) age at maturity and generation time and (b) age at maturity and total body length. Populations increase in size when $\lambda > 1$ and decrease when $\lambda < 1$. (After Cortes, 2002.)

Fig. 3.7 Historical distribution of four locally extinct skates in the northwest and northeast Atlantic: (a) barndoor skate *Dipturus laevis*; (b) common skate *D. batis*; (c) white skate *Rostroraja alba*; (d) long-nose skate *D. oxyrhinchus*. Key: e, area of local extinction; p, present in recent fisheries surveys; ?, no knowledge of status; e?, possible local extinction. (e) Frequency distribution of skate body size – the four locally extinct species are black. (After Dulvy & Reynolds, 2002.)

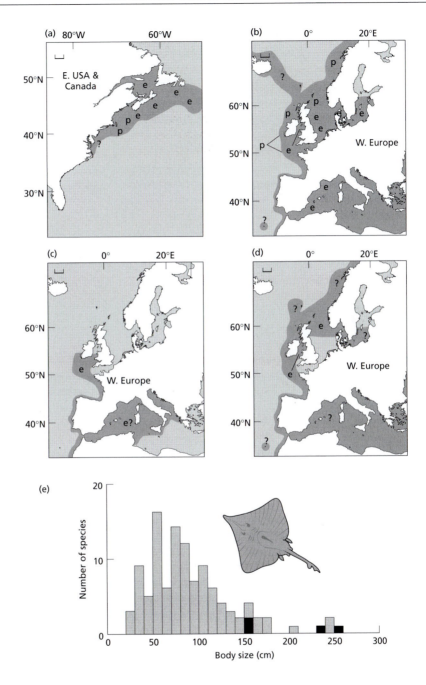

Within the sharks and skates, then, it is the *K*-strategists that are most vulnerable to overexploitation. King and McFarland (2003) provide a broader perspective in their analysis of life-history traits of a diverse array of marine fish (42 species) that includes skates and sharks. Now we see all skates and sharks as relatively *K*-selected, being among the largest of fish and showing a high degree of parental investment.

In stark contrast are the small 'opportunists' (or *r*-selected species), such as clupeids (e.g. sardines) and smelts (e.g. *Thaleichthys pacificus*). And intermediate between the *K*- and *r*-species are a group of medium-sized fish, including the rockfishes and mudfishes, that are long-lived and slow growing, but with high fecundity. Some general management principles could be developed on the basis of the properties of these groupings, for use even in new fisheries where the details of species' ecologies are unknown. (I deal with harvest management in Chapter 7.) Certainly, species at the *K* end of the continuum can clearly only withstand modest harvest levels.

Other human impacts may also be felt by different species in different ways because of contrasts in their life-history traits. Prinzing et al. (2002), for example, exposed forest soil to a single application of an insecticide used to control gypsy moth (*Lymantria dispar*) and other forest pests. Diflubenzuron interferes with chitin synthesis, disturbing the maturation and reproduction of insects, crustaceans, and also the soil mites that are the focus of Prinzing's study. The mite taxa that are least affected by the disturbance are those with shorter generation times and therefore able to recover more quickly. Cultivation activities also have impacts in freshwater bodies adjacent to the land. Doledec et al. (2006) studied the relationships between species traits of stream invertebrates and intensity of agricultural development in the stream's catchment area. Traits associated with population resilience, including short generation time (many generations per year, known as plurivoltine) and hermaphroditic reproduction, become more prevalent with more intense land use (Figure 3.8), reflecting increases in the intensity and frequency of disturbances to stream chemistry. This pattern mirrors that of the forest mites. There is also a shift away from laying unattached eggs at the water surface and a decrease in gill respiration, reflecting the increasing likelihood of smothering by sediment introduced to the stream as a result of grazing animals disturbing the soil and banks.

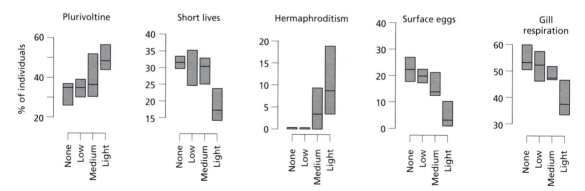

Fig. 3.8 Relationships between selected species traits of stream invertebrates and agricultural intensity in the stream's catchment area. There were four levels of land-use intensity: none, ungrazed native tussock grassland; low, grazed tussock grassland; medium, extensively grazed pasture; high, dairy or deer farming. Key: plurivoltine, % individuals in the community with more than one generation per year; life duration, % individuals with short lives (10–30 days); hermaphroditism, % individuals reproducing hermaphroditically (one individual possessing both sexes); surface, % individuals laying unattached eggs at the stream surface; gill respiration, % individuals with external gills. Boxes show the 25%, 50% and 75% distribution values. (From Doledec et al., 2006).

3.4.3 *When competitiveness matters – CSR theory, grazing and habitat fragmentation*

The herbaceous species of grasslands differ in the extent to which they can withstand the pressure of grazing by vertebrates. In grazing trials in Argentina and Israel, Diaz et al. (2001) asked whether grazing vulnerability could be predicted on the basis of species traits. If so, managers in a new area with different species could predict (and manage) their grasslands without the need to conduct their own time-consuming grazing trials. Diaz and his colleagues found that plant species that decreased in abundance under increased grazing pressure ('decreasers') were characterized by tallness, perennial life history, large leaves and low specific leaf area ($cm^2_{leaf} g_{leaf}^{-1}$). The 'increasers' possessed the opposite traits. If grazers attack plants from above, as they do where the vegetation is relatively dense, tall plants are most vulnerable because they are the first ones encountered and, by their presence, smaller species are protected to some degree. Other things being equal, large-leaved species are doubtless preferable to grazers because they offer bigger mouthfuls; small-leaved species require more bites per unit leaf material consumed and the bites probably include more unwanted stem material. Perennial species, in comparison to annuals, remain exposed to grazing for longer and also allocate less to reproduction, reducing their ability to compensate for grazing mortality. Similarly, species with low specific leaf areas may be relatively more affected by intensive grazing because they do not have such a capacity for rapid regrowth.

So we see a number of predictable responses to grazing, some that can be related to the *r*/*K* continuum as in Section 3.4.2 (small annuals with large reproductive allocations do better under grazing pressure), and others to relative competitive status in the face of grazing pressure (small species gain a greater share of resources if larger species protect them by their presence). These species seem to combine Grime's competitive and ruderal strategies (CR) – competing for resources best when disturbances are relatively frequent, whereas the vulnerable species lack these traits. Such generalizations will be helpful to managers, but Vesk et al. (2003) warn against extrapolating too far. They note that in more arid rangelands, with their more open, patchy vegetation, simple predictions based on species traits do not hold so well.

The importance of sensitivity to disturbance and competitive status receives further support from a variety of studies of tropical trees in forest fragments (e.g. Metzger, 2000). For example, shade-tolerant species typical of dense forest are vulnerable because, in comparison to 'edge' species, they have higher mortality, lower growth rate and lower dispersal capability. This results in a competitive disadvantage under the modified light and increased disturbance regimes that dominate a larger percentage of the area of smaller forest fragments. Poor dispersal ability (flightlessness) is also linked to extinction risk of beetles of woodland in Australia, which has become increasingly fragmented through conversion to agriculture (Driscoll & Weir, 2005).

Summary

Life cycles and species traits
To understand why one species is successful in a particular location while another is not, we need to focus on differences in their life cycles – birth, juvenile period, reproductive period, and possibly a post-reproductive period. Some species have one generation each year (annuals), but others (perennials) have life cycles that last for several or many years. Many can spend a substantial time lying dormant (e.g. as seeds or eggs). Individuals of different species may be small or large, invest little or heavily in their offspring, grow fast or slowly, and be more or less vulnerable to

various sources of mortality. Some species are very specialized in their requirements, while others are generalists. Species also vary in their competitiveness for limited resources, their tolerance of physically stressful conditions or of environmental disturbances. Different conceptual schemes link particular groups of traits to particular kinds of environment (e.g. *r/K* and CSR schemes).

Species traits and restoration

Effective restoration needs to be based on knowledge of which species will do well. By monitoring the results of grassland restoration attempts, ecologists have determined the species traits that are linked with greater probability of establishment – including a generalist trait, good powers of colonization, the ability to grow vegetatively and the longevity of the soil seed-bank. In a similar manner, the success of certain forest restoration attempts has been linked to seed size, ability to compete and shade tolerance.

Species traits and invasion success

The success of some invasive species has a strong element of predictability. Among pine tree species, for example, successful invaders are those with small seeds, high growth rates, a short interval between successive large seed crops and a short juvenile period. In the case of birds, on the other hand, the ability of young to feed themselves, a large brain size (linked to behavioral adaptability) and a broad diet help to predict invader success. For freshwater fish, a broad diet and a wide range of physiological tolerance seem important. But exceptions to these 'rules' are common and there are many cases where no relationships have been found. Commonly, the best predictor of invasion success is previous success as an invader elsewhere.

Species traits and extinction risk

Conservation managers need to know if there are suites of species traits that 'protect' native species in the face of extinction threats or, alternatively, that make them more likely to succumb to the pressures. Again the answer varies among taxonomic groups but some generalizations emerge. For many taxa it is the ecological specialists that seem more at risk, while those showing greater flexibility (in terms of physiological tolerance ranges or diet) often fare better. And a pattern that has repeatedly emerged is that extinction risk tends to be highest for species with a large body size.

The final word

Carl (Section 3.1) takes some comfort from our increasing knowledge of the species traits that predict invasion risk. '*I still feel bad about accidentally introducing some harmful meadow flowers in my commercial wild flower mixes, but such an event must be a lot less likely now.*' Or is it?

Make a list of five or more exotic herbaceous flowering plants that have become invasive in your area. What problems do they cause? How did they arrive? Check out your national biosecurity precautions: do you consider that these have eliminated the probability of arrival of further invasive flowers?

References

Angermeier, P.L. (1995) Ecological attributes of extinction-prone species: loss of freshwater fishes of Virginia. *Conservation Biology* 9, 143–158.

Begon, M., Townsend, C.R. & Harper, J.L. (2006) *Ecology: from individuals to ecosystems*, 4th edn. Blackwell Publishing, Oxford.

Cardillo, M. & Bromham, L. (2001) Body size and risk of extinction in Australian mammals. *Conservation Biology* 15, 1435–1440.

Cassey, P., Blackburn, T.M., Russell, G.J., Jones, K.E. & Lockwood, J.L. (2004) Influences on the transport and establishment of exotic bird species: an analysis of the parrots (Psittaciformes) of the world. *Global Change Biology* 10, 417–426.

Cortes, E. (2002) Incorporating uncertainty into demographic modeling: application to shark populations and their conservation. *Conservation Biology* 16, 1048–1062.

Craig, M.D. (2002) Comparative ecology of four passerine species in Jarrah forests used for timber production in southwestern Western Australia. *Conservation Biology* 16, 1609–1619.

Cronk, Q.C.B. & Fuller, J.L. (1995) *Plant Invaders*. Chapman & Hall, London.

Díaz, S., Noy-Meir, I. & Cabido, M. (2001) Can grazing response of herbaceous plants be predicted from simple vegetative traits? *Journal of Applied Ecology* 38, 497–508.

Doledec, S., Phillips, N., Scarsbrook, M., Riley, R.H. & Townsend, C.R. (2006) A comparison of structural and functional approaches to determining land-use effects on grassland stream communities. *Journal of the North American Benthological Society* 25, 44–60.

Driscoll, D.A. & Weir, T. (2005) Beetle responses to habitat fragmentation depend on ecological traits, habitat condition and remnant size. *Conservation Biology* 19, 182–194.

Dulvy, N.K. & Reynolds, J.D. (2002) Predicting extinction vulnerability in skates. *Conservation Biology* 16, 440–450.

Gibson, D.J., Spyreas, G. & Benedict, J. (2002) Life history of *Microstegium vimineum* (Poaceae), an invasive grass in southern Illinois. *Journal of the Torrey Botanical Society* 129, 207–219.

Grime, J.P. (1974) Vegetation classification by reference to strategies. *Nature* 250, 26–31.

Grime, J.P., Hodgson, J.G. & Hunt, R. (1988) *Comparative Plant Ecology: a functional approach to common British species*. Unwin-Hyman, London.

Grotkopp, E., Rejmanek, M. & Rost, T.L. (2002) Toward a causal explanation of plant invasiveness: seedling growth and life-history strategies of 29 pine (*Pinus*) species. *American Naturalist* 159, 396–419.

Haig, S.M., Belthoff, J.R. & Allen, D.H. (1993) Population viability analysis for a small population of red-cockaded woodpeckers and an evaluation of enhancement strategies. *Conservation Biology* 7, 289–301.

Henle, K., Davies, K.F., Kleyer, M., Margules, C. & Settele, J. (2004) Predictors of species sensitivity to fragmentation. *Biodiversity and Conservation* 13, 207–251.

Hooper, E., Condit, R. & Legendre, P. (2002) Responses of 20 native tree species to reforestation strategies for abandoned farmland in Panama. *Ecological Applications* 12, 1626–1641.

King, J.R. & McFarland, G.A. (2003) Marine fish life history strategies: applications to fishery management. *Fisheries Management and Ecology* 10, 249–264.

Kolar, C.S. & Lodge, D.M. (2001) Progress in invasion biology: predicting invaders. *Trends in Ecology and Evolution* 16, 199–204.

MacArthur, R.H. & Wilson, E.O. (1967) *The Theory of Island Biogeography*. Princeton University Press, Princeton, NJ.

Mack, R.N., Simberloff, D., Lonsdale, W.M., Evans, H., Clout, M. & Bazzaz, F.A. (2000) Biotic invasions: causes, epidemiology, global consequences and control. *Ecological Applications* 10, 689–710.

Marchetti, M.P., Moyle, P.B. & Levinbe, R. (2004) Invasive species profiling: exploring the characteristics of exotic fishes across invasion stages in California. *Freshwater Biology* 49, 646–661.

Metzger, J.P. (2000) Tree functional group richness and landscape structure in a Brazilian tropical fragmented landscape. *Ecological Applications* 10, 1147–1161.

Pianka, E.R. (1970) On *r*- and *k*-selection. *American Naturalist* 104, 592–597.

Prinzing, A., Kretzler, S., Badejo, A. & Beck, L. (2002) Traits of oribatid mite species that tolerate habitat disturbance due to pesticide application. *Soil Biology & Biochemistry* 34, 1655–1661.

Pysek, P., Sadlo, J., Mandak, B. & Jarosik, V. (2003) Czech alien flora and the historical pattern of its formation: what came first to Central Europe? *Oecologia* 135, 122–130.

Pywell, R.F., Bullock, J.M., Roy, D.B., Warman, L., Walker, K.J. & Rothery, P. (2003) Plant traits as predictors of performance in ecological restoration. *Journal of Applied Ecology* 40, 65–77.

Rejmanek, M. & Richardson, D.M. (1966) What attributes make some plant species more invasive? *Ecology* 77, 1655–1661.

Richardson, D.M. & Rejmanek, M. (2004) Conifers as invasive aliens: a global survey and predictive framework. *Diversity and Distributions* 10, 321–331.

Sol, D. & Lefebvre, L. (2000) Behavioral flexibility predicts invasion success in birds introduced to New Zealand. *Oikos* 90, 599–605.

Vesk, P.A., Leishman, M.R. & Westoby, M. (2003) Simple traits do not predict grazing response in Australian dry shrublands and woodlands. *Journal of Applied Ecology* 41, 22–31.

Williamson, M. (1999) Invasions. *Ecography* 22, 5–12.

Wittenberg, R. & Cock, M.J.W. (2001) *Invasive Alien Species: a toolkit of best prevention and management practices*. CAB International, Oxford.

4 Dispersal, migration and management

All organisms move, whether as seeds, by colonial growth or the active relocation of animals. But they vary in the distance, directionality and function of their movements. Plans for conservation areas or marine reserves, and for animal recovery or vegetation restoration, may be doomed to failure in the absence of an understanding of dispersal and migration behavior. Equally, biosecurity strategies against potential invaders must recognize patterns of movement along the world's trade and tourist routes. Less obvious is the need to take animal movement into account when assessing the environmental sustainability of industry.

Chapter contents

4.1 Introduction – why species mobility matters 82
4.2 Migration and dispersal – lessons for conservation 84
    4.2.1 For whom the bell tolls – the surprising story of a South American bird 84
    4.2.2 The ups and downs of panda conservation 85
    4.2.3 Dispersal of a vulnerable aquatic insect – a damsel in distress 86
    4.2.4 Designing marine reserves 88
4.3 Restoration and species mobility 89
    4.3.1 Behavior management 89
    4.3.2 Bog restoration – is assisted migration needed for peat's sake? 89
    4.3.3 Wetland forest restoration 91
4.4 Predicting the arrival and spread of invaders 92
    4.4.1 The Great Lakes – a great place for invaders 92
    4.4.2 Lakes as infectious agents 94
    4.4.3 Invasion hubs or diffusive spread? 95
    4.4.4 How to manage invasions under globalization 96
4.5 Species mobility and management of production landscapes 97
    4.5.1 Squirrels – axeman spare that tree 97
    4.5.2 Bats – axeman cut that track 97
    4.5.3 Farming the wind – the spatial risk of pulverizing birds 100
    4.5.4 Bee business – pollination services of native bees depend on dispersal distance ... 103

> **Key concepts**
> In this chapter you will
>
> recognize the distinction between dispersal – the spreading of individuals away from each other – and the mass directional movement of migration
> appreciate that the mobility traits of many species link diverse geographical areas that may be beyond the scope of any single nature reserve

> understand that marine reserves need to be big enough for replenishment by local dispersal of propagules and/or close enough to neighboring reserves for propagule dispersal among them
>
> realize that the necessary steps to vegetation restoration vary according to the dispersal powers of key species and whether a seed-bank is still present in the soil
>
> note that countless invasive plants, animals and microorganisms travel the world's trade and tourist routes on people and their goods
>
> appreciate why sustainable power generation, forestry and agriculture may depend on our understanding of animal dispersal and migration

4.1 Introduction – why species mobility matters

Maria is a businesswoman who wants to turn wind power into electricity. She's angry. *'You'd think with all the concern about global climate change that the public would be keen to see wind farms developed. Wind power promises a much cleaner and, just as important, more sustainable supply of electricity than the fossil-fuel power stations that spew out carbon dioxide.'* Indeed, this is why some international wildlife organizations are in favor of wind power development. But there are plenty of people who find something to object to. Maria comes from the Netherlands, famous for its historic landscape of windmills, and finds the notion of wind farms rather romantic, but others think a horizon dominated by wind turbines will be ugly. And she points out: *'People living in the area we want to develop are suffering from a bad case of NIMBY syndrome (Not in My Back Yard!).'* Then again, local conservationists are worried about migration routes and endangered birds getting chewed up by the turbines. Maria's not a happy woman.

A useful distinction can be made between dispersal (the spreading of individuals away from each other) and migration (a mass directional movement – Box 4.1). Local bird populations, with their daily dispersal movements, as well as wider-ranging species that follow set migration routes may both be at some risk from wind farms. But movement is not just critical to birds. Even an oak tree (*Quercus robur*) is where it is because an acorn was moved to that location before it germinated. On a very different scale, an arctic tern (*Sterna paradisaea*) migrates from its breeding site in the Arctic to the Antarctic and back again each year. Movement, then, ranges from passive to active, short to long range, and from random to purposeful. Maria will have to wait until Section 4.5 to read about the significance of understanding movements of animals and plants for the management of power generation (as well as other production industries – forestry, agriculture). First, I will deal with the application of knowledge about dispersion and migration to problems of conservation (Section 4.2), restoration (Section 4.3) and invasion control (Section 4.4).

I have already discussed species traits in Chapter 3. The timing, location and distance involved in dispersal and migration vary from species to species and are themselves species traits. Individuals of some migratory species are more likely to go extinct because of a complex need for different types of habitat at different times in their life cycle. Similarly, some species are more likely to arrive under their own steam in restoration projects, or as unwanted invaders, because of the details of their dispersal or migration traits. This suite of traits deserves the prominence of a chapter of its own because of the particular significance of species mobility in applied ecology.

Box 4.1 Dispersal and migration – the conceptual framework

Dispersal

Dispersal is a spreading of individuals away from others, such as plant seeds or crab larvae moving away from each other and their parents, or squirrels shifting from one area of a forest to another, leaving residents behind and possibly with a reverse flow of squirrels in the other direction, or bees buzzing from one flower patch to another in an 'archipelago' of flower patches.

Dispersal can be active or passive. The seeds of many plants, whether wind-dispersed or explosively ejected from seedpods, are passively dispersed in air currents and most come to earth quite close to their parents. Whether they fall on stony ground, or in more hospitable locations, is simply a matter of chance. Passive dispersal can also occur by means of an active agent, probably with an increased chance of arriving somewhere desirable. Thus, seeds may have spines that increase the probability of dispersal in the coats of animals, and fleshy fruits attract bird dispersers, which eat them and later deposit viable seeds in their feces.

Some species can also disperse by clonal growth. The bracken fern (*Pteridium aquilinum*), for example, spreads by means of underground rhizomes – an individual eventually fragments into independent plants that continue their clonal spread. Note that the world's major aquatic weed invaders multiply as clones but fragment and fall to pieces as they grow.

Young spiders, on gossamer threads, are passively dispersed in air currents like seeds, and when the wind is blowing hard enough many insects and even birds may be passively dispersed to new locations. However, there is often an element of active control in these cases. For example, aphids may have no say about prevailing wind direction, but they can control when they take off and, to some extent, where they land, increasing the chances of shifting to somewhere better from a location that has become unfavorable. Like animal-dispersed seeds, many mites hitch a ride from dung pat to dung pat, or from one piece of carrion to another, by climbing onto dung or carrion beetles.

Of course, most animals actively explore, visiting different locations before returning to the most favored.

Migration

Migration, in contrast to dispersal, is a mass directional movement of large numbers of a species from one location to another. The term applies to the classic long-distance migrations of arctic terns, but also to the to-and-fro movements of shore animals following the tidal cycle.

Individuals of many species move from one habitat to another and back again repeatedly during their life, on a timescale of hours, days, months or years. Crabs move with the advance and retreat of the tides to maintain themselves in essentially the same habitat. Some planktonic algae and animals make daily movements between very different habitats – hiding from predators in deep, dark water during the day but migrating to resource-rich surface waters at night. Other species track feeding habitats that vary seasonally: giant pandas (*Ailuropoda melanoleuca*), for example, move up in elevation each summer. Then there are the seasonal migrations of frogs, toads and newts between aquatic breeding habitat in spring and the adult's terrestrial feeding habitat for the remainder of the year.

The most dramatic migrations, however, are those that involve extremely long distances. These are triggered by an external seasonal phenomenon like changing day length and are often preceded by profound physiological changes such as accumulation of body fat. Many birds migrate from temperate to tropical locations during the winter, moving between habitats that experience alternating periods of glut and famine – taking advantage, it need hardly be said, of the periods of glut in each. Similarly, baleen whales of the Southern Hemisphere move south in summer to feed in the food-rich waters of the Antarctic but return north to breed in tropical and subtropical waters. The arctic tern goes one stage further, alternating between consecutive summers near the north (midsummer in June) and south poles (midsummer in December). In all these cases an individual may make the return journey several times. But some long-distance migrants have just one return ticket. Eels (*Anguilla* spp.) breed in the sea and move into freshwater to spend most of their lives, while pacific salmon (*Oncorhynchus nerka*) do it the other way round. Red admiral butterflies (*Vanessa atalanta*), on the other hand, have a strictly one-way ticket – the species breeds at both ends of its migration – in summer in Great Britain and in autumn in the Mediterranean region – but each adult flies to and breeds in only one of the locations. The American monarch butterfly (*Danaus plexippus*), already discussed in Section 2.3.1, also holds a one-way ticket.

Dormancy – migration through time

Habitats are not always favorable and, by moving at the appropriate time, migrant animals can take advantage of conditions and resources that suit them. This option is not available to plants. However, many plants (and some animals) achieve the same ends by means of time travel, spending a period of delay in dormant form, most often as a seed (or an egg). The time travel afforded by seasonal dormancy is very common, but there are also cases of dormancy for much longer periods, analogous to long-distance migration in space. You have already come across the phenomenon of seed-banks in the soil (Box 3.1; Section 3.2.1). The species whose seeds persist for longest in the soil are often short-lived opportunists waiting for a favorable opening for germination that may take years or decades to arrive – the same species often lack features that would disperse them far through space.

4.2 Migration and dispersal – lessons for conservation

It is no easy matter deciding the best place for a nature reserve. But the task is doubly difficult when the species to be protected moves during its lifetime through a variety of habitats that are more or less remote from each other. Here I describe the cases of a tropical bird, a charismatic mammal and a diminutive stream insect (Sections 4.2.1–4.2.3) before considering the related matter of how best to design marine reserves in light of the long-distance dispersal behavior of many marine species (Section 4.2.4).

4.2.1 *For whom the bell tolls – the surprising story of a South American bird*

Although the three-wattled bellbird (*Procnias tricarunculata*) is one of Costa Rica's largest fruit-eating species, little was known about its migratory behavior and habitat requirements. Costa Rica has an excellent record of setting aside protected natural areas, but this has not always been done on the basis of detailed ecological knowledge, and Powell and Bjork (2004) wanted to find out whether the nation's nature reserves can adequately protect this species. A number of birds were captured in the Tilaran Mountain Range, close to the Monteverde Nature Reserve (Figure 4.1a), tagged with small radio transmitters and their locations determined successively through the year.

The birds breed between March and June at mid-elevations (1000–1800 m) in moist forest on the Atlantic slopes of the mountains before their first migratory move (step 1) westward over the continental divide to spend several months feeding on forest fruits at similar elevations but on the mountains' Pacific slopes (Figure 4.1a). These western forest habitats occur mostly in small, isolated fragments on private farms, and are poorly represented in Costa Rica's system of protected areas. During September and October the bellbirds migrate eastward (step 2) to lowland Atlantic forest in Costa Rica and Nicaragua, habitats that are also under threat. Next they move south (step 3) to heavily modified and fragmented forest, mainly on private property, along the Pacific coast of Costa Rica (Figure 4.1b). In March they return north (step 4) to their breeding area.

The complex nature of this bird's migratory behavior poses some big challenges for managers. While breeding habitat seems to be sufficiently protected, the bellbird's situation remains precarious because of habitat decline in three other migration destinations whose importance was unknown before this study. Perhaps the species will show some flexibility in habitat use as prime areas are fragmented and destroyed, but this cannot be taken for granted. The bellbird is just one among many species whose ecology links diverse geographical areas that are beyond the scope of

Fig. 4.1 (a) Locations of radio-tagged bellbirds during the breeding season (circles) (March–June), primarily on the Atlantic slope of the Tilaran mountain range in north-central Costa Rica (white area), and post-breeding season (triangles) (July–September), primarily on the Pacific slope of the mountain range (shaded area). The square in (b) outlines the whole of the area shown in (a). (b) Locations of radio-tagged bellbirds during September–November (circles) and December–March (triangles). The arrows in (a) and (b) show the four annual migration steps. (After Powell & Bjork, 2004.)

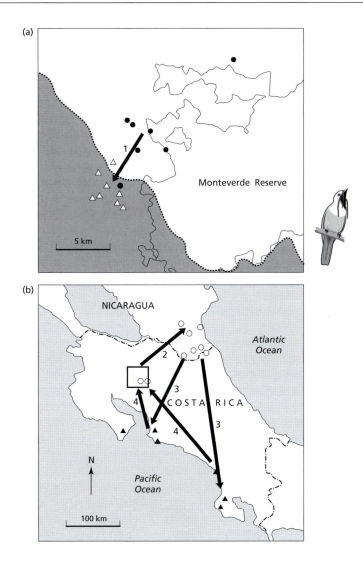

any single nature reserve. Powell and Bjork conclude that protected areas should not be considered in isolation but as pieces of an integrated system.

4.2.2 The ups and downs of panda conservation

The Qinling Province in China is home to approximately 220 giant pandas, representing about 20% of the wild population of one of the world's most imperilled mammals. The pandas in this region are elevational migrants – in other words, the population needs both low and high elevation habitats to persist. This is something that existing nature reserves did not cater for.

Pandas are extreme dietary specialists, primarily consuming a few species of bamboo. From June to September the pandas eat *Fargesia spathacea*, which grows from 1900 to 3000 m. But as colder weather sets in, they travel to lower elevations and between October and May feed primarily on *Bashania fargesii*, a bamboo species that grows from 1000 to 2100 m. Loucks et al. (2003) used a combination of satellite imagery and field work to identify a landscape to meet the long-term needs of the species. The protocol for selecting potential habitat first excluded areas lacking giant pandas, forest block areas that were smaller than 30 km² (the minimum area needed to support a pair of giant pandas over the short term), forest with roads or settlements, and plantation forests. Figure 4.2 maps summer habitat (1900–3000 m; *F. spathacea* present), autumn/winter/spring habitat (1400–2100 m; *B. fargesii* present) and a small amount of year-round habitat (1900–2100 m, both bamboo species present) and identifies four areas of 'core' panda habitat (A–D) that provide for the migrational needs of the pandas.

Superimposed on Figure 4.2 are the nature reserves that existed at the time of this study; disturbingly, they covered only 45% of the core habitat. Loucks et al. (2003) recommended that their four core habitat areas should be incorporated into a reserve network. They also noted the importance of promoting linkage between the zones, because extinction in any one area (and in all combined) is more likely if the populations are isolated from each other. Their plan identified two important linkage zones for protection, between areas A and B where steep topography means few roads exist, and between B and D across high-elevation forests. Most of these recommendations have now been acted upon by the provincial government.

4.2.3 *Dispersal of a vulnerable aquatic insect – a damsel in distress*

The bellbird and panda migrations took advantage of patterns of food availability that vary seasonally, but which recur predictably from year to year. Some animals, on the other hand, have to contend with patterns in the distribution of suitable

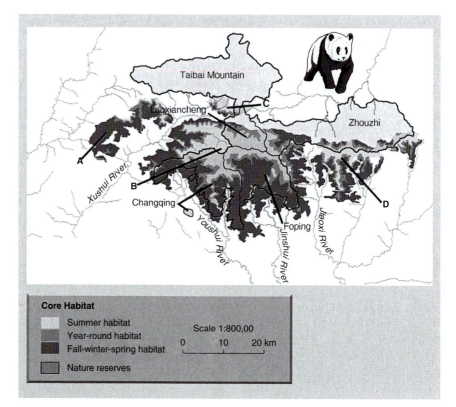

Fig. 4.2 Core panda habitats A, B, C and D, each of which caters for the year-round needs of the elevational migration of giant pandas in China's Qinling Province. Superimposed are current nature reserves (in gray) and their names. (From Begon et al., 2006, after Loucks et al., 2003.)

habitat that are short lived and unpredictable. Take, for example, the southern dam-selfly, *Coenagrion mercuriale*, which has flying adults but aquatic eggs and larvae. The species is restricted in Britain to the highly fragmented habitat of small streams and ditches associated with ancient water meadows that have long been managed by periodic grazing or cutting (Purse et al., 2003). Within these habitats, the aquatic larvae are further restricted to unshaded flowing water with perennial herbaceous vegetation. Such sites have recently been grazed or cut but will alter with time (through a process called ecological succession – Chapter 8) to a shady state unfavorable to the damselflies. Persistence of the species within the shifting mosaic of favorable habitat patches depends on adult colonization of newly created patches, counterbalancing the loss of subpopulations from patches as they become unsuitable. But to what extent do the poor powers of flight of adults, which only live for 7 or 8 days, limit dispersal to new suitable patches?

Purse's team captured adults in some water meadows in Wales and released them again after marking on the forewing and thorax so they could be individually identified when subsequently recaptured. As in our bellbird example, this allowed the net lifetime movements of individuals to be determined (Figure 4.3) – you can see that the damselflies operate over a markedly smaller range than do the birds. Purse's team concluded that management effort should be focused on locations where a cluster of habitable sites occur within 1–3 km of each other. In other locations, where there are only a few, relatively isolated sites, suitable habitat is so fragmented that reintroduction will not lead to the establishment of stable populations. The authors also suggest that hedges should be removed between any existing population and empty, but suitable sites, to facilitate stepping-stone dispersal movements. This is analogous, but again on a much smaller scale, to the corridors linking core panda habitats.

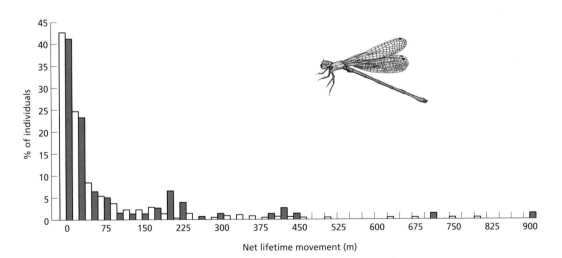

Fig. 4.3 Percentage distribution of net lifetime movements for male (open bars) and female damselflies (closed bars) at a site in Preseli, Wales. Most recaptured individuals dispersed over short distances but a few moved up to 1 km. (After Purse et al., 2003.)

4.2.4 *Designing marine reserves*

In contrast to the adult damselflies in Section 4.2.3, the main dispersal phase of organisms that live on the seabed is an immature stage. Thus, many marine species, including algae, invertebrates and fish, release spores, eggs or larvae that spend time in the water column, moving in ocean currents, before settling and maturing on the bed. Viable populations need a sufficient rate of replenishment by offspring produced by local adults or by distant adults whose 'propagules' (spores, etc.) are brought on ocean currents. When designing marine reserves, self-replenishment can be achieved if a reserve is big enough to contain a substantial amount of local dispersal. Alternatively, in a system of reserves, self-replenishment can occur if neighboring reserves are at appropriate distances for transport of propagules from one to another. In either case, management decisions require knowledge about dispersal distances.

Shanks et al. (2003) reviewed a considerable body of literature to obtain estimates of the time propagules spend in the water column, and the distance they disperse, to provide information of value when designing reserves. Duration in the floating phase ranges from less than 2 minutes to 293 days, and estimated dispersal distance ranges from less than 1 m to 4400 km. For the most part, there is a convincing relationship between dispersal distance and propagule duration (Figure 4.4), but some species travel less far, given their time in the water column, and fall below the regression line in the figure. This applies to spores, eggs and larvae that are negatively buoyant or that swim toward the bottom, spending their time close to the bed where current speeds are lower.

Figure 4.4 reveals an intriguing pattern. There appears to be a bimodal distribution of dispersal distances, most being less than 1 km or more that 20 km. The short-distance dispersers are spores or nonfeeding larvae of a variety of plants and animals. Long-distance species, on the other hand, have propagules that feed while dispersing and require longer periods of larval development before they are ready to settle. The lack of propagules competent to travel between about 1 and 20 km has significant implications for reserve design. Shanks and his team suggest that reserves should be at least 4–6 km in diameter to contain the short-distance dispersing propagules, and should be spaced from 10 to 20 km apart to capture long-distance dispersers released from neighboring reserves.

Fig. 4.4 Dispersal distance (km) plotted against propagule duration in the water column (hours) for a variety of marine plants and animals (note the log scales used). The points labeled A–F (for a coral, a seaweed, a shrimp, a fish, an abalone and a giant kelp, respectively) fall below the regression line, dispersing less far than would be predicted from their propagule duration; these spend their time close to the bed where currents are slower. (After Shanks et al., 2003.)

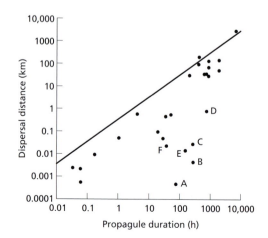

4.3 Restoration and species mobility

Managers sometimes want to restore populations of endangered species to locations where they have gone extinct, or to restore whole communities of plants and animals. Once again, without detailed knowledge of migratory and dispersal behavior, efforts to manage endangered animals may be doomed to failure (Section 4.3.1). And when it comes to plant community restoration, an intimate knowledge of the dispersal ability of seeds (whether through space or time) can be equally critical (Sections 4.3.2, 4.3.3).

4.3.1 Behavior management

Attempts to restore populations of migratory species must, of course, take their migratory behavior carefully into account. Europe's lesser white-fronted goose (*Anser erythropus*) moves between widely separated locations, and Sutherland (1998) tells a fascinating story where knowledge of migratory behavior has proved useful for management. A scheme was devised to alter the migration route of the lesser white-fronted geese from southeastern Europe, where they tend to get shot, to spend their winters in the Netherlands, where hunting is not a problem. A population of captive barnacle geese (*Branta leucopsis*) breeds in Sweden at Stockholm Zoo but overwinters in the Netherlands. Some were taken to the wilds of Swedish Lapland where they nested and were given lesser white-fronted goose eggs to rear. The young geese then flew with their adoptive parents to the Netherlands for the winter, but next spring the lesser white-fronted geese returned to Lapland and bred there, subsequently returning again to the Netherlands. (Saved from the hunters, one wonders whether the geese might meet an equally unpleasant end in one of Maria's wind farm turbines?)

Another example where restoration effort has benefited from an intimate knowledge of animal behavior involves reintroduction of captive-reared *Phascogale tapoatafa*, a small carnivorous marsupial that has disappeared from part of its range in Australia. Soderquist (1994) found that if males and females were released together, the males quickly dispersed from the area and females could not find a mate. Much more successful was a 'ladies first' release scheme; this allowed the females to establish a home range before males came and joined them.

4.3.2 Bog restoration – is assisted migration needed for peat's sake?

Peat is a valuable commodity, for fuel and garden supplies. Many *Sphagnum* peatbogs in eastern Canada, as elsewhere, are exploited for peat for several decades before they are abandoned, leaving flat surfaces up to 5 km² in area that are intersected by dense networks of ditches. A relatively thin organic peat substrate, 1 m or more deep, usually remains, but this is acid and poor in nutrients. There is no persistent seedbank in these highly disturbed sites, so 'migration through time' is not possible (Box 4.1). However, unexploited peatbog habitat will usually be found at the edges of the disturbed areas. Subsequent recolonization of exploited sites from these healthy fragments is very slow, partly because of unfavorable physicochemical conditions, but also because of restricted powers of dispersal of some of the peatbog plants. Campbell et al. (2003) wanted to distinguish between these two causes of poor recovery so that restoration management could be focused for maximum effect.

They assessed the wind-dispersal ability of a wide range of peatbog plants, including mosses, herbaceous plants, shrubs and trees. For each species, three components of wind dispersal were determined for plants growing in the wild, namely 'propagule release height' (those released from greater height are generally exposed to higher wind velocities), 'fall time' (a longer fall time increases the likelihood of being carried away on the wind) and 'wing loading' (mass per unit surface area – predicts horizontal dispersal distance). Note the parallels with marine propagule dispersal discussed in Section 4.2.4. Species with a combination of high release height, long fall time and small wing loading should be the best wind dispersers. With this in mind, Campbell's team calculated a composite 'dispersal ability by wind' as the mean of the three variables, each scaled to values between 0 and 1 (Figure 4.5). They then carried out one more data manipulation to derive an index of 'immigration potential by wind'. This recognizes that immigration potential does not depend on dispersal ability alone, but also on fecundity (the number of propagules produced by adult plants) and the relative abundance of the plants adjacent to the exploited peatbog.

Fig. 4.5 Indexes of wind dispersal ability and of immigration potential for a variety of species of moss, herb, shrub and tree that are associated with peatbog communities in eastern Canada. (After Campbell et al., 2003.)

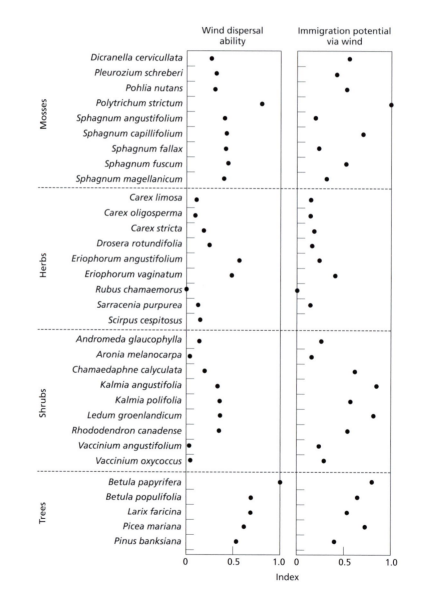

Again they calculated a mean of the three components and scaled between 0 and 1 to provide a relative index (Figure 4.5).

Not all plants have wind-dispersed propagules: some are capable of dispersing by water (assessed experimentally as the proportion of seeds still floating after 72 hours) or via animals (those with fleshy fruits). Once again, immigration potential of wind- and animal-dispersed plants was estimated by also taking into account fecundity and relative plant abundance adjacent to abandoned peat workings. Figure 4.6 plots the percentage occurrence of plant species in abandoned peatbog workings – a measure of their ability to recolonize naturally – against estimated immigration potential for plants in all the dispersal categories.

The recolonization of many vascular plants is quite well predicted by their estimated immigration potential – those with a dispersal index above 0.9 predictably occur in abandoned bogs. For these species, it seems that physicochemical conditions in the abandoned bogs are adequate. On the other hand, mosses have high immigration potential but are very poorly represented, strongly suggesting that adverse physicochemical conditions are limiting their successful re-establishment. Put another way, the physicochemical conditions of the abandoned peatbog workings encompass the fundamental niches of many of the vascular plants but not of the mosses (Box 2.1 defines fundamental niche). Campbell's team concludes that restoration efforts should concentrate on actively introducing species with low immigration potential. In addition, microenvironments should be created that are suitable for the establishment of the mosses and other species with relatively high immigration potential but poor representation in abandoned bogs.

4.3.3 *Wetland forest restoration*

Next I turn to a case of restoration that contrasts with the peatbogs because extensive seed-banks remain and many species (but not all) can recolonize by means of 'time travel'.

A very large proportion of the world's forested wetlands were lost when people drained and converted them for agriculture. Ironically, however, market changes have led to abandonment of vast areas of the wettest of this agricultural land, especially in Europe and North America. In the southeastern USA, for example, 89,000 hectares of floodplain land is due for reforestation in the early years of this century (King & Keeland, 1999). Restoration faces two major problems. The first is that some

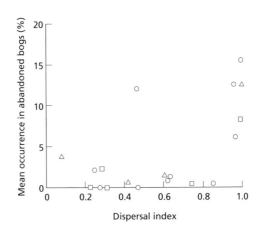

Fig. 4.6 Relationship between mean occurrence of study species in abandoned peatbog workings in eastern Canada and their immigration potential by wind (circles), water (squares) or animal dispersal (triangles). (After Campbell et al., 2003.)

species that play a critical role in the functioning of wetland forest ecosystems have restricted powers of recolonization. Just as important, the nature of these systems depends on floods that periodically inundate the land. These were engineered away to permit agricultural production. Now, human settlement in the floodplains limits the opportunity to re-engineer the appropriate flood dynamics.

Middleton (2003) set out to examine the natural restoration potential of farmland that had originally been baldy cypress swamp (*Taxodium distichum*) along the Cache River in Illinois, USA. She compared the seed-banks of 51 ex-swamps that had been farmed for up to 50 years with those from nine intact swamps, collecting soil samples for germination under laboratory conditions. Herbaceous species proved to be common in the seed-banks of both farmed (207 species) and intact (173 species) swamps. On the other hand, neither farmed nor intact sites were well represented by seeds of woody species, including those of the baldy cypress itself.

Baldy cypress seeds are short lived and are naturally dispersed to seed-banks by water during a flood (after winter rain or snowmelt), germinating once the flood-waters recede. Middleton argues that the re-establishment and maintenance of the dominant tree species will be best served by hydrological engineering (e.g. removing levees and restoring the linkage between the river and its floodplains). This will re-establish the supply of tree seeds to seed-banks. Although it is good news that many swamp herbs remain abundant in the seed-bank, not all will prosper unless the hydrological regime is restored to regain the necessary physicochemical conditions. Moreover, some herbaceous species rely on vegetative organs for dispersal (e.g. *Heteranthera dubia* and *Lemna minor*) and are absent from seed-banks: these will require active reintroduction. Middleton looks forward to the possibility of restoration of a natural flood regime in newly acquired conservation land in the Cache River floodplain.

4.4 Predicting the arrival and spread of invaders

Human travel and commerce have opened up novel 'dispersal' routes for thousands of invaders. Just like the seeds and mites that hitch a ride on animal dispersers, countless plants, animals and microorganisms travel the world's trade and tourist routes on people and their goods.

The best way to deal with invaders is to prevent their arrival in the first place and this involves identifying these novel 'dispersal' pathways. If we know the major invader routes, such as hitchhiking in mail or cargo, on aircraft and in ships, we can plan to manage the associated risks (Wittenberg & Cock, 2001). Biosecurity precautions are now routine in many parts of the world, including the screening and cleaning of cargo at the point of embarkation and on arrival, and inspection of camping gear and boots of incoming passengers. As noted before (Section 3.3.3), however, it is not just the likelihood of invader arrival that must be attended to, but also the probability of spread from the point of entry. The Great Lakes of North America provide an excellent case study of both biosecurity to prevent invasion and the prediction of subsequent spread.

4.4.1 The Great Lakes – a great place for invaders

The Great Lakes of North America have been invaded by a motley collection of more than 145 alien species, including fish, mussels, amphipods, cladocerans and snails. Many of these arrived in ships' ballast water taken aboard at the other end of an important trade route in the Black and Caspian Seas (Ricciardi & MacIsaac, 2000). An ocean freighter before taking on cargo in the Great Lakes might discharge

3 million liters of Caspian ballast water, containing a diversity of plants and animals (even the cholera bacterium *Vibrio cholerae* has been found in ballast water). One management solution is to make it compulsory (rather than voluntary) to dump ballast water in the open ocean – this is now the case for the Great Lakes. Other possible methods involve filter systems when loading ballast water, and on-board treatment by ultraviolet irradiation or waste heat from the ship's engines.

The most damaging invaders are not simply those that arrive in a new part of the world – the subsequent pattern and speed of their spread also matters. Zebra mussels (*Dreissena polymorpha*) have had a devastating effect (Section 1.2.5) since arriving in North America via the Caspian Sea/Great Lakes trade route. The expansion of their range occurred quickly through all commercially navigable waters, but overland dispersal into inland lakes, mainly attached to recreational boats, has been much slower (Kraft & Johnson, 2000). Geographers have developed a method to predict human dispersal patterns based on distance to and attractiveness of destination points, and Bossenbroek et al. (2001) adopted their modeling approach to predict the spread of zebra mussels through the inland lakes of Illinois, Indiana, Michigan and Wisconsin (364 counties in all). The model, which we need not consider in detail, has three steps dealing respectively with the probability of a boat traveling to a zebra mussel source and inadvertently picking up mussels, the probability of the same boat making a subsequent outing to an uncolonized lake, and the probability of zebra mussels becoming established in the uncolonized lake.

Figure 4.7a shows the number of lakes colonized by zebra mussels in each county as predicted by the model. This can be a fraction of one, because of the probabilistic

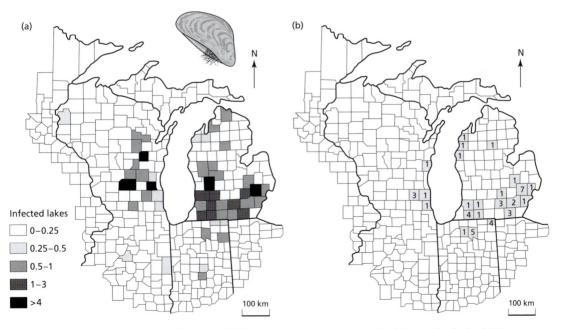

Fig. 4.7 (a) The predicted distribution (based on 2000 iterations of a computer model of dispersal) of inland lakes colonized by zebra mussels in 364 counties in the USA; the large lake in the middle is Lake Michigan, one of the Great Lakes of North America. (b) The actual distribution of colonized lakes as of 1997. The numbers refer to the number of lakes in each county predicted to be invaded (a) or actually invaded (b). (After Bossenbroek et al., 2001.)

nature of the predictions, while the maximum number colonized per county ranged up to four or more. The predictions proved to be highly correlated with the pattern of colonization that actually occurred until 1997, indicating that the model is realistic. Note that parts of central Wisconsin and western Michigan were expected to be colonized, but no mussel colonies were documented. Bossenbroek's team suggest that invasion may be imminent in these locations, which should therefore be the focus of biosecurity efforts and education campaigns.

4.4.2 *Lakes as infectious agents*

Each new lake that is colonized by zebra mussels represents a separate invasion from a neighboring lake, analogous to the original invasion of North America from Eurasia but on a smaller scale. Muirhead and MacIsaac (2005) were interested to know whether all lakes in the landscape possess equal potency as sources for invasion of other lakes. Or are some created more equal than others in their capacity to infect? Lakes in a landscape have the characteristics of a network, not unlike the human brain or the internet. The question is: are some lakes likely to serve as invasion hubs, and others as dead ends?

The spiny waterflea, *Bythotrephes longimanus*, was discovered in Lake Ontario in 1982, from where it dispersed among the interconnected Great Lakes as well as to adjacent inland lakes. The waterflea produces highly resistant resting eggs that become entangled in fishing gear, resist the desiccation associated with road-travel of boats between lakes and subsequently detach and hatch in new lakes (MacIsaac et al., 2004). Muirhead and MacIsaac surveyed recreational boaters at boat ramps and marinas to determine which invaded and uninvaded lakes were visited by boaters departing from particular invaded lakes. It turns out that most invaded lakes connect to only a few uninvaded counterparts (0–5), while some lakes connect to many others (16–25) and may act as invasion hubs (Figure 4.8).

Fig. 4.8 Frequency histogram of the number of links between invaded source lakes and invaded and uninvaded destination lakes from recreational boater surveys in 2003. It is the links to uninvaded lakes that really matter in terms of invasive spread through the landscape. (After Muirhead & MacIsaac, 2005.)

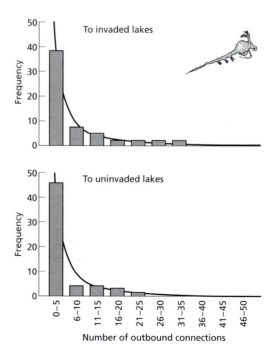

Lake Muskoka was the first inland lake to be invaded by the waterflea as a result of regular boat traffic from Lake Huron, one of the Great Lakes (Figure 4.9). It quickly developed into a regional hub for invader spread for two reasons. First, all of its outbound traffic to inland lakes was to uninvaded lakes and, second, the total amount of traffic leaving was high (1452 people towed boats on trailers from the lake each year). As lakes become invaded, the number of uninvaded lakes decreases while the vulnerability of remaining uninvaded lakes changes. Thus, Lake Muskoka is unlikely to continue to serve as a hub because most of its outbound traffic is now to other lakes that have already been invaded. On the other hand, the heavy outbound traffic from Lakes Simcoe (3774 annual movements) and Kashagawigamog (1840 movements) is predominantly to uninvaded lakes. These two lakes are developing into the most important invasion hubs (Figure 4.9). Lake Panache, which is located in northern Ontario away from the current cluster of invaded lakes, supports a large recreational fishery and Muirhead and MacIsaac's study suggests it may well be an important hub of the future. Thousands of lakes are still uninvaded and managers need to know how the network functions so they can focus effort in the most effective way to educate and/or regulate boaters.

You may be wondering whether a tiny waterflea is worth all the bother. Certainly, it does not cause such economic or ecological harm as the zebra mussel. Its importance here lies in an ability to act as a general indicator of the means and pattern of spread of aquatic organisms from lake to lake.

4.4.3 *Invasion hubs or diffusive spread?*

Lakes in a landscape are obvious examples of 'discontinuous' habitat. It is because assorted lakes are differentially linked into the network that invasion hubs can be identified for special management treatment. Other patchily arranged habitats – such as oceanic islands, mountain ranges and forest remnants – can be expected to

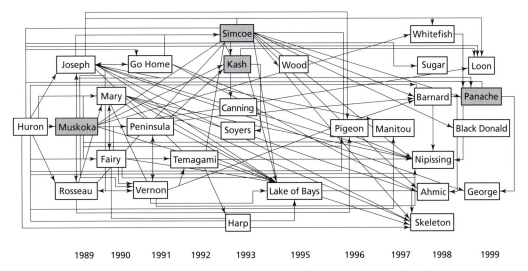

Fig. 4.9 Diagram of network traffic from previously to recently invaded lakes based on records of first reporting. Stages in the development of the ever-increasing invasion network are indicated by the year in which lakes were invaded (from left to right). Shaded boxes represent existing or developing network hubs. Thousands of lakes (not shown) are so far uninvaded. (Lake Kashagawigamog is abbreviated as Kash.) (After Muirhead & MacIsaac, 2005.)

function in a similar way, unless the organisms' powers of dispersal are so strong that the habitats are functionally continuous. Thus, the timing and pattern of invasive spread depends on both the geography of the discontinuous habitat and an organism's dispersal potential.

Then again there are habitats that are 'continuous', including large expanses of grassland (prairies, steppes, pampas, agricultural land) or forest (rainforest, boreal forest, plantations). In these cases the spread of invaders can be envisaged as a progressive wave of diffusion across the landscape, at a rate that depends on the species' intrinsic rate of population growth and its 'coefficient of diffusion'. A high coefficient of diffusion may be associated with the ability to hitch a ride with humans or other animals or on the wind, or the capacity to walk or fly quickly through the landscape.

The red fire ant (*Solenopsis invicta*) has spread rapidly through much of the southern USA with dramatic economic consequences (Section 1.2.5). The species, which originated in Argentina, occurs in two distinct social forms. The single-queened (monogyne) form and the multiple-queened (polygyne) form differ in patterns of reproduction and modes of dispersal. Queens from monogyne colonies take part in mating flights and create new colonies, whereas queens from polygyne colonies are adopted into already existing nests after mating. As a result, the monogyne populations spread a thousand times faster than their polygyne counterparts (Holway & Suarez, 1999). The ability of managers to predict potentially problematic invaders and devise strategies to counter their spread depends on a thorough understanding of dispersal behavior.

4.4.4 How to manage invasions under globalization

We now have some power to predict the species that pose the greatest risk as invaders, and know the major invasion routes; national biosecurity strategies are based on this information. However, as Perrings et al. (2005) point out, protecting national borders is difficult because those whose actions result in invasions usually bear no legal responsibility and do not have to pay the costs associated with invaders. In economic terms, invasion costs are an 'externality' of global trade – an unintended side effect whose cost is not reflected in the market price of the goods that pose the risk. Invasions are a form of biological pollution; the same externality problem occurs if the costs of chemical pollution are borne by society in general rather than by those who pollute. It is now more generally accepted that polluters should pay for the damage they cause – the 'polluter pays principle'. This is an example where what was an externality (cost of pollution) becomes internalized (pollution costs reflected in the market price of the industrial products). Can the same thing be done for invasion costs?

You saw in Section 3.3.3 that the probability of a successful invasion by a parrot species was positively correlated with the development of the international parrot market. There are many other examples among both animals and plants. What is needed, according to Perrings' team, is a measure that confronts exporters with the costs of their actions – the introduction of invasion risk-related tariffs. Import tariffs can be expected to reduce export activities that are particularly risky, which is desirable, but such tariffs could disproportionately hurt the economies of poor countries. Thus, the team suggests that tariffs should be coupled with international financial support for low-income countries that adopt biosecurity-enhancing measures and, thus, confer a worldwide benefit.

4.5 Species mobility and management of production landscapes

Human industry and resource exploitation inevitably have consequences for natural ecosystems. Given the undeniable need to exploit forest for timber (Sections 4.5.1, 4.5.2), to generate energy (Section 4.5.3), and to grow agricultural crops (Section 4.5.4), the aim of resource managers is twofold. First, to designate areas where this might be done with minimal consequences for the natural world and, second, to ensure that production landscapes are managed in a sustainable manner. I will come to the question of the zoning of production and conservation across landscapes and waterscapes in Chapter 10, because this is the domain of community and ecosystem ecology. Here, I will confront the question of sustaining individual species in exploited landscapes, with continuing emphasis on the relevance of species mobility for conservation.

4.5.1 Squirrels – axeman spare that tree

The flying squirrel *Pteromys volans* in Finland has declined dramatically because of habitat loss, habitat fragmentation and reduced habitat connectivity caused by intensive forestry. This nocturnal squirrel favors spruce-dominated forest (*Picea abies*) with a marked component of deciduous trees such as aspen (*Populus tremula*), birch (*Betula* spp.) and alder (*Alnus* spp.). Aspens are probably the most important of these, providing both food for the squirrels and shelter in the form of woodpecker cavities. Areas of natural forest, critical to the squirrels, are now separated by clear-cut and regenerating areas.

The core breeding habitat of a flying squirrel only occupies a few hectares, but individuals, particularly males, move to and from this core for temporary stays in a much larger 'dispersal' area (1–3 km^2), and juveniles permanently disperse away from their parents within this range. Reunanen et al. (2000) compared the landscape structure around known flying squirrel home ranges (63 sites) with randomly chosen areas (96 sites) to determine the forest patterns that the squirrels need. They first established that the landscape could be divided into optimal breeding habitat (mixed spruce-deciduous forests), dispersal habitat (pine (*Pinus sylvestris*) and young forests) and unsuitable habitat (young sapling stands and open habitats). Figure 4.10 shows the spatial arrangement of breeding and dispersal habitat for a typical flying squirrel site and a random forest site. Overall, flying squirrel landscapes contain three times more suitable breeding habitat and 23% more dispersal habitat than random landscapes. But, most notably, squirrel dispersal habitat is much better connected (fewer fragments per unit area) than random landscapes.

Reunanen and his colleagues recommend that forest managers should restore and maintain a deciduous mix of trees for optimal breeding habitat. And of particular significance in the context of dispersal behavior, they need to ensure good physical connectivity between optimal breeding and dispersal habitat.

4.5.2 Bats – axeman cut that track

In stark contrast to the flying squirrels, bats can sometimes be favored by certain forestry practices! Bats vary in wing shape and echolocation calls in a way that influences their operation in different habitats. For example, those with long thin wings, which make them less maneuverable, or with calls of relatively low frequency, find it difficult to negotiate cluttered habitats. As vegetation regenerates after logging, stem density and vegetation clutter (structural complexity) increase. With this in mind, Law and Chidel (2002) investigated the effects of logging (15 years later) in a *Eucalyptus* forest in New South Wales, Australia. Prior to logging, the tall, wet forest was dominated by two tree species, Sydney blue gum (*Eucalyptus saligna*)

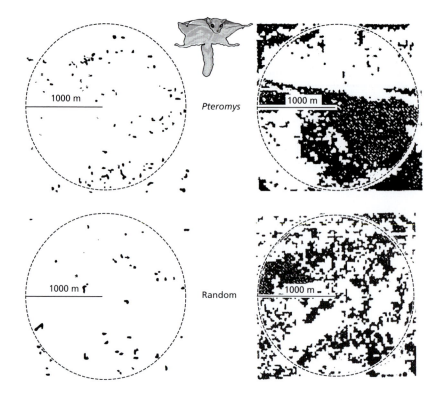

Fig. 4.10 The spatial arrangement of patches (dark) of breeding habitat (left hand panels) and breeding plus dispersal habitat (right hand panels) in a typical landscape patch containing flying squirrels (top panels) and a random forest location (bottom panels). The flying squirrel patch contains 4% breeding habitat and 52.4% breeding plus dispersal habitat, compared with 1.5% and 41.5% for the random landscape. Dispersal habitat in the squirrel landscape is much more highly connected than in the random landscape. (After Reunanen et al., 2000.)

and silvertop stringybark (*Eucalyptus laevopinea*). By recording bats using ultrasonic detectors, the researchers were able to document the number of passes made by a variety of species that could be separately identified from their calls. They compared bat activity both in logged and unlogged forest patches, and in different locations in each case – along forest tracks, off track in mid-forest and along unlogged 'riparian' margins beside forest streams.

Surprisingly there was no statistically significant difference in bat activity between unlogged and logged locations. What really mattered was the presence of forest tracks as dispersal pathways for feeding activity. In both logged and unlogged situations, much higher activity was recorded on forest tracks (183 and 196 bat movements per night in logged and unlogged forest respectively) than off track (5 and 36 movements per night). The stream riparian areas, which like forest tracks provide linear pathways for the bats, showed intermediate activity (55 and 26 movements per night). Bat species richness was actually highest on forest tracks in the logged areas.

The least maneuverable bats would be predicted to do well in the uncluttered flight paths provided by tracks. This was indeed the case (Figure 4.11a–d), but in addition the highly maneuverable *Rhinolophus megaphyllus* made use of the pathways (Figure 4.11e), perhaps because insect prey density was higher there, or because of enhanced hunting success or simply as a 'commuter' route.

An index of clutter was calculated for every recording site to integrate percentage vegetation cover in each of four strata – groundcover, shrub, understorey trees (including *Eucalyptus* regrowth in the logged sites) and large eucalypts emerging into the upper canopy. Overall bat activity declined significantly with the increasing

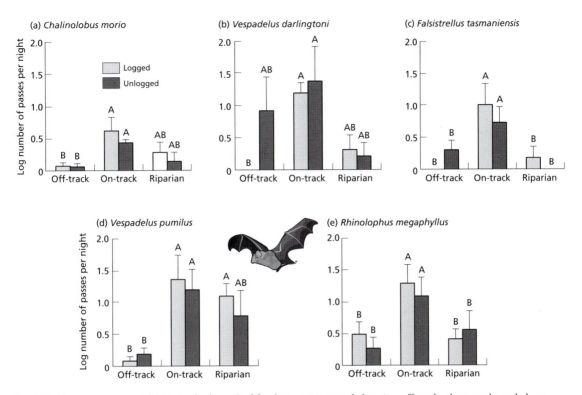

Fig. 4.11 Mean counts per night (+ standard error) of five bat species recorded at sites off track, along tracks and along stream riparian corridors. Histograms that do not share the same letter are significantly different. (a)–(d) Species considered to be clutter-sensitive because of their morphology and behavior; (e) a clutter-tolerant species. (After Law & Chidel, 2002.)

Fig. 4.12 Relationship between total number of recorded bat movements and an index of forest clutter for all logged and unlogged sites combined. (After Law & Chidel, 2002.)

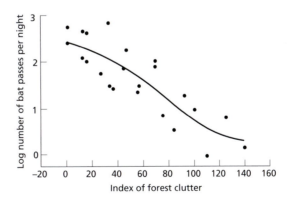

clutter provided by regrowth in logged forest and the amount of understorey euca-lypts in unlogged forest (Figure 4.12). This indicates that low bat activity away from tracks and stream corridors was related to clutter.

While the opening up of forest tracks has clear benefits as feeding dispersal routes for many species of bats, and the bat community recovers well within 15 years of logging, it would be unsafe to conclude that logging is a blessing in disguise. Bird diversity, for example, takes 30–50 years to recover after logging (Williams et al., 2001). And even for the bats, essential maternity roosts occur in the hollows of large mature trees that would take more than 100 years to grow. Thus, sustainable forestry practice in Australia requires that five such habitat trees are retained per hectare. The principles of sustainable harvesting will be considered in more detail in Chapter 7.

4.5.3 Farming the wind – the spatial risk of pulverizing birds

For many species, conservation involves leaving important structures in place, such as trees of particular species (flying squirrels, Section 4.5.1) or age (bats, Section 4.5.2). With wind farms, on the other hand, it is similarly large structures that are the problem.

Wind farms can be built on land or at sea, each posing its own threats to migrating and dispersing birds. On land, soaring birds such as falcons and vultures are at particular risk of colliding with the turbines (up to 100 m above the ground), particularly because the engineers often select their locations for the same wind-related reasons that birds select their routes. Many wind farms are also planned for marine settings – in Europe, for example, more than a hundred applications have been submitted (Garthe & Huppop, 2004). Each may consist of as many as 1000 turbines, up to 150 m tall, as far offshore as 100 km and in water as deep as 40 m. The turbines may pose risks to migrating birds (from the smallest of songbirds to cranes and birds of prey) as well as seabirds dispersing locally to find food.

What information is needed before we can get to grips with the risks that wind farms pose for wildlife? First, there is a need for behavioral information about migration routes, daily and seasonal dispersal patterns, flying height and maneuverability. However, the vulnerability of individual birds does not necessarily translate into a significant risk of population decline or extinction. To assess this, information is also needed about the population size and conservation status of each species. The race is on to expand the wind-power sector, often in the absence of much detailed ecological information. Next I describe two studies, one on land and the other at sea, that begin to build the necessary database.

Two land-based wind farms have already been installed in Gibraltar close to what has been described as an important migration 'bottleneck' where many soaring birds funnel through the Straits of Gibraltar en route between Africa and Europe. Both farms consist of rows of turbines along the north–south ridges of mountains and hills where they can most effectively harness the prevailing east–west winds (Figure 4.13). The first, along the ridge of the Sierra de Enmedio, has 66 turbines in two

Fig. 4.13 Location of two wind farms in the Campo de Gibraltar region of the Iberian Peninsula (inset). The contours are at 100-m intervals. (After Barrios & Rodriguez, 2004.)

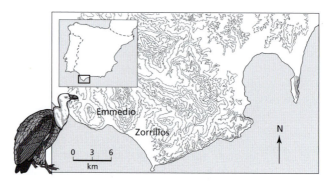

rows at an altitude of 420–550 m above sea level and with steep slopes below – the vegetation consists of dense shrubland less than 1 m in height. The other, with 190 turbines in seven rows, is at lower altitude (80–300 m) in the less steeply sloping forested landscape of the Dehesa de los Zorrillos. The rotors of both, which are 18–23 m in diameter on towers 21–36 m high, operate at wind speeds between 4.5 and 24 m s^{-1}.

Barrios and Rodriguez (2004) removed old bird carcasses before commencing twice-weekly surveys of randomly selected turbines to retrieve all birds killed by collision. They chose this frequency because small bird carcasses were likely to be scavenged if left for longer; the carcasses of larger birds persisted for months. Observations were also made of the timing and location of near misses when birds were either flying straight, at constant altitude, or circling to gain height using the lift generated on windward slopes.

During the 1-year study, 68 birds died as a result of collisions with the turbines under investigation. Griffon vultures (*Gyps fulvus* – 30 individuals of all ages) and kestrels (*Falco tinnunculus* – 12 juvenile individuals) were most frequently killed. Griffon vultures were more likely to die in autumn/winter. These birds need vertical air currents to gain height and, in summer, thermals from valley bottoms are readily generated in the high temperatures. However, lower temperatures in winter make thermals scarcer and this means the vultures have to make do with slope updrafts, whose force is often insufficient to lift them well above the ridgeline – hence the increased likelihood of a collision with a turbine, especially at lower wind speeds. Kestrels, which do not use thermals, were most at risk in their post-fledging period in summer when they were young and inexperienced.

Mortality rate differed markedly between the two wind farms, with 0.15 vultures and 0.19 kestrels killed per turbine per year at Zorrillos, but only 0.03 and zero, respectively, for the Enmedio farm. In the case of the vultures this is because the gentler slopes at Zorrillos generated updrafts that were too weak for them to clear the turbines. The explanation is different for the kestrels, which hover rather than soar. They were more likely to be killed at Zorrillos because their hunting habitat of open areas in forest was better represented there.

Some of the vultures that died may have been migrating individuals rather than locals. In general, however, migrating 'soaring' birds, including the white stork (*Ciconia ciconia*), were not badly affected, partly because the farms were placed away from the main migration routes and partly because their flights were at sufficient altitude to miss the turbines. While few data are available on vulture and kestrel population densities and the consequent risk of their decline or extinction, the affected birds have each been listed as threatened or vulnerable in Spain. Barrios and Rodriguez make two recommendations: to suspend operation of turbines associated with high mortality at wind speeds where deaths are likely to occur and, more generally, to undertake detailed behavioral studies at locations of planned farms before they are built.

According to German development plans, thousands of square kilometers of the marine environment will be developed for wind farming by 2030, yet almost nothing is known about likely impacts on bird populations. Garthe and Huppop (2004) took a two-pronged approach to predict the possible consequences. First they developed a species sensitivity index (SSI) for 26 seabirds, ranking them on a 1 to 5 scale for each of a range of properties and integrating all the scores into a single value. The properties related mainly to dispersion behavior, including flight maneuverability (less agile species, with high scores, are more likely to collide with turbines), flight altitude (species flying at 50–200 m, scoring 5, are more vulnerable to turbines than low flyers at altitudes of 0–5 m, scoring 1), percentage time spent flying (those in the air for more of the time score higher: 5 for >80%, 1 for >20%). But also included were regional population size (bigger populations being less vulnerable: score 5 for <10,000, score 1 for > 3000,000) and conservation status (score 5 for 'vulnerable', 4 for 'declining', etc.). The most sensitive species included the nonmaneuverable and 'vulnerable' black-throated diver (*Gavia arctica*: SSI = 44.0) and the maneuverable but 'declining' sandwich tern (*Sterna sandvicensis*) that flies almost constantly and at perilous altitudes (SSI = 25.0). The least sensitive species were the black-legged kittiwake (*Rissa tridactyla*: SSI = 7.5) and northern fulmar (*Fulmarus glacialis*: SSI = 5.8).

The SSI for each species was then coupled with density distribution data to produce vulnerability maps for the German area of the North Sea. The so-called WSI (wind farm sensitivity index) is essentially the product of local density for each seabird and its SSI. In areas where the WSI was less than the overall average, species were considered to be of 'less concern'. At the other extreme, map grid cells with the highest 20% of WSI values were considered to indicate 'major concern'. Between these extremes, WSI values were considered to be of 'concern'. There was, not surprisingly, seasonal variation in the WSI configurations, but the overall annual pattern indicates the areas where most care is needed (Figure 4.14). Garthe and Huppop suggest that this information could act as a basis for the selection of marine wind farm locations.

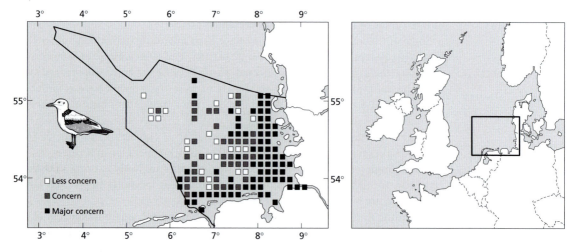

Fig. 4.14 Areas in the German sector of the North Sea (inset) where wind farm development is considered to be of 'less concern', 'concern' or 'major concern' on the basis of a wind farm sensitivity index (WSI) that takes into account both bird density patterns and species specific sensitivity indexes (SSIs). Areas not studied in at least one of the four seasons of the year are left blank. (After Garthe & Huppop, 2004.)

4.5.4 *Bee business –
pollination services
of native bees
depend on dispersal
distance*

A large proportion of the world's crops depend on insect pollinators and bees play a predominant role. Farmers often rely on the ancient relationship with domesticated honeybees (*Apis mellifera*), many importing hives when their crops are in bloom. However, a multitude of native, unmanaged bee species also provide pollination services for various crops, and these are likely to become even more important as populations of domesticated and feral honeybees decline.

Kremen et al. (2004) studied the role played by native bees in 26 watermelon (*Citrullus lanatus*) fields on farms in northern California (Figure 4.15). The farms, some of which were organic (meeting set standards for pesticide and fertilizer use) and others conventional (greater use of pesticides and fertilizer), varied in the proportion of native and other habitats found nearby. Satellite imagery was used to determine the amount of native upland habitat (woodland and chaparral), riparian woodland (beside streams) and highly modified land classes (agriculture, grassland dominated by non-native species, urban land) in the vicinity of each field. They gathered information on pesticide use at the same scale and also recorded the total number of plant species on and around each watermelon field.

Each watermelon plant requires 500–1000 pollen grains to be deposited by insect vectors to produce marketable fruit. The 30 native bee species in the area deposit from 4 to 197 grains each time they visit a flower, whereas honeybees deposit 21 grains per visit. Kremen's team recorded on a number of sunny days over a 2-year period a total of 3349 visits by native bees to watermelon flowers and 7023 by honeybees. They estimated the number of pollen grains deposited at each visit.

In an analysis to determine the importance of all environmental variables (including local habitats, pesticide use, organic vs conventional farms, and plant species richness) only the proportion of upland native habitat within 1–2.5 km of the fields was related to pollen deposition by native bees (Figure 4.16a). This accords with the maximum foraging dispersal distances of about 2.2 km for native bees that nest in

Fig. 4.15 Map of northern California showing the locations of farms and surrounding habitat cover. Circles denote organic and triangles conventional watermelon farms. Gray, upland (woodland plus chaparral) habitat; black, riparian (streamside) habitat; white, agricultural or other human-dominated land type. The town of Davis and Lake Berryessa are indicated, and the scale is in kilometers. (After Kremen et al., 2004.)

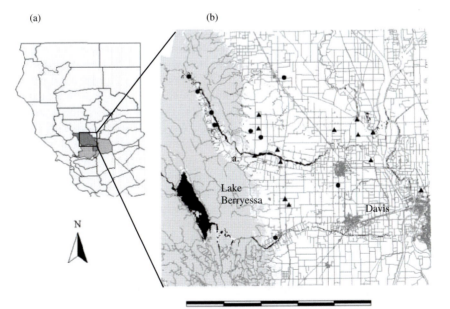

(a)

(b)

Fig. 4.16 (a) Mean
number of pollen grains
deposited by native
bees per 10-minute
observation period in
relation to the
proportion of upland
habitat within 2.4 km of
the watermelon field.
Circles denote organic
farms and triangles
conventional farms. (b)
Estimated total pollen
deposition per day by
native bees. Each
watermelon plant
requires 500–1000
pollen grains to set
marketable fruit. (After
Kremen et al., 2004.)

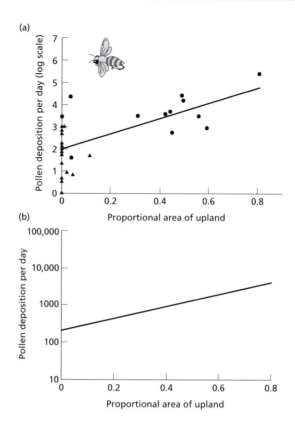

these natural habitats. None of the environmental variables was related to honeybee
pollination – remember that many of these bees would have been from hives
imported to the farms for this purpose.

Kremen and her colleagues converted their records of pollen deposited per
10-minute observation period to the total per day. This allowed them to determine
the proportion of surrounding land that must consist of upland native habitat in
order to yield the 500–1000 pollen grains required per melon plant (Figure 4.16b).
About 40% of natural habitat within 2.4 km of a field is sufficient to provide for
melon pollination needs. Their results provide a strong economic argument for
conserving these natural habitats. For farms that are far from natural habitat, active
restoration with native plants in hedgerows and ditches and around fields, barns
and roads, might allow a target of about 10% native habitat to be achieved. According to Figure 4.16b this would equate to 20–40% of watermelon pollination needs
being met by native bees.

Pollination is just one of the many economically important 'ecosystem services'
provided by native plants and animals. (Others include climate regulation, waste
treatment, provision of health-enhancing products, and even spiritual refreshment
– they will be discussed in detail in Chapter 9.) To achieve the economic benefit of
native bees, it turns out that attention must be paid to the mobility of the species
in relation to the mosaic of their habitat patches in the landscape. Once again, dispersal traits help us understand how to manage our natural resources.

Summary

Dispersal and migration

Dispersal, the spreading of individuals away from others, can be passive (e.g. wind-dispersed seeds) or active (e.g. squirrels dispersing away from their birth site). By contrast, migration is a mass directional movement of large numbers of a species from one location to another. The term applies to the classic long-distance migrations of Arctic terns, but also to the to-and-fro movements of shore animals following the tidal cycle. Some species migrate not through space but through time. Many plants (and some animals) spend a period in dormant form, restarting their development when favorable conditions return. Plans for conservation areas, for animal recovery or vegetation restoration, may be doomed to failure in the absence of an understanding of dispersal and migration behavior.

Species mobility and conservation

Deciding the best place for a nature reserve is especially difficult when the species to be protected moves during its lifetime through a variety of habitats that are more or less remote from each other. Some species have larval stages that disperse remotely from their birth place, others migrate among several areas in a geographical region, some move up and down mountains as the seasons change, and others live in ephemeral habitats that disappear and reappear elsewhere. In every case, powers and patterns of movement will influence the success of conservation strategies.

Species mobility and restoration

When managers try to restore populations where a species has gone extinct, or to restore whole communities of plants and animals, they need to take account of species' mobility. Attempts to restore populations of migratory species must, of course, take their migratory behavior carefully into account. But when it comes to plant community restoration, an intimate knowledge of the dispersal ability of seeds can be equally critical. Some species, because of limited powers of movement, need to be carried (as seeds, young plants or animals) to their newly created havens.

Predicting the arrival and spread of invaders

The best way to deal with invaders is to prevent their arrival in the first place, and this involves identifying the novel 'dispersal' pathways opened up by human travel and commerce. Biosecurity precautions include the screening and cleaning of cargo at the point of embarkation and on arrival, and inspection of camping gear and boots of incoming passengers. Once arrived, however, knowledge of dispersal power is needed to predict patterns of spread – so that suitable precautions can be taken.

Species mobility and management of production landscapes

Human industry and resource exploitation inevitably have consequences for natural ecosystems, and managers need to confront the problem of sustaining species in exploited landscapes; species mobility is again an important consideration. Forestry management can take account of movement patterns, such as those of flying squirrels and bats, to minimize adverse effects on biodiversity. Similar considerations apply to the siting of wind turbines, to minimize risk to migrating and dispersing birds in their vicinity. And judicious intermingling of native forest habitat with agricultural production can be managed to maximize the 'free' crop pollination services performed by native bees that move back and forth from their forest home.

The final word

Maria, our wind farm proponent, is certainly not against controls on development for the good of the natural world. '*Business may be my focus, but I appreciate the spiritual refreshment of visiting natural landscapes as much as any greeny. And as for the wild bee story, how could a businessperson not be happy with a service that enhances profitability but is essentially free?*' But as far as her proposed wind farm development goes, and having learnt about the land-based and marine-based wind farm studies, Maria is now even angrier. '*You don't need to be a rocket scientist to know that some birds will fly into turbines, just as they do into plate-glass windows, aeroplane jets and power lines. The question is whether these losses, regrettable as they are, would actually result in declines or extinctions of any bird species at all. Don't forget that these vague risks need to be balanced against things like the gainful employment of building and running the wind farms. Not to mention the environmental benefit of wind power as opposed to the burning of fossil fuel.*'

Consider the two wind farm studies in Section 4.5.3. Do you judge their results to be of use in deciding where wind farm development should be allowed? Outline some studies that might be performed to allow stronger recommendations to be made.

References

Barrios, L. & Rodriguez, A. (2004) Behavioral and environmental correlates of soaring-bird mortality at on-shore wind turbines. *Journal of Applied Ecology* 41, 72–81.

Begon, M., Townsend, C.R. & Harper, L. (2006) *Ecology: from individuals to ecosystems*, 4th edn. Blackwell Publishing, Oxford.

Bossenbroek, J.M., Kraft, C.E. & Nekola, J.C. (2001) Prediction of long-distance dispersal using gravity models: zebra mussel invasion in inland lakes. *Ecological Applications* 11, 1778–1788.

Campbell, D.R., Rochefort, L. & Lavoie, C. (2003) Determining the immigration potential of plants colonizing disturbed environments: the case of milled peatlands in Quebec. *Journal of Applied Ecology* 40, 78–91.

Garthe, S. & Huppop, O. (2004) Scaling possible adverse effects of marine wind farms on seabirds: developing and applying a vulnerability index. *Journal of Applied Ecology* 41, 724–734.

Holway, D.A. & Suarez, A.V. (1999) Animal behavior: an essential component of invasion biology. *Trends in Ecology and Evolution* 14, 328–330.

King, S.L. & Keeland, B.D. (1999) Evaluation of reforestation in the Lower Mississippi River Alluvial Valley. *Restoration Ecology* 7, 348–359.

Kraft, C.E. & Johnson, L.E. (2000) Regional differences in rates and patterns of North American inland lake invasions by zebra mussels (*Dreissena polymorpha*). *Canadian Journal of Fisheries and Aquatic Sciences* 57, 993–1001.

Kremen, C., Williams, N.M., Bugg, R.L., Fay, J.P. & Thorp, R.W. (2004) The area requirements of an ecosystem service: crop pollination by native bee communities in California. *Ecology Letters* 7, 1109–1119.

Law, B. & Chidel, M. (2002) Tracks and riparian zones facilitate the use of Australian regrowth forest by insectivorous bats. *Journal of Applied Ecology* 39, 605–617.

Loucks, C.J., Zhi, L., Dinerstein, E., Dajun, W., Dali, F. & Hao, W. (2003) The giant pandas of the Qinling Mountains, China: a case study in designing conservation landscapes for elevational migrants. *Conservation Biology* 17, 558–565.

MacIsaac, H.J., Borbely, J., Muirhead, J. & Graniero, P. (2004) Backcasting and forecasting biological invasions of inland lakes. *Ecological Applications* 14, 773–783.

Middleton, B.A. (2003) Soil seed-banks and the potential restoration of forested wetlands after farming. *Journal of Applied Ecology* 40, 1025–1034.

Muirhead, J.R. & MacIsaac, H.J. (2005) Development of inland lakes as hubs in an invasion network. *Journal of Applied Ecology* 42, 80–90.

Perrings, C., Dehnen-Schmutz, K., Touza, J. & Williamson, M. (2005) How to manage biological invasions under globalization. *Trends in Ecology and Evolution* 20, 212–215.

Purse, B.V., Hopkins, G.W., Day, K.J. & Thompson, D.J. (2003) Dispersal characteristics and management of a rare damselfly. *Journal of Applied Ecology* 40, 716–728.

Powell, G.V.N. & Bjork, R.D. (2004) Habitat linkages and the conservation of tropical biodiversity as indicated by seasonal migrations of three-wattled bellbirds. *Conservation Biology* 18, 500–509.

Reunanen, P., Monkkonen, M. & Nikula, A. (2000) Managing boreal forest landscapes for flying squirrels. *Conservation Biology* 14, 218–227.

Ricciardi, A. & MacIsaac, H.J. (2000) Recent mass invasion of the North American Great Lakes by Ponto-Caspian species. *Trends in Ecology and Evolution* 15, 62–65.

Shanks, A.L., Grantham, B.A. & Carr, M.H. (2003) Propagule dispersal distance and the size and spacing of marine reserves. *Ecological Applications* 13(1) Supplement, S159–S169.

Soderquist, T.R. (1994) The importance of hypothesis testing in reintroduction biology: examples from the reintroduction of the carnivorous marsupial *Phascogale tapoatafa*. In: *Reintroduction Biology of Australian and New Zealand Fauna* (M. Serena, ed.), pp. 156–164. Beaty & Sons, Chipping Norton, UK.

Sutherland, W.J. (1998) The importance of behavioral studies in conservation biology. *Animal Behavior* 56, 801–809.

Williams, M.R., Abbott, I., Liddelow, G.L., Vellios, C., Wheeler, I.B. & Mellican, A.E. (2001) Recovery of bird populations after clearfelling of tall open eucalypt forest in western Australia. *Journal of Applied Ecology* 38, 910–920.

Wittenberg, R. & Cock, M.J.W. (2001) *Invasive Alien Species: a toolkit of best prevention and management practices*. CAB International, Oxford.

5 Conservation of endangered species

Species at risk of extinction, by virtue of their rarity, are subject to various forms of uncertainty in their population dynamics and, as a crisis discipline, conservation decisions need to be made in the absence of a detailed understanding of these dynamics. Small populations also face risks associated with their genetic status, something that at best is only ever partially documented. This chapter is about making the most of limited information to reduce the likelihood of extinction of threatened populations.

Chapter contents

5.1 Dealing with endangered species – a crisis discipline	109
5.2 Assessing extinction risk from correlational data	113
5.3 Simple algebraic models of population viability analysis	117
5.3.1 The case of Fender's blue butterfly	117
5.3.2 A primate in Kenya – how good are the data?	118
5.4 Simulation modeling for population viability analysis	119
5.4.1 An Australian icon at risk	120
5.4.2 The royal catchfly – a burning issue	122
5.4.3 Ethiopian wolves – dogged by disease	123
5.4.4 How good is your population viability analysis?	126
5.5 Conservation genetics	127
5.5.1 Genetic rescue of the Florida panther	128
5.5.2 The pink pigeon – providing a solid foundation	128
5.5.3 Reintroduction of a 'red list' plant – the value of crossing	129
5.5.4 Outfoxing the foxes of the Californian Channel Islands	130
5.6 A broader perspective of conservation – ecology, economics and sociopolitics all matter	130
5.6.1 Genetically modified crops – larking about with farmland biodiversity	131
5.6.2 Diclofenac – good for sick cattle, bad for vultures	133

Key concepts

In this chapter you will

understand that the population dynamics of threatened species are governed by high levels of uncertainty

appreciate that population viability can be gauged by methods that vary in the amount of data required and in the confidence that can be placed in their conclusions

understand that attempts to define the minimum viable population are not so much aimed at precise estimation of extinction probability, but to allow managers to compare the likely outcomes of alternative management scenarios

---

recognize that small populations may also be at risk because of low genetic diversity and inbreeding
   depression
realize that every conservation question also has an economic angle and usually a sociopolitical
   one too

5.1 Dealing with endangered species – a crisis discipline

The population of white-backed vultures (*Gyps bengalensis*) in India has declined by more than 99% in less than a decade. You would be lucky to see one today where just a few years ago thousands were carrying out their daily refuse disposal business. Vultures may not be your favorite birds, but to Jijanji they are a vital part of everyday life. '*Nowadays, dear sir, it is not unusual to see a dead buffalo rotting into a stinking mess close to where children are playing. Contamination of nearby wells and the spread of disease by flies must surely be more likely now that dead animals are not quickly picked clean by the vultures.*' Jijanji fears that the white-backed vulture, together with its long-billed (*G. indicus*) and slender-billed cousins (*G. tenuirostris*), could be extinct within a few years. This is a crisis whether you are a conservationist or part of the local community that depends on the vultures for waste disposal. But to Jijanji the crisis is even more personal. He is a Parsee, a descendant of Persian refugees who settled long ago in western India. '*According to my religion dead people should not be buried, burnt or thrown into river or ocean, because of the sacredness of earth, fire and water. Instead, within a few hours of death a Parsee is taken in daylight to a special tower (dakhma) where the naked body is laid on a platform in full view of vultures. The corpse is completely stripped within an hour or two, according to our rites, but only if the vultures are there.*'

In the previous section of the book (Chapters 2–4), my focus was the application of ecological theory at the level of the ecology of *individuals* (niche theory, life-history theory, dispersal/migration theory). The individual organisms of a single species are collectively known as a population. In this chapter and the two following I shift attention to the application of ecological theory at the level of *populations*. When confronted with a declining population, such as that of the white-backed vulture, managers need to invoke the theory of population dynamics, which seeks to understand what determines population size and the way this varies through time. In simple terms, a population's dynamics depend on the interplay of processes that increase (birth and immigration) or decrease population size (death and emigration) (Box 5.1). You will discover in the next two chapters that the theory of population dynamics is also the key to success in pest control operations (Chapter 6) and the management of wild harvests (Chapter 7). The 'theory' boxes of all three chapters (Boxes 5.1, 6.1 and 7.1) can be read together as an overview of population dynamics theory.

Population dynamics theory is one of the most mature of ecological disciplines, but only relatively recently has attention been turned from the fundamental understanding of population dynamics to the application of theory to populations at risk of extinction. Extinction has always been a fact of life – the fossil record tells us that the vast majority of species that ever existed became extinct long ago. But the arrival on the scene of humans injected novelty into the list of causes of extinction. Overexploitation by hunting was probably the first, but more recently many others have been brought to bear, including habitat destruction, introduction of invaders and pollution. I will return to the likely causes of the demise of Jijanji's vultures in Section 5.6.

It should be noted that many, probably most, species on earth are naturally rare. In the absence of human interference there is no reason to expect that the rarer types would be substantially more at risk of extinction. However, while some species are born rare, others have rarity thrust upon them. The actions of humans have undoubtedly reduced the abundance and range of many species (including naturally rare species) and the probability of their extinction may be enhanced for reasons related to the population dynamics of small populations (Box 5.1) and/or their population genetics (Box 5.2).

Given the environmental circumstances and life-history characteristics of a species of concern, what is the chance it will go extinct in a specified period? Alternatively, how big must its population be to reduce the chance of extinction to an acceptable level? These are frequently the crunch questions in conservation management.

So how are managers to decide what constitutes the 'minimum viable population'? Three approaches will be highlighted in this chapter: a simple correlational approach that seeks to identify easily measured factors that are correlated with extinction risk (Section 5.2); the use of general algebraic population models when detailed population information is lacking (Section 5.3); and the use of specific population viability analyses, involving simulation models designed for particular species at risk (Section 5.4). All the approaches have their limitations, which I will explore by looking at particular examples. But first it should be noted that attempts to define the minimum viable population, because of inherent constraints, are not so much aimed at the precise estimation of extinction probability or the predicted time to extinction, but to allow managers to compare the likely outcomes of alternative management scenarios. In Section 5.5, I discuss how our knowledge of the genetics of small populations can assist in conservation efforts. Then in Section 5.6, I broaden the scope to consider not only the ecological aspects of population viability, but also the economic and social aspects of actions based on the ecological theory.

Box 5.1 Population dynamics theory 1

Some basics

In simple terms, the way that the number of individuals changes in a population is given by:

$$N_{now} = N_{then} + B - D + I - E$$

In other words, the number now (N_{now}) equals the number there previously (N_{then}; a year ago, say), plus the number of births between then and now (B), minus the number of deaths (D), plus the number of immigrants (I), minus the number of emigrants (E). Each of these components may be expressed as absolute numbers or as densities (numbers per unit area). Note also that the processes that add (B, I) or subtract individuals from the population (D, E) may usefully be expressed as rates (per head of population per year).

Put another way:

$$N_{now} = \lambda N_{then}$$

Where λ is the *fundamental net per capita rate of increase*. Clearly, the population will increase when $\lambda > 1$, and decrease when $\lambda < 1$. A value of $\lambda = 2$, for example, means that on average every individual in the population will give rise to two individuals in the next generation (either by producing one surviving offspring and staying alive itself, or by dying but producing two surviving offspring).

Finally, note that birth, death, migration rates and λ are likely to be influenced by a variety of environmental factors: physicochemical conditions (such as temperature, humidity, pH), resource availability (light or nutrients for plants, food or nest sites for animals) and the presence of enemies (predators, parasites or species that compete for the same resources).

Why are some species common and others rare? Why does a species occur at low density in one place and at high density in another? Why does one species fluctuate dramatically in density, while another remains constant? These are the central questions of population dynamics. Their answers depend on knowing how the environmental factors affect B, D, I and E and the consequences of these effects for the mean and variance of population size.

Population regulation and determination – density-dependent and density-independent factors

Population regulation is the tendency of a population to decrease in size when it is above a particular level, but to increase in size when below that level. This occurs as a result of density-dependent processes that act on rates of birth and/or death and/or migration. A population may be regulated at a certain density because the resources are sufficient to support only a certain number of animals: thus, the per capita death rate due to starvation may be higher when density is high than when low. Similarly, a limit on potential nest sites may cause the per capita birth rate to be lower when density is high. A density-dependent regulatory effect might equally apply because predators or parasites cause a higher death rate in the affected population when it is at high density.

It is important to understand that while the processes discussed above *may* act in a density-dependent way, they do not necessarily do so. In addition, physicochemical factors affect populations in a density-independent manner. For example, a blizzard can be expected to kill the same proportion of baby reindeer whether the population is at high or low density.

In contrast to *regulation*, the actual density of individuals in a population is *determined* by the combined effects of all the processes that affect it, whether they are dependent or independent of density. Figure 5.1 illustrates this in a simple way for three hypothetical populations of a plant species. The birth rate is density-dependent (because the number of seeds per individual is reduced at high density as a result of competition for limited soil nutrients) but the death rate is density-independent and depends on temperature, which differs in three locations. There are three equilibrium populations (N^*_1, N^*_2, N^*_3), corresponding to the three death rates that, themselves, correspond to the three sets of physical conditions. Density dependence is operating in all three populations, but the density around which they are regulated differs. Thus, we must turn to the *determination* of density if we want to understand why a particular population has a particular density in a particular place, and not some other density.

Models of population dynamics

The development of population dynamics theory has primarily involved the formulation of general equations that incorporate, as realistically as possible, the various processes that influence population size. We need not go into detail (see Begon et al., 2006, for a comprehensive treatment), but it will be useful to give a flavor of the approach and to introduce some key parameters that will crop up later in the chapter.

The speed (dN/dt) at which a population increases in size (N) as time (t) progresses can be written as:

$$dN/dt = rN$$

where r is the population's *intrinsic rate of natural increase*. As long as r is greater than zero the population will increase exponentially. (Note in passing that r is related to λ, described earlier, by $r = \ln \lambda$.)

However, real populations do not continue to increase indefinitely in this way. We can incorporate density dependence (resulting from competition for resources among population members) by introducing the concept of the carrying capacity (K):

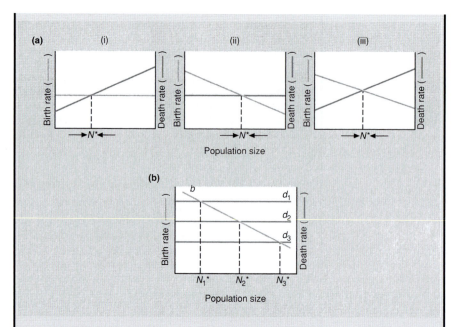

Fig. 5.1 (a) Population regulation with (i) density-independent death; (ii) density-dependent birth and density-independent death; (iii) density-dependent birth and death. Population size increases when the birth rate exceeds the death rate and decreases when the death rate exceeeds the birth rate. N^* is therefore a stable equilibrium population size. The actual value of the equilibrium population size is seen to depend on both the magnitude of the density-independent rate and the magnitude and slope of any density-dependent process.
(b) Population regulation with density-dependent birth, b, and density-independent death, d. Death rates are determined by physical conditions which differ in three sites (death rates d_1, d_2 and d_3). Equilibrium population size varies as a result (N_1^*, N_2^*, N_3^*). (From Townsend et al., 2003.)

$$dn/dt = rN\{1 - (N/K)\}$$

This is known as the logistic equation. Now the rate at which the population increases declines as the population size N approaches the carrying capacity K. When N and K are the same, the population no longer increases but stabilizes at the carrying capacity. The carrying capacity can be formally defined as the population size that the environment can just support (carry) without a tendency to increase or decrease. Look again at Figure 5.1 and note that the equilibrium population sizes (N^*) can equally be described as carrying capacities (K).

Modelers have made the logistic equation (and others like it) progressively more realistic and complex by incorporating elements that account for competition with individuals of other species, mortality caused by predators, changes to reproduction and survivorship caused by parasites, the way the population is arranged in space, the effects of random environmental changes, and so on.

A different approach has been to build simulation models that can be run on a computer for many generations, allowing us to follow the consequences for population dynamics of changing the schedules of rates of birth and death of the various age classes that make up a real population, or of incorporating density dependence or elements of environmental unpredictability.

Both algebraic and simulation approaches figure in the conservation management of endangered populations.

Uncertainty and the risk of extinction – the population dynamics of small populations

While the general theory of population dynamics has focused on central tendencies such as *mean* population size, species at risk of extinction are almost always rare and more weight needs to be placed on *variability* and the prediction of extreme values (such as extinction). The dynamics of small populations are governed by a high level of uncertainty, whereas large populations can be described as being governed by the law of averages (Caughley, 1994). Three kinds of uncertainty are of particular importance to the fate of small populations.

1 Demographic uncertainty: random variations in the number of individuals that are born male or female, or in the number that happen to die or reproduce in a given year, or in the genetic quality of offspring in terms of survival/reproductive capacities can matter very much to the fate of small populations. Suppose a breeding pair produces a clutch consisting entirely of males – such an event would go unnoticed in a large population but would be the last straw for a species down to its last pair.

2 Environmental uncertainty: unpredictable changes in environmental factors, whether large scale (such as floods, storms or droughts of a magnitude that occurs very rarely) or small scale (year to year variation in average temperature or rainfall), can also seal the fate of a small population. Even where the average rainfall of an area is known accurately from historical records, we cannot predict whether next year will be average or extreme, nor whether we are in for a number of years of particularly wet conditions. A small population is more likely than a large one to be reduced by adverse conditions to zero (extinction), or to numbers so low that recovery is impossible (quasi-extinction).

3 Spatial uncertainty: many species consist of subpopulations that occur in more or less discrete patches of habitat. Since the subpopulations are likely to differ in terms of demographic uncertainty, and the patches they occupy in terms of environmental uncertainty, the patch dynamics of extinction and local recolonization can be expected to have a large influence on the chance of extinction of the population as a whole (the so-called 'metapopulation').

Thus, chance events play a particularly large role for small populations. As a result, the models that have been developed to predict their behavior often feature important elements of random change or *stochasticity* (expressed as variation in reproduction and survival). This will become clear in the various examples discussed later in the chapter.

One thing to note before proceeding is that the extinction of a population does not necessarily signal the extinction of a species (global extinction). Global extinction occurs when the last surviving local population goes extinct.

5.2 Assessing extinction risk from correlational data

When dealing with species at risk and in the absence of detailed *demographic* data (birth and death rates, etc.), a simple approach might be to search for correlations between environmental factors and presence/absence or species density. If individuals are associated with habitat X but not habitats Y or Z, managers might decide to protect or enhance habitat X. This is a risky assumption, however, because habitat use of a species in decline may not equate to optimal niche conditions, as you saw in Section 2.3.2 for a New Zealand bird, the takahe. A better approach would be to look for correlations between environmental factors and *population decline* rather than current distribution. The corncrake *Crex crex*, a bird from eastern Africa that breeds in summer in the UK, has been declining for more than a century. Stowe et al. (1993) identified sites occupied by corncrakes during censuses in 1978–79 that had retained or had lost singing males by 1988. Their analysis showed that locations retaining corncrakes were hay meadows with tall vegetation. It seems that an agricultural switch to early-season mowing had been responsible for reduced breeding success. Management recommendations have since been made to reduce the impact of mowing operations and some populations have started to recover (Green & Gibbons, 2000).

Box 5.2 Genetics of small populations

What is genetic variation?

Each gene can exist in a number of forms or 'alleles'. Recall that sexually reproducing organisms receive one copy of each gene from each parent. These may be identical (the same allele derives from both parents) or different. In the latter case, one form of the allele is usually 'dominant' and expressed in the individual; the 'recessive' allele is unexpressed, but is passed on to some of the offspring. Now consider, for example, the gene for flower color in a particular plant species, or the gene for a particular enzyme in an animal species. If all individuals in a population have the same form of the gene in question (i.e. all possess an identical allele), the population has low diversity for that gene – all the flowers are the same color and all the animals operate optimally under the same physicochemical conditions. If, on the other hand, a population contains individuals that possess one of a number of alleles of the gene (and of other genes), the population has high genetic diversity.

Genetic variation in a population is determined by the joint action of 'natural selection' (where the frequency of an allele in a population is related to the evolutionary advantage it confers) and 'genetic drift' (where the frequency of an allele is determined simply by chance). Geneticists have powerful molecular tools (such as DNA fingerprinting) to determine genetic variation.

Loss of genetic variation in small populations

Box 5.1 explained how small populations are subject to increased demographic risks. Population genetics theory tells us to also beware genetic problems in small populations, which arise through loss of genetic diversity. The influence of genetic drift is greater in small isolated populations, which as a consequence are expected to lose genetic variation. And populations that some time in their history have been reduced to just a few individuals (a 'bottleneck'), or that have arisen from a few 'founder' migrant individuals, will have particularly low genetic diversity, even if they subsequently increase in numbers. Migrants can be important in another respect too – populations where immigration is a common event are likely to be genetically more diverse because of alleles contributed by the migrants.

Greater prairie-chickens (*Tympanuchus cupido pinnatus*) provide a good example of the relationship between population size and genetic diversity. These birds were once widespread throughout the tall-grass prairies of Midwestern North America, but with the loss and fragmentation of this habitat many populations have become small and isolated. Johnson et al. (2003) used molecular (DNA) techniques to measure genetic diversity in both large (from 1000 to more than 100,000 individuals) and small prairie-chicken populations (fewer than 1000 individuals). The mean number of alleles (per region of DNA) ranges from 7.7 to 10.3 in the large populations, but is only 5.1 to 7.0 in the small populations. It seems that prairie-chicken populations were once linked by the 'gene flow' provided by migrants, which kept genetic diversity high, but current populations are isolated in their habitat fragments.

Why might loss of genetic diversity be a problem? Rare alleles that confer no immediate advantage might turn out to be well suited to changed environmental conditions in the future. Small populations that have lost rare alleles may therefore have less potential to adapt, increasing their risk of extinction in the long term. Consider two populations – in the first, all individuals possess the same allele for a particular enzyme, but in the second many alleles are represented. An increase in temperature (associated with an unusual sequence of hot years or human-induced global warming) might lead to uniformly poor enzyme performance and a high risk of extinction of the first population. In the second population, on the other hand, some individuals with an allele that confers good enzyme activity at higher temperatures may now be at a selective advantage and prosper (passing their alleles on to offspring) – with a consequently reduced likelihood of extinction.

Inbreeding depression

A more immediate potential problem is inbreeding depression. When populations are small there is a tendency for related individuals to breed with one another. All populations carry recessive alleles that can be lethal to individuals when homozygous (when the alleles provided by the mother and father are identical). Individuals that breed with close relatives are more likely to produce offspring where the harmful alleles are derived from both parents – so the deleterious effect is expressed. Domestic and zoo breeders have long been aware that inbred individuals may show reductions in fertility, survivorship, growth rates and resistance to disease. Evidence is also accumulating of a deleterious effect of inbreeding depression in wild populations, although this is not invariably the case. For example, there is some evidence that island populations, which may be naturally small, can persist despite high levels of inbreeding. Ironically, frequent inbreeding may allow the more lethal of recessive deleterious alleles to be exposed to natural selection so that they are 'purged' from the population.

It has been suggested that an effective population size above 50 individuals would be unlikely to suffer from inbreeding depression, whilst 500–1000 might be needed to maintain high genetic diversity and thus long-term evolutionary potential (Franklin & Frankham, 1998). But such rules of thumb should be applied cautiously. Conservation management action should ideally be based on specific genetic information. But the objectives are usually straightforward, whether managers are concerned with captive rearing programs or translocations of individuals from vulnerable to safe habitats – maximize genetic diversity and minimize the risks of inbreeding depression. You will see in Section 5.5 how conservation managers set out to achieve these objectives.

The relative importance of genetic and demographic risks for small populations

Despite the genetic risks, no example of extinction due to genetic problems has been reported. Perhaps inbreeding depression has occurred undetected as part of the 'death rattle' of some declining populations (Caughley, 1994). Thus, a population may have been reduced to a very small size by one or more of the processes illustrated in Figure 5.2, leading to an increased frequency of matings among relatives and the expression of deleterious alleles in offspring, leading to reduced survivorship and fecundity, and causing the population to become smaller still – the so-called extinction vortex.

Another reason for the dearth of examples of genetics-related extinctions could be that the potential pathways to extinction (demographic or genetic) operate at contrasting rates. Habitat loss or the introduction of an exotic predator might, through demographic changes, drive a population extinct within years. By contrast, severe inbreeding problems might conceivably lead (or contribute) to extinction within a few generations. Then again, any effect of loss of genetic diversity, via a reduction in the ability to adapt to changed circumstances, might take decades or centuries to become obvious – if and when a dramatic change in environmental circumstances takes place. Thus, genetic problems in small populations follow a very long and winding road (Jamieson, in press), while demographic problems are more akin to a high-speed crash on the freeway. But this does not mean that we can afford to ignore the warnings from population genetics theory.

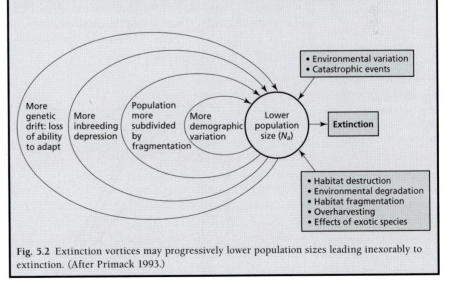

Fig. 5.2 Extinction vortices may progressively lower population sizes leading inexorably to extinction. (After Primack 1993.)

Long-term historical data sets often provide valuable insights. For example, records have been kept for up to 70 years in the case of populations of bighorn sheep (*Ovis canadensis*) in various desert areas in North America. By grouping these into classes according to their population sizes at the commencement of record keeping, it becomes clear that the smaller the population, the greater the risk of extinction (Figure 5.3). Let's set an arbitrary definition of the necessary minimum viable population, as conservation managers often do, as one that will give at least a 95% probability of persistence for 100 years. Figure 5.3 can be explored to provide an approximate answer. Populations of fewer than 50 individuals all went extinct within 50 years while only 50% of populations of 51–100 sheep lasted for 50 years. Evidently, for our minimum viable population we require a population with more than 100 individuals: such populations demonstrated close to 100% success over the maximum period studied of 70 years.

A similar analysis of long-term records of various species of birds on the Channel Islands off the Californian coast indicates a minimum viable population of between 100 and 1000 pairs of birds (Thomas, 1990). These unusual data sets are available because of the extraordinary interest people have in hunting (bighorn sheep) and ornithology (Californian birds). Their value for conservation, however, is limited because they deal with species that are generally not at risk. It is at our peril that we use them to produce recommendations for management of endangered species. There might be a temptation to report to a manager 'if you have a population of more than 100 pairs of your bird species you are above the minimum viable threshold'. But this would only be a (reasonably) safe recommendation if the species of concern and those in the study had very similar life-history characteristics and if the environmental regimes were similar, something that it would rarely be safe to assume.

One final point in the context of lessons for conservation managers from correlational data concerns the species traits I discussed in Section 3.4.2. There you saw that extinction risk is higher for large-bodied, *K*-selected species because of their low reproductive rates. Thus, a link becomes clear between individual life-history traits and population processes. The message for managers is clear – pay particular attention to large-bodied species at risk.

Fig. 5.3 The percentage of populations of bighorn sheep that persist over a 70-year period is lower when the initial population size is smaller. (After Berger, 1990.)

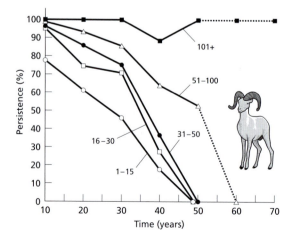

5.3 Simple algebraic models of population viability analysis

If the fundamental net per capita rate of increase λ is constant and greater than 1 (or equivalently, if the intrinsic rate of population increase r is constant and greater than zero) a population will grow indefinitely. Conversely, if λ is constant and less than 1 (or r is less than 0) the population will shrink to extinction. When λ equals 1 (or r equals 0) the population size remains unchanged. In fact, of course, environmental variation causes survival and reproduction to vary from year to year, and thus population growth rate also varies, and usually in an unpredictable manner. Thus, we must view population growth rates as variable, each population having a growth rate with a mean and a variance about the mean. Sometimes managers are faced with populations at risk for which little detailed demographic information is available. At best, they may only have a run of several years of population density estimates. However, such census data can be used to roughly calculate the mean and variance of population growth rate and, given a range of simplifying assumptions, to derive an estimate of the probability of extinction over a particular time period.

At its simplest, the likely persistence time of a population, T, can be expected to increase with population size N (as you saw in Figure 5.3), to increase with average population growth rate (r or λ), and to decrease with an increase in variance in growth rate V resulting from temporal variation in environmental conditions.

5.3.1 The case of Fender's blue butterfly

Schultz and Hammond (2003) performed an analysis based on this approach for the rare Fender's blue butterfly (*Icaricia icarioides fenderi*) in the USA, using population data from surveys lasting at least 8 years in 12 different local populations (Figure 5.4). The butterfly lives exclusively in the few surviving patches of prairie grassland in Oregon that support plants used as food by its caterpillars – Kincaid's lupine (*Lupinus sulphureus kincaidii*) and spur lupine (*Lupinus arbustus*). They followed a procedure described by Morris et al. (1999) to estimate the mean and variance of growth rate for each population, and then to determine the probability that each population would persist for 100 years. The key assumptions of the method were: (i) that the data represent a complete count of individuals or a constant fraction of the population; (ii) that variability between years is true environmental variability and not influenced by observer error; (iii) that there are no extreme catastrophes or bonanza years in the data set; (iv) that population growth rate is independent of density; and (v) that current environmental conditions will persist for at least a century. The probability of population persistence for 100 years varied from less than 0.01 (i.e. lower than a 1% chance of survival) at Gopher Valley to 0.92 (92% chance of survival) at Butterfly Meadows. You can see in Figure 5.4 that Gopher Valley has a small population that follows a particularly erratic pattern (high variance in population growth rate) while the population at Butterfly Meadows is both large and relatively constant through time. Schultz and Hammond estimated that given the average observed variation in population growth rate ($V = 0.79$) and an initial population size of 300 butterflies, a minimum average population growth rate λ of 1.83 is needed for a 95% probability that a population will survive 100 years. Regrettably, none of the populations look set to achieve this. To provide for a 95% probability of persistence of a minimum of one population, they estimate that three independent populations will need to be managed, or restored, each to have a population growth rate λ of at least 1.55.

Fig. 5.4 Fender's blue butterfly population dynamics recorded for populations with a mean population size of (a) more than 25 and (b) less than 25 butterflies. FR, Fern Ridge; WC, Willow Creek. (After Schultz & Hammond, 2003.)

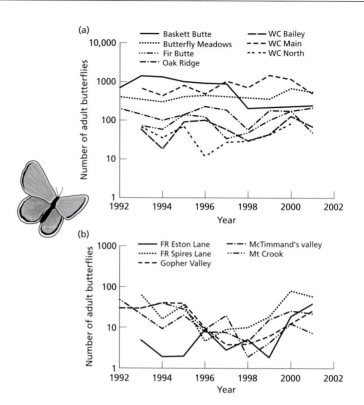

5.3.2 *A primate in Kenya – how good are the data?*

In their study of the Tana River crested mangabey (*Cercocebus galeritus galeritus*) in Kenya, Kinnaird and O'Brien (1991) used a similar approach to estimate the population size needed to provide a 95% probability of persistence for 100 years. This endangered mammal is confined to the floodplain forest of a single river where it declined in numbers from 1200 to 700 between 1973 and 1988 despite the creation of a reserve. Its naturally patchy habitat has become progressively more fragmented because of agricultural expansion. The key parameters were estimated from just a few years of data to be an average population growth rate of $r = 0.11$ with variance $V = 0.20$. This corresponds to a minimum viable population of 8000. Given the amount of habitat actually available, Kinnaird and O'Brien concluded that the mangabeys could not attain such a population size. In fact, they think it unlikely that this naturally rare and restricted species has ever been so abundant. The data were probably deficient; for example, environmental variation in r may be smaller than estimated if the mangabeys are able to undergo dietary shifts in response to habitat change.

While these simple census-based approaches have the advantage of requiring a minimum of information, the results must be treated with caution. Lotts et al. (2004) analyzed a series of published data sets and concluded that ideally at least two decades of census data are needed, because predictions based on short time series produce unreliable estimates of variance in growth rate – as with the mangabeys. They noted, however, that the approach can be useful to rank populations according to extinction risk, the approach taken with the butterflies in Section 5.3.1. It is certainly not without some value.

5.4 Simulation modeling for population viability analysis

Simulation models provide an alternative, more specific way of gauging viability. These encapsulate survivorships and reproductive rates in age-structured populations. To see what is involved let's consider the example of a seabird population (Morris et al., 1999). The first step is to arrange the available information on birth and death rates into a matrix where the columns represent three different age classes (from left to right – juveniles, subadults and adults) and the rows represent the probability over 1 year that an individual in one age class will progress to the next. Thus, in the matrix M below there is a probability of 0.665 that a juvenile will survive to be a subadult (i.e. 43.5% of juveniles die as juveniles), a probability of 0.724 that a subadult will survive to be an adult and a probability of 0.95 that an adult will still be alive the following year (i.e. only 5% of adults die each year). The first row contains the crucial birth rate information: only adults have babies, and each adult only produces 0.054 surviving juveniles per year.

$$
\begin{array}{c}
\text{From:} \\
\begin{array}{ccc}
\text{Juveniles} & \text{Subadults} & \text{Adults}
\end{array} \\
\text{TO:} \begin{array}{c} \text{Juveniles} \\ \text{Subadults} \\ \text{Adults} \end{array}
\begin{bmatrix}
0 & 0 & 0.054 \\
0.665 & 0 & 0 \\
0 & 0.724 & 0.95
\end{bmatrix} = M
\end{array}
$$

Next we construct an initial population 'vector', which consists simply of numbers in each age class in year 1 (N_t): for our seabird population this is 40 juveniles, 50 subadults and 60 adults. Now the matrix M is multiplied by the vector, as shown below, to give us new numbers for each age class in the following year (N_{t+1}; 3.2 juveniles, 26.6 subadults and 93.2 adults). This new population vector is then multiplied by the matrix M to give numbers in the third year, and so on. Some taxa, plants for example, may be better represented by size classes rather than age classes, and there may be from two to many classes in the model, but the procedure is the same.

$$
\begin{array}{ccc}
M & \times\ N_t\ = & N_{t+1}
\end{array}
$$

$$
\begin{bmatrix}
0 & 0 & 0.054 \\
0.665 & 0 & 0 \\
0 & 0.724 & 0.95
\end{bmatrix}
\times
\begin{bmatrix}
40 \\ 50 \\ 60
\end{bmatrix}
=
\begin{bmatrix}
(0\times40)+(0\times50)+(0.054\times60) \\
(0.665\times40)+(0\times50)+(0\times60) \\
(0\times40)+(0.724\times50)+(0.95\times60)
\end{bmatrix}
=
\begin{bmatrix}
3.2 \\ 26.6 \\ 93.2
\end{bmatrix}
$$

Ultimately the simulated population will take on a constant rate of growth (λ) reflecting the birth and death rates of age classes in the population in question. But to better represent reality, random variations in the rates in the matrix are introduced to denote the impact of environmental variation. This can be done each year, for example, by drawing each matrix entry from a continuous range of possible values whose mean and variance have been estimated from the available data. The models can be extended to include disasters with a specified frequency (one 'random' year in each century, for example) and intensity (reflected in the extent to which birth and death rates are affected). Density dependence can be introduced where required, as can population harvesting or supplementation. In the more sophisticated models, every individual is treated separately in terms of the probability, with its imposed uncertainty, that it will survive or produce a certain number of offspring

Table 5.1 Values used as inputs for simulations of koala populations at Oakey (declining) and Springsure (secure). Values in brackets are standard deviations due to environmental variation; the model procedure involves the selection of values at random from the range. Catastrophes are assumed to occur with a certain probability; in years when the model 'selects' a catastrophe, reproduction and survival are reduced by the multipliers shown (e.g. in a year with a catastrophe, reproduction is reduced to 55% of what it would otherwise have been). (After Penn et al., 2000.)

Variable	Oakey	Springsure
Maximum age	12	12
Sex ratio (proportion male)	0.575	0.533
Litter size of 0 (%)	57.00 (±17.85)	31.00 (±15.61)
Litter size of 1 (%)	43.00 (±17.85)	69.00 (±15.61)
Female mortality at age 0	32.50 (±3.25)	30.00 (±3.00)
Female mortality at age 1	17.27 (±1.73)	15.94 (±1.59)
Adult female mortality	9.17 (±0.92)	8.47 (±0.85)
Male mortality at age 0	20.00 (±2.00)	20.00 (±2.00)
Male mortality at age 1	22.96 (±2.30)	22.96 (±2.30)
Male mortality at age 2	22.96 (±2.30)	22.96 (±2.30)
Adult male mortality	26.36 (±2.64)	26.36 (±2.64)
Probability of catastrophe	0.05	0.05
Multiplier for reproduction	0.55	0.55
Multiplier for survival	0.63	0.63
% of males in breeding pool	50	50
Initial population size	46	20
Carrying capacity, K	70 (±7)	60 (±6)

in the current time period. The program is run many times, each giving a different population trajectory because of the random elements involved. The outputs include estimates of population size each year and the probability of extinction during the modeled period (the proportion of simulated populations that go extinct).

I will present three case studies to illustrate how simulation modeling can be applied to species at risk. The account of Australian koalas serves to highlight that different populations of a single species can have significantly different demographic features (Section 5.4.1). Plants have their own problems when it comes to simulation modeling, as you will see for the royal catchfly (Section 5.4.2). Finally, disease can be a particular risk factor, and I explore how the epidemiology of rabies has been introduced into simulation modeling of Ethiopian wolf populations (Section 5.4.3).

5.4.1 *An Australian icon at risk*

Koalas (*Phascolarctos cinereus*) are regarded as potentially threatened nationally, with populations in different parts of Australia varying from secure to vulnerable or extinct. The primary aim of the nation's national management strategy is to retain viable populations throughout their natural range (ANZECC, 1998). Penn et al. (2000) used a widely available simulation modeling package, known as VORTEX (Lacey, 1993), to model two populations in Queensland, one thought to be declining (at Oakey), the other secure (at Springsure).

Koala breeding commences at 2 years old in females and 3 years old in males; these values were used for both population models. The other demographic values used in the two analyses were derived from extensive databases from the two populations and are shown in Table 5.1. Note how the Oakey population has somewhat higher annual female mortality and fewer females producing young each year. The Oakey population was modeled from 1971 and the Springsure population from 1976 (when first estimates of density were available) and the trends in the modeled populations were indeed declining and stable, respectively. Over the modeled period (shown in Figure 5.5), the probability of extinction of the Oakey population is 0.380

Fig. 5.5 Observed koala population trends (solid circles) compared with predicted population performance (open circles, ±1 SD) based on 1000 repeats of the VORTEX modeling procedure at (a) Oakey and (b) Springsure. Real population censuses were not performed every year. (After Penn et al., 2000.)

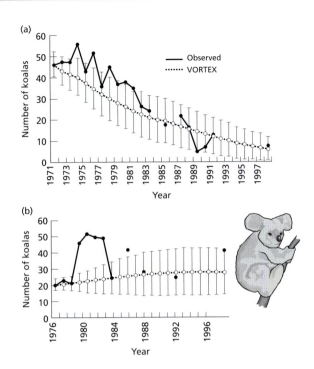

Table 5.2 Sensitivity analysis of the effect of varying fertility or mortality at different life stages on the average population growth rate λ and the probability of extinction (E) during the modeled period (1000 repeats). The 'Standard' values are for the scenarios shown in Table 5.1 and Figure 5.5. (From Penn et al., 2000.)

	Oakey		Springsure	
Scenario	λ	E	λ	E
Standard	0.930	0.464	1.034	0.084
−10% fertility	0.915	0.608	1.015	0.117
+10% fertility	0.943	0.321	1.052	0.046
−10% newborn mortality	0.936	0.386	1.042	0.053
+10% newborn mortality	0.923	0.504	1.026	0.089
−10% juvenile mortality	0.933	0.418	1.038	0.068
+10% juvenile mortality	0.927	0.488	1.031	0.086
−10% adult mortality	0.936	0.323	1.041	0.042
+10% adult mortality	0.923	0.526	1.028	0.099

(i.e. 380 out of 1000 population models with the stated demographic values went extinct), while that for Springsure is only 0.063.

Next, a *sensitivity analysis* was performed to explore the effects of small alterations to demographic values (mortality, % breeding, etc.) (Table 5.2). This allows managers to gauge the likely effects of measurement errors, or to estimate consequences of a management change designed to alter a demographic parameter in a beneficial way. The declining Oakey population, as you would expect, has an average population growth rate λ of less than 1. The population growth rate of the Springsure population is slightly greater than 1 for all scenarios but note that this does not necessarily lead to a zero risk of extinction, as would be expected if λ were greater than 1 and constant. This is because the random elements in the model mean that some of the modeled populations will have been subject to a run of deleterious values for reproduction or mortality, reflecting the reality, for example, that a sequence of very dry or very wet years will occur occasionally in nature. The probability of extinction is most favorably affected by a 10% decrease in adult mortality in the Springsure population (halving extinction probability from 0.084 to 0.042). On the other hand, extinction probability of the Oakey population is most sensitive to a 10% increase in fertility (reducing extinction probability from 0.464 to 0.321).

Managers concerned with critically endangered species do not usually have the luxury of monitoring populations to check the accuracy of their predictions. But Penn et al. (2000) were able to check their predictions because the koala populations have been continuously monitored since the 1970s. The predicted population trends are close to the actual recorded trends, particularly for the Oakey population, and this gives added confidence to the modeling approach (Figure 5.5).

How can such modeling be put to management use? Local authorities in New South Wales are obliged both to prepare comprehensive koala management plans and to ensure that developers survey for potential koala habitat when a building application affects an area greater than 1 hectare. Penn et al. (2000) argue that population viability modeling can be used to determine whether any effort made to protect habitat is likely to be rewarded by a viable population.

5.4.2 *The royal catchfly – a burning issue*

The life histories of plants present particular challenges for simulation modeling, particularly where there is seed dormancy, highly periodic recruitment or clonal growth (Menges, 2000). However, as with endangered animals, different management scenarios can be simulated in population viability analyses. The royal catchfly, *Silene regia*, is a long-lived prairie perennial whose range has shrunk dramatically. Menges and Dolan (1998) collected demographic data for up to 7 years from 16 populations in the Midwest of the USA. The populations, whose total adult numbers ranged from 45 to 1302, were subject to different management regimes. This species, whose seeds do not show dormancy, has high survivorship and frequent flowering, but successful germination is very episodic – most populations in most years fail to produce seedlings. Population projection matrices, such as the one illustrated in Table 5.3, were produced for each population in each year (using another simulation

Table 5.3 An example of a projection matrix for a particular *Silene regia* population from 1990 to 1991, assuming successful germination of seedlings. Numbers represent the proportion changing from the stage in the column to the stage in the row (bold values represent plants remaining in the same stage). 'Alive undefined' represents individuals with no size or flowering data, usually as a result of mowing or herbivory. Numbers in the top row are seedlings produced by flowering plants. The finite rate of increase λ for this population is 1.67. The site is managed by regular burning. (After Menges & Dolan, 1998.)

	Seedling	Vegetative	Small flowering	Medium flowering	Large flowering	Alive undefined
Seedling	–	–	5.32	12.74	30.88	–
Vegetative	0.308	**0.111**	0	0	0	0
Small flowering	0	0.566	**0.506**	0.137	0.167	0.367
Medium flowering	0	0.111	0.210	**0.608**	0.167	0.300
Large flowering	0	0	0.012	0.039	**0.667**	0.167
Alive undefined	0	0.222	0.198	0.196	0	**0.133**

modeling package: RAMAS-STAGE). Multiple simulations, each lasting 1000 years, were then run for every matrix to determine the population growth rate λ and the probability of extinction. Sensitivity analyses, similar to those performed on the koala populations, showed that population performance was not more sensitive to any particular demographic parameter, indicating that a combination of growth, fecundity and survival all contribute to population performance.

Figure 5.6 shows the median population growth rate λ for the 16 populations, grouped into cases where particular management regimes were in place. This was done both for years when recruitment of seedlings occurred and for years when seedling recruitment did not occur. All sites where λ is greater than 1.35 when recruitment took place are managed by burning and some by mowing as well; none of these was predicted to go extinct during the modeled period. On the other hand, populations with no management regime, or whose management does not include fire, have lower values for λ and all except two have predicted extinction probabilities (over 1000 years) of from 0.10 to 1.00.

The obvious management recommendation is to use prescribed burning to provide opportunities for seedling recruitment. Low establishment rates of seedlings may be due to rodents or ants eating fruits or to competition for light with other plants – burnt areas probably reduce one or both of these negative effects. While management regime was by far the best predictor of persistence, it is also worth noting that populations with higher genetic diversity also had higher median values for λ.

5.4.3 *Ethiopian wolves – dogged by disease*

It is not unusual for disease to contribute significantly to the population dynamics of a species but epidemiology has rarely been incorporated into population viability simulations. An overly simplistic approach would be to incorporate disease simply as an additional mortality factor. However, epidemiology theory makes clear that the size of epidemics and the intervals between them are acutely dependent on

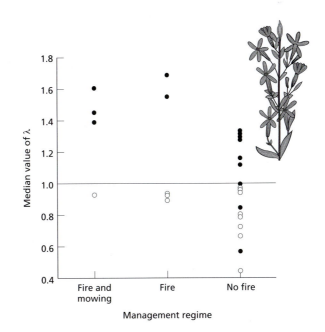

Fig. 5.6 Median rates of population increase (λ) of *Silene regia* populations in relation to management regime, for years with seedling recruitment (black circles) and without (open circles). Unburned management regimes include just mowing, herbicide use or no management. (After Menges & Dolan, 1998.)

population size, age-class structure and immunological status of individuals in the population, so these need to be incorporated if simulation models are to be worthwhile.

The Ethiopian wolf (*Canis simensis*) is critically endangered, now restricted in the Ethiopian highlands to just seven fragmented and isolated populations ranging in size from 15 to 250 individuals, with a grand total of about 500. The habitat patches are surrounded by agricultural land where domestic dogs are a reservoir for rabies. We know something about the risk associated with disease because a rabies outbreak in the early 1990s reduced the biggest of the wolf populations by two thirds and only after 10 years did pack numbers approach pre-epidemic levels. As a result, dog vaccination was instigated in 1996. The question of whether other disease management strategies are needed, and where these should be applied, was addressed by Haydon et al. (2002) using a standard population simulation model approach but with an additional epidemiological component.

The wolves live in close-knit, male-biased packs consisting of an average of six adults, one to six yearlings, and one to seven pups. Only dominant females breed. Dispersal is limited by the scarcity of unoccupied habitat: males stay in the pack where they were born, but two thirds of females disperse at 2 years old and become 'floaters' until a breeding vacancy becomes available.

The simulation models were constructed on the basis of published information or realistic biological assumptions. Carrying capacity of each occupied patch was set at one individual per km^2. Each pack (of which there were several in each habitat patch) had a maximum size of 13 adults, of which only two could be female. Each year packs with at least one male and one female gave birth with a probability of 0.63 to litters of pups of from one to six (with average probabilities for each litter size ranging from 0.05 to 0.32). Age classes (pups 0–1 years, juveniles 1–2 years, adults >2 years) were updated before breeding, and immediately prior to breeding a pack could undergo realistic events such as loss of a randomly selected female to the 'floater' pool, recruitment of a female from the floater pool or splitting of large packs into two. Adults had an average 0.15 probability of dying each year but for younger animals this probability was 0.45 for males and 0.55 for females. Together with random elements (stochasticity) in the precise birth and death probabilities applied, this is the normal bread and butter of individual-based population viability models.

To account for disease, each individual was also subject to a susceptible-infectious-recovered (S-I-R) process. Susceptible individuals may become infected (and for a period be infectious), followed by death or recovery. The models incorporated realistic probabilities for all these epidemiological elements. Infection probabilities from dog to wolf differed for pack members and floaters, and there were different probabilities again from wolf to wolf among members of a pack, between different packs and between pack members and floaters. Recovered individuals are immune to further infection. Population simulations were performed for various sized areas (25–250 km^2), to encompass the known sizes of habitat patches occupied by the wolves, for a range of dog disease incidence (1–140% of a realistic baseline incidence) and for a range of anti-rabies wolf vaccination efforts.

In disease-free simulations (1000 for each scenario), the probability of populations persisting in the different sized patches is generally 100%; only in the case of the smallest habitat patch (25 km^2, with its correspondingly small population – carrying

Fig. 5.7 (a) Proportion of 1000 wolf population simulations that went extinct in 50 years using baseline demographic parameters in habitat patches of different areas, exposed to different levels of disease incidence in the adjacent domestic dog population. The likelihood of extinction increases with disease incidence but is smaller for populations in larger areas. (b) Effect of percentage of wolves vaccinated on population persistence in models with baseline disease incidence. Vaccination of just 20–40% of wolves seems sufficient to prevent the largest epidemics. (After Haydon et al., 2002.)

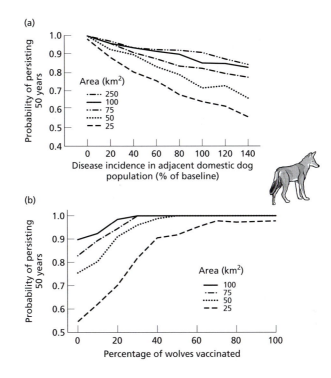

capacity 25 individuals) did 2% of simulated populations go extinct in 50 years (see probabilities of persistence when disease incidence is zero in Figure 5.7a).

When exposed to the baseline disease incidence (i.e. 100% on the horizontal axis of Figure 5.7a), the 50-year probability of extinction for the largest populations increases to 0.08–0.09, with populations showing one or two crashes down to as low as 33% of carrying capacity. However, under this scenario, populations remain reasonably viable until habitat size drops below 100 km². For the 25-km² case, extinction probability increases to 0.46. As disease incidence in the dog population declined from 140% of baseline to 20%, the 50-year extinction probabilities also fall, in the case of the 250-km² habitat from 0.13 to 0.04.

The potential value of management intervention to vaccinate wolves is illustrated in Figure 5.7b. With 20% vaccination in the 250-km² population, percentage population extinction declines from 0.10 to zero. In the smallest population, on the other hand, 40% vaccination is needed to reduce extinction risk from 0.46 to 0.10. Haydon's team concluded that direct vaccination of as few as 20–40% of wolves might be sufficient to eliminate the largest epidemics and to protect populations against the very low densities that make recovery unlikely.

Finally, when they applied a sensitivity analysis to the 250-km² case by looking at the effects of slightly perturbing various parameters, an intriguing result emerged. Population viability proved particularly sensitive to the process of recruitment of females to packs. When females cannot be recruited because none or few are present as floaters, pack extinction rates double. Clearly occasional female recruitment is essential, but paradoxically when the size of the floater pool is large this imparts an additional source of disease transmission and overall population viability declines.

This result provides a very useful reminder that understanding the details of individual behavior can make a difference to conservation management.

5.4.4 How good is your population viability analysis?

In an ideal world a population viability analysis would enable us to produce a specific and reliable recommendation for an endangered species of the population size that would permit persistence for a given period with a given level of probability. Conservation biologists know this is not possible in practice, but it is important to know how good our predictions are likely to be. Brook et al. (2000) addressed this by conducting retrospective tests using 21 long-term ecological studies (involving birds, mammals, a reptile and a fish). The demographic parameters for simulation modeling were estimated on the basis of the first half of each data set, while the second half was used to test the accuracy of predictions. Brook's team scrutinized five commonly used population simulation packages, each with slightly different model structures and underlying assumptions, some of which you have come across in this chapter. The results provide comfort to managers because they show a close relationship between model predictions (whichever package was used) and the historical behavior of the 21 populations (Figure 5.8).

On the other hand, we shouldn't be lured into a false sense of security – no model is better than the data upon which it is based. Within the inevitable constraints of lack of knowledge and lack of time and opportunity to gather data, the model-building exercise is no more nor less than a rationalization of the problem and a careful quantification of ideas. Common sense tells us to trust the results only in a qualitative fashion. Nevertheless, the examples I have discussed show how models can be constructed to make the very best use of available data and provide the confidence to choose between various possible management options and to identify the relative importance of factors that put a population at risk (Reed et al., 2003). The sorts of management interventions that may then be recommended include translocating individuals to augment target populations, restricting unwanted dispersal by fencing, restoring habitat, creating larger reserves, raising carrying capacity by augmenting resources, providing foster care for young, reducing mortality by controlling predators or poachers, and vaccinating against influential pathogens.

The most widely recognized system for ranking species at risk is the IUCN 'Red List' of the World Conservation Union (Box 1.1). (The puzzling acronym IUCN arises because the World Conservation Union was once called the International Union for

Fig. 5.8 Plots of the proportion of populations predicted to decline below a critical threshold (an index of extinction risk) versus the actual number of populations declining below that threshold. For each of the five population viability analysis (PVA) software packages, a perfect fit with reality lies on the 45 degree line shown. You have come across two of these packages already in this chapter: VORTEX in Section 5.4.1 and RAMAS-STAGE in Section 5.4.2. (After Brook et al., 2000, who give details of the other software packages.)

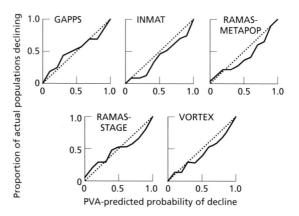

Fig. 5.9 Levels of threat as a function of time and probability of extinction. Note that the time axis is an a log scale. (After Akçakaya, et al., 1999. Reproduced with permission of *Applied Biomathematics.*)

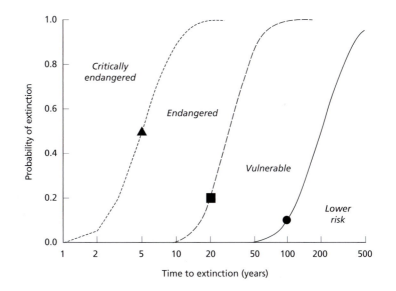

the Conservation of Nature!) From the point of view of extinction probability a species is categorized *vulnerable* if there is considered to be a 10% probability of extinction within 100 years, *endangered* if the probability is 20% within 20 years or five generations, whichever is longer, and *critically endangered* if within 10 years or three generations the risk of extinction is at least 50% (Rodrigues et al., 2006) (Figure 5.9). However, because assessment of extinction risk is only one of the Red List criteria, the correlation between population viability predictions and Red List classes is relatively weak (but positive – O'Grady et al., 2004). Moreover, population viability models have only been used for a minority of species recognized to be at risk. Morris et al. (2002) believe that the value of population modeling needs to be more widely recognized by managers. They recommend that conservation agencies routinely use population viability analyses in their future recovery planning.

5.5 Conservation genetics

You saw in Box 5.2 that small populations can be expected to have lower genetic diversity so that their ability to adapt to environmental change may be compromised over the long term. This may provide another reason, beyond the demographic vagaries discussed in Section 5.4, for an increased risk of extinction when populations have been driven to small numbers. In the shorter term, the genetic risk with the greatest potential to ease a small population over the edge to oblivion is inbreeding depression. In Sections 5.5.1–5.5.3 I present case studies of a mammal, a bird and a plant, all of which depend on a process of genetic mixing to reduce inbreeding effects and thus increase the likelihood of persistence of the population. In Section 5.5.4 you will see how molecular technology can be used to uncover patterns of genetic variation within a species, identifying evolutionary significant units (ESUs) that are worthy of conservation effort. Now the focus is on keeping genetically distinct populations apart because they represent meaningfully distinct genetic units that could generate new species.

5.5.1 *Genetic rescue of the Florida panther*

The last remaining population of the Florida panther (*Puma concolor coryi*) became so small that genetic variation was remarkably low and deleterious alleles occurred at high frequency (or were 'fixed' – showing no variation at all). A suite of traits, doubtless a result of this genetic makeup and rarely seen in other subspecies of the *Puma* genus, includes unilateral undescended testes, a kinked tail, a 'cowlick' pattern on the back, and the poorest semen quality of any cat species (Roelke et al., 1993). Managers decided to translocate individuals from another subspecies, the Texas cougar (*Puma concolor stanleyana*), in an attempt to eliminate deleterious variants and restore more normal levels of genetic variation.

Eight female Texas cougars were introduced in 1995, and five of these produced offspring with resident Florida panther males. Some of the offspring have also mated with resident panthers and now about 20% of Florida panthers have some cougar ancestry. Individuals with cougar ancestry show dramatic reversals in the frequency of undesirable traits – kinked tails reduced from 88% to 7%, the cowlick trait reduced from 93% to 24% and abnormal testes development from 68% to zero (Land et al., 2001). Only one animal with cougar descent has been tested for semen quality, but this is at least as good as the average for cougars.

It cannot be certain that the Florida panther would have gone extinct without this genetic rescue, or that the population is now completely safe. But the signs are good that the probability of extinction has been reduced. Under less urgent circumstances, managers would have shied away from diluting the Florida panther gene pool with that of a different subspecies. But in desperate times, this genetically desperate measure can be fairly justified.

5.5.2 *The pink pigeon – providing a solid foundation*

The pink pigeon (*Columba mayeri*), once widespread on the island of Mauritius, recovered from only nine or ten birds in 1990 to 355 free-living individuals (plus more in captivity) by 2003. The recovery program depended on captive-breeding on Mauritius and overseas. In captivity, the aim has been to manage matings to retain high levels of genetic diversity and to minimize inbreeding. The captive population was originally descended from just 11 founder individuals but this was augmented during 1989–94 by adding to the captive gene pool 12 further founder individuals (offspring from 12 of the 13 remaining wild individuals).

Once captive-reared birds are released into the wild the incidence of inbreeding depression is not easy to control – the tactic of releasing a large number of individuals probably provides the greatest chance of success. Between 1987 and 1997, 256 birds were reintroduced as three subpopulations on Mauritius – wherever possible selecting birds with minimal inbreeding (based on records in breeding 'stud books') and releasing them in groups with good representation of the different founder ancestries. All birds were banded for unique identification.

The genetics and ecological success of both captive and wild populations have been carefully monitored. This database means we can evaluate the impact of inbreeding on survival and reproduction under the controlled situation of captive rearing and also in the more risky circumstances of the wild. Inbreeding reduces egg fertility, and survival of nestlings (up to 30 days post-hatching; Figure 5.10), as well as juveniles and adult birds, but effects are only strongly marked in the most inbred birds. Highly inbred birds are generally more dramatically affected in the wild than in captivity, where threats can be more carefully controlled.

Fig. 5.10 Effect of inbreeding on probability of survival to 30 days of age of pink pigeon nestlings (a) in captivity and (b) in the wild population. Inbreeding is expressed as an index derived from known ancestry in relation to 23 'founder' individuals. The fewer founders in a bird's ancestry, the higher the index of inbreeding. Birds are grouped into three classes – non-inbred, moderately inbred and highly inbred. Only highly inbred birds show a powerful effect of inbreeding. (After Swinnerton et al., 2004.)

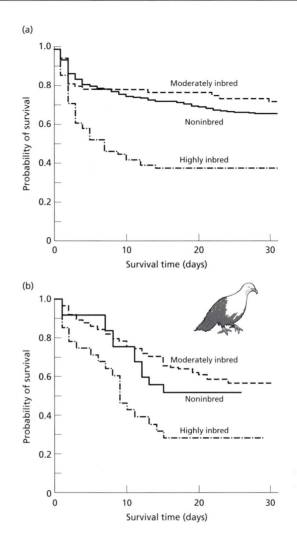

This reintroduction success story has the added benefit of providing a rare quantification of the value of avoiding inbreeding when managing endangered populations.

5.5.3 *Reintroduction of a 'red list' plant – the value of crossing*

A rare variety of *Silene douglasii* (var. *oraria*) is a white-flowered perennial plant that lives in coastal grassland but is now threatened throughout the state of Oregon, USA, and endangered globally. During a reintroduction attempt in previously grazed habitat within its historical range, the opportunity was taken to compare the success of individuals that varied in terms of inbreeding. Over a 3-year period plants were derived from seeds from self-pollinated flowers (inbred, because both 'parents' are the same plant) and cross-pollinated plants (outbred – different parents). A third class of plants was 'open-pollinated' by insects, having been naturally pollinated and without the experimental manipulation of forced self- or cross-pollination.

Progeny of the outbred, cross-pollinated flowers show significantly greater survival than offspring of self-pollinated or open-pollinated flowers. The outbred progeny also produce more reproductive stems and flowers (Figure 5.11). The similarity in performance of self-pollinated and open-pollinated indicates that self-pollination was the general rule at least in the circumstances of this experiment. The results again reinforce the point that inbreeding depression can reduce the chances of a successful conservation management program.

5.5.4 *Outfoxing the foxes of the Californian Channel Islands*

DNA analysis can also be used to identify the closeness of the relationships among different species or subspecies, producing information of use when managers need to decide which populations are deserving of particular attention because of their genetic distinctiveness. Figure 5.12 shows the relationships among subspecies of the island fox *Urocyon littoralis*. Individual foxes are shown as lines, and those whose DNA makeup is most similar branch closest together. For each island population, and the mainland population, individual foxes are most closely related to others in the same population. Among the island populations, those of San Miguel and Santa Rosa are genetically very similar, while populations from the southern islands are readily distinguishable from those in the northern islands. Because of this distinctiveness, the northern and southern populations should not be mixed. On the other hand, in response to the recent extinction of the fox population on San Miguel (due to predation by a non-native eagle species), managers can reasonably decide to translocate individuals from the closely related Santa Rosa population.

5.6 **A broader perspective of conservation – ecology, economics and sociopolitics all matter**

You have seen how population viability analyses or genetic information can reveal the likely benefits for population persistence of particular management actions. However, in conservation, as in every area of life, action costs money. Some populations at greatest risk of extinction require huge recovery efforts with little chance of success, while less threatened species might be secured relatively cheaply. Conservation managers have limited funds, so the process of prioritization for action is critical, yet most often agencies simply allocate resources to species at highest risk (Possingham et al., 2002). In desperate times, painful decisions have to be made about priorities. Thus, wounded soldiers arriving at field hospitals in the First World War were subjected to a *triage* evaluation: Priority 1 – those who were likely to survive but only with rapid intervention; Priority 2 – those who were likely to survive without rapid intervention; Priority 3 – those who were likely to die with or without intervention. Conservation managers are often faced with the same kind

Fig. 5.11 Percentage of surviving *Silene* plants that produced at least one open flower during a 3-year study. 'Outbred' individuals derived from artificially performed cross-pollination always performed better than their more inbred counterparts derived from self-pollination. Open-pollinated plants, derived from unmanipulated parents that had been naturally pollinated by insects, perform more like the artificially self-pollinated plants, indicating that self-pollination is naturally more common. (After Kephart, 2004.)

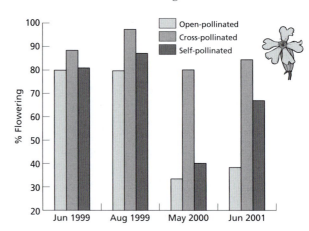

Fig. 5.12 Genetic relationship tree of individuals of fox subspecies inhabiting different Channel Islands and the Californian mainland. In this branching display, individuals that are genetically very similar branch close together. Populations on the southern islands are distinct from those on the northern islands, while the populations of adjacent islands contain individuals that are more similar to each other. (After Wayne & Morin, 2004, based on Aguilar et al., 2004.)

of choices and need to demonstrate some courage in giving up on hopeless cases, and prioritizing those species where something can be done within the limits of available resources.

Whenever economics and ecology are considered together, as they have to be when a change in human behavior is required for the sake of sustainability, there is inevitably a sociopolitical dimension too. In Section 5.6.1, I discuss a case where the environmental cost of novel agrotechnology depends not only on the ecology of affected species but also on the way farmers behave in the new situation.

Whether people are prepared to change their behavior depends on their perceptions of the costs and benefits involved. Until recently, the value of conserving biodiversity hardly figured in the equation and, even now, the public, and the governments that serve them, are often unaware of the gains to be made by taking care of the natural world. When is public education needed to make clear the costs and benefits of action (or inaction)? How should government take account of questions of sustainable behavior – bottom-up, by encouraging different sector groups to work out sustainable strategies together or, top-down, by regulating behavior through new laws? There are no easy answers. In Section 5.6.2, I return to the plight of India's vultures. You will see from a population viability analysis that the decline of the species, and of the immensely valuable (but free) ecosystem services it performs, is linked to a pharmaceutical product. In this case, government has acted to ban its use.

5.6.1 *Genetically modified crops – larking about with farmland biodiversity*

The progressive intensification of agriculture, and particularly increases in mechanization, field size and pesticide use, have been linked to declines in farmland biodiversity. Genetic modification (GM) of crops is the most recent of these pressures and Watkinson et al. (2000) ask how genetic modification of sugar beet (*Beta vulgaris*) might impact on birds that depend on the seeds of cropland weeds. This is not a traditional population viability analysis. Rather, Watkinson's team combines the correlation approach described in Section 5.2 (correlating bird density with the seeds upon which they feed) with a detailed knowledge of the behavior of individual birds. The final ingredient is to test the consequence of assumptions about the way farmers will behave in relation to the new technology.

Sugar beet has been genetically modified to be resistant to the broad-spectrum herbicide glyphosate. This allows the herbicide to be used to effectively control the weeds that usually compete with the crop. Fat hen (*Chenopodium album*), one weed that will be affected adversely, produces seeds that are important winter food for farmland birds, including the skylark (*Alauda arvensis*). Watkinson and his colleagues take advantage of the fact that the population biology of both fat hen and skylarks is well studied and can be readily incorporated into a model of the impacts of GM sugar beet.

In a typical 5-year crop rotation in eastern England, sugar beet is grown every fifth year with cereals in intervening years. Fat hen can only establish in year 5, when sugar beet is grown, persisting between beet crops as dormant seed banks. Survival from germination to flowering depends on weed control regimes (whether traditional or involving GM sugar beet and glyphosate), whilst seed production depends on competition for resources between weed and crop plants.

Skylarks do not feed only on fat hen, but the weed is a suitable 'model' species to generate predictions of how GM sugar beet might affect the weed–bird interaction. Figure 5.13 shows how skylarks aggregate in fields in response to seed density. This

Fig. 5.13 (a) The relationship between density of skylarks (per hectare) in fields in Norfolk, England and weed seed density per square meter near the soil surface. (From Robinson & Sutherland 1999.) (b) Frequency distributions of mean seed densities across farms before the introduction of GM sugar beet (solid line), and in two situations where the technology has been adopted: where the technology is preferentially adopted on farms where weed density is currently high (dotted line) and where it is currently low (dashed line). (After Watkinson et al., 2000.)

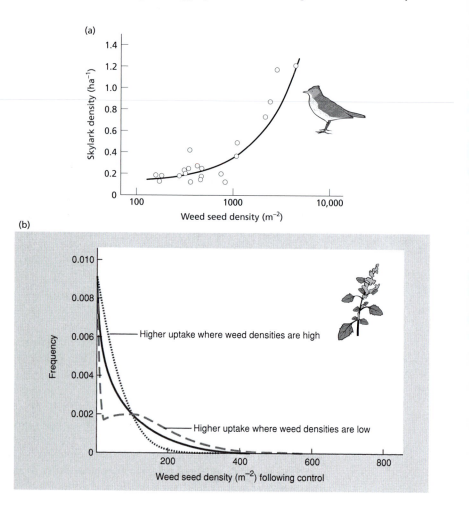

Fig. 5.14 The relative density of skylarks in fields in winter (vertical axis; unity indicates field use before the introduction of GM crops) in relation to ρ (horizontal axis; positive values mean farmers are more likely to adopt GM technology where seed densities are currently high, negative values where seed densities are currently low) and to the approximate reduction in weed seed bank density due to the introduction of GM crops (Γ, third axis; realistic values are those less than 0.1). (After Watkinson et al., 2000.)

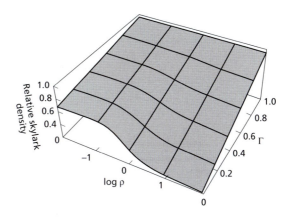

means that the impact of GM sugar beet will depend critically on the extent to which high-density patches of weeds are affected.

The steps in building their model of GM sugar beet/fat hen/skylark populations are as follows. First, they built a general model of the population dynamics of fat hen in a sugar beet crop. Next they modified this to simulate the effect of the introduction of GM sugar beet on the weed population (densities of fat hen are expected to be reduced by about 90%). Then they incorporated the predicted use of fields by skylarks according to weed seed density (as shown in Figure 5.13a).

Farmer behavior was also modeled. Before the introduction of GM technology most farms had a relatively low density of weed seeds, but a few had very high densities (solid line in Figure 5.13b). The probability of a farmer adopting GM crops is related in the model to seed bank density through a parameter ρ. Positive values of ρ mean that farmers are more likely to adopt the technology where seed densities are currently high and there is the potential to reduce yield losses to weeds. This leads to an increase in the relative abundance of low-density fields (dotted line in Figure 5.13b). Negative values of ρ indicate that farmers are more likely to adopt the technology where seed densities are currently low (intensively managed farms), perhaps because a history of effective weed control is correlated with a willingness to adopt new technology. This leads to a decreased frequency of low-density fields (dashed line in Figure 5.13b). The way that farmers will respond is not self-evident and Watkinson's team argued that this needs to be included in the model.

It turns out that the relationship between current weed levels and uptake of the new technology (ρ) is as important to bird populations as the direct impact of the technology on weed abundance (Γ), because small positive or negative changes to ρ give quite different skylark densities (Figure 5.14). In other words, predicting the impact of GM technology on biodiversity requires not only an understanding of how the ecological system operates but also of how the farming community will respond.

5.6.2 Diclofenac – good for sick cattle, bad for vultures

In early 1997 white-backed vultures began dropping from their perches to die shortly afterwards. Local people in India and other parts of South Asia were quick to notice dramatic declines in vulture numbers, but the reasons perplexed scientists. A conference called in Mumbai, India, in August 1999 to consider the possible causes was eerily reminiscent of a similar meeting held 34 years earlier in

Table 5.4 Modeled percentages of ungulate carcasses with lethal levels of diclofenac required to cause population declines at rates observed for long-billed vultures (LBW) or white-backed vultures (WBW) in India and Pakistan between 2000 and 2003. For each population, results are given for four feasible adult survival rates (S_0) in the absence of diclofenac, and three values of the interval between vulture feeding bouts in days, F. This approach has much in common with the sensitivity analyses described in Sections 5.4.1–5.4.3 (From Green et al., 2004.)

		Percentage of carcasses with lethal level			
	F	$S_0 = 0.90$	$S_0 = 0.95$	$S_0 = 0.97$	$S_0 = 0.99$
LBV India	2	0.132	0.135	0.137	0.138
	3	0.198	0.202	0.205	0.208
	4	0.263	0.271	0.273	0.277
WBV India	2	0.339	0.347	0.349	0.350
	3	0.508	0.521	0.526	0.533
	4	0.677	0.693	0.699	0.700
WBV Pakistan	2	0.360	0.368	0.372	0.376
	3	0.538	0.551	0.558	0.564
	4	0.730	0.734	0.743	0.751

Wisconsin, USA, to discuss the plight of the peregrine falcon (*Falco peregrinus*) (Risebrough, 2004). In their case, it turned out that DDE (a derivative of the agricultural insecticide DDT) had moved through the food web, accumulated in falcon tissue and caused eggshells to become thin and easily crushed during nesting, causing populations in Britain and the USA to plummet.

What about the decline of the vultures? It took a few years to determine a common element in the deaths of otherwise healthy birds – each had suffered from visceral gout (accumulation of uric acid in the body cavity) followed by kidney failure. Furthermore, vultures dying of visceral gout were found to contain residues of the drug diclofenac, whereas birds dying of other, known, causes did not (Oaks et al., 2004); and carcasses of domestic animals treated with diclofenac proved lethal to captive vultures. Diclofenac, a nonsteroidal anti-inflammatory drug developed for human use in the 1970s, came into common use in the past decade as a veterinary medicine in Pakistan and India. Thus a drug with evident agricultural benefit to domestic mammals proved lethal to birds that fed upon them.

Given the relatively small numbers of dead mammals that had been treated with diclofenac, was the associated vulture mortality sufficient explanation for the population crashes? This is the question addressed by Green et al. (2004) by means of a simulation population model. On the basis of population surveys and knowledge of demographic rates for both white-backed and long-billed vultures (see Section 5.1), they built models to predict population behavior in a similar way to the studies I discussed in Section 5.4. But Green's team also posed the specific question: what proportion of carcasses would have to contain lethal doses of diclofenac to cause the observed population declines?

Table 5.4 provides the answers for a range of feasible scenarios. Pre-diclofenac adult annual survival rates are unknown, but information for related species suggests that annual survival will lie somewhere in the modeled range of 0.90–0.99 (i.e. 90–99% of adults survive each year). Moreover, vultures vary in feeding frequency and this has a bearing on the outcome: models were run for intervals between feeding bouts of 2, 3 and 4 days.

You can see that at a maximum (for the Pakistani white-backed vultures when adult survival is set at 0.99 and feeding interval is 4 days) only 0.751% or, in other words, 1 in 133 carcasses need to be dosed with diclofenac to cause the observed population decline. At a minimum (for Indian long-billed vultures when adult survival is 0.90 and feeding interval is 2 days) only 0.132% or 1 in 757 carcasses are needed. Proportions of vultures found dead or dying in the wild with signs of diclofenac poisoning were similar to the proportions of deaths expected from the model if the observed population decline was due entirely to diclofenac poisoning.

The researchers recommended urgent action to prevent the exposure to livestock carcasses contaminated with diclofenac and we can note, for example, that the Punjab government has now banned its use. Green and his colleagues also highlighted the need for research to identify alternative drugs that are effective in livestock and safe for vultures. Clearly, on the one hand, the banning of a veterinary option can be expected to have consequences for the agricultural economy. On the other hand, the unmeasured economic benefit of ecosystem services provided by vultures is likely to be many times as great as the economic loss of banning this particular drug. The social cost of vulture decline is the heightened disease risk associated with slow decomposition of large mammal carcasses, while the cultural cost to Parsees has been the disruption of funeral rites (Section 5.1). Finally, given the depths to which the vulture populations have sunk, Green's team point to the importance of holding and breeding vultures in captivity until diclofenac is under control. This is a sensible precaution to ensure the long-term survival of the threatened species and to provide for future reintroduction programs.

Summary

Population dynamics

The individual organisms of a single species are collectively known as a population. When confronted with a declining population, managers need to call on population dynamics theory, which seeks to understand what determines population size and the way this varies through time. A population's dynamics depend on the interplay of processes that increase (birth and immigration) or decrease population size (death and emigration). Some processes act on rates of birth, death and migration in a density-dependent manner (bigger percentage effect at higher density). However, the abundance we observe is determined by the combined effects of all the processes that affect the population, whether they are dependent or independent of density.

The population dynamics of small populations

Many species are naturally rare, but others have had rarity thrust upon them by human actions. The probability of their extinction is enhanced because of features of the population dynamics of small populations. The behavior of small populations is governed by a high level of uncertainty, whereas large populations can be described as being governed by the law of averages. Three kinds of uncertainty are of particular importance to small populations: *demographic uncertainty* (e.g. random variations in the number of individuals that are born male or female, or in the number that happen to die or reproduce in a given year), *environmental uncertainty* (unpredictable changes in environmental factors such as storms, floods or droughts) and *spatial uncertainty* (the patchy dynamics of extinction and local recolonization of populations that exist as a number of relatively discrete subpopulations).

Using an understanding of population dynamics, managers often need to decide what constitutes the 'minimum viable population'. There are three basic approaches: (i) a simple correlational approach to identify easily measured factors that are correlated with extinction risk; (ii) the use of general algebraic population models when detailed population information is lacking; and best of all (iii) the use of specific population viability analyses, involving simulation models designed for particular species at risk. Population viability analyses appear to produce good predictions of population performance, but the results need to be used cautiously. They are not so much aimed at precise estimation of extinction probability or the predicted time to extinction, but to allow managers to compare the likely outcomes of alternative management scenarios.

The population genetics of small populations

Populations vary in the amount of genetic variation they contain, and in small populations this can have profound implications for their persistence. Genetic variation in a population is determined by the joint action of *natural selection* (where the frequency of a particular form of a gene in a population is related to the evolutionary advantage it confers) and *genetic drift* (where the frequency is determined simply by chance). The influence of genetic drift is greater in small isolated populations, which as a consequence are expected to lose genetic variation. This may prove to be a problem because rare genetic forms that confer no immediate advantage might turn out to be well suited to changed environmental conditions in the future. A more immediate potential problem is *inbreeding depression*. When populations are small there is a tendency for related individuals to breed with one another, increasing the likelihood of expression of harmful forms of genes.

The objectives of managers, whether they are concerned with captive rearing programs or translocations of individuals from vulnerable to safe habitats, are straightforward – maximize genetic diversity and minimize the risks of inbreeding depression. This usually involves a process of genetic mixing, based on knowledge of relatedness among the breeding population, to reduce inbreeding effects and thus increase the likelihood of persistence of the population.

A benefit of modern molecular technology is that it can be used to uncover patterns of genetic variation within a species, identifying evolutionary significant units (possibly subspecies) that are worthy of special conservation focus.

The extinction vortex

There may be an interaction between the population dynamics and population genetics of declining populations. Thus, a population may be reduced to a very small size by habitat loss, pollution, disease or overexploitation, leading to an increased frequency of matings amongst relatives and the expression of deleterious forms of genes in offspring, leading to reduced survivorship and birth rates, and causing the population to become smaller still – the so-called extinction vortex.

The final word

It's a great relief to Jijanji (Section 5.1) that the reason for the vulture collapse in India is now understood and that action has begun to be taken. It has occurred to Jijanji that it might be helpful if the economic value of an ecosystem service, like that performed by vultures, could actually be calculated. 'Wouldn't this enable our government to view the wonderful service provided by vultures in the wider context of economic activities that put the service at risk?'

Check out the website http://ecovalue.uvm.edu/evp/doc_economic_methods.asp to discover the various ways that ecosystem services might have a value put on them.

Now outline how you would go about assessing the value of the services performed by Jijanji's vultures.

References

Aguilar, A., Roemer, G., Debenham, S. et al. (2004) High MHC diversity maintained by balancing selection in an otherwise genetically monomorphic mammal. *Proceedings of the National Academy of Sciences, USA* 101, 3490–3494.

Akçakaya, H.R. (1992) Population viability analysis and risk assessment. In: *Proceedings of Wildlife 2001: Populations* (D.R. McCullough, ed.), pp. 148–158. Elsevier, Amsterdam.

Akçakaya, H.R., Burgman, M.A. & Ginzburg, L.R. (1999) *Applied Population Ecology: principles and computer exercises using RAMAS EcoLab 2.0*, 2nd edn. Sinauer Associates, Sunderland, Massachusetts.

ANZECC (Australia and New Zealand Environment and Conservation Council) (1998) *National Koala Conservation Strategy*. Environment Australia, Canberra.

Begon, M., Townsend, C.R. & Harper, J.L. (2006) *Ecology: from individuals to ecosystems*, 4th edn. Blackwell Publishing, Oxford.

Brook, B.W., O'Grady, J.J., Chapman, A.P., Burgman, M.A., Akcakaya, H.R. & Frankham, R. (2000) Predictive accuracy of population viability analysis in conservation biology. *Nature* 404, 385–387.

Caughley, G. (1994) Directions in conservation biology. *Journal of Animal Ecology* 63, 215–244.

Franklin, I.R. & Frankham, R. (1998) How large must populations be to retain evolutionary potential. *Animal Conservation* 1, 69–73.

Green, R.E. & Gibbons, D.W. (2000) The status of the corncrake *Crex crex* in Britain in 1998. *Bird Study* 47, 129–137.

Green, R.E., Newton, I., Shultz, S. et al. (2004) Diclofenac poisoning as a cause of vulture population declines across the Indian subcontinent. *Journal of Applied Ecology* 41, 793–800.

Haydon, D.T., Laurenson, M.K. & Sillero-zubiri, C. (2002) Integrating epidemiology into population viability analysis: managing the risk posed by rabies and canine distemper to the Ethiopian wolf. *Conservation Biology* 16, 1372–1385.

Jamieson, I.G. (in press) Has the debate over genetics and extinction of island endemics truly been resolved? *Animal Conservation*.

Johnson, J.A., Toepfer, J.E. & Dunn, P.O. (2003) Contrasting patterns of mitochondrial and microsatellite population structure in fragmented populations of greater prairie-chickens. *Molecular Ecology* 12, 3335–3347.

Kephart, S.R. (2004) Inbreeding and reintroduction: progeny success in rare *Silene* populations of varied density. *Conservation Genetics* 5, 49–61.

Kinnaird, M.F. & O'Brien, T.G. (1991) Viable populations for an endangered forest primate, the Tana River crested mangabey (*Cerocebus galeritus galeritus*). *Conservation Biology* 5, 203–213.

Lacey, R.C. (1993) VORTEX – a model for use in population viability analysis. *Wildlife Research* 20, 45–65.

Land, D., Cunningham, M., Lotz, M. & Shindle, D. (2001) *Florida Panther Genetic Restoration and Management. Annual Report 2000–2001*. Florida Fish and Wildlife Conservation Commission, Tallahassee, FL.

Lotts, K.C., Waite, T.A. & Vucetich, J.A. (2004) Reliability of absolute and relative predictions of population persistence based on time series. *Conservation Biology* 18, 1224–1232.

Menges, E.S. (2000) Population viability analyses in plants: challenges and opportunities. *Trends in Ecology and Evolution* 15, 51–56.

Menges, E.S. & Dolan, R.W. (1998) Demographic viability of populations of *Silene regia* in midwestern prairies: relationships with fire management, genetic variation, geographic location, population size and isolation. *Journal of Ecology* 86, 63–78.

Morris, W., Doak, D., Groom, M. et al. (1999) *A Practical Handbook for Population Viability Analysis*. The Nature Conservancy, Arlington, Virginia.

Morris, W.F., Bloch, P.L., Hudgens, B.R., Moyle, L.C. & Stinchcombe, J.R. (2002) Population viability analysis in endangered species recovery plans: past use and future improvements. *Ecological Applications* 12, 708–712.

Oaks, J.L., Gilbert, M., Virani, M.Z. and ten others (2004) Diclofenac residues as the cause of vulture population decline in Pakistan. *Nature* 427, 629–633.

O'Grady, J.J., Burgman, M.A., Keith, D.A. et al. (2004) Correlations among extinction risks assessed by different systems of threatened species categorization. *Conservation Biology* 18, 1624–1635.

Penn, A.M., Sherwin, W.B., Gordon, G., Lunney, D., Melzer, A. & Lacy, R.C. (2000) Demographic forecasting in koala conservation. *Conservation Biology* 14, 629–638.

Possingham, H.P., Andelman, S.J., Burgman, M.A., Medellin, R.A., Master, L.L. & Keith, D.A. (2002) Limits to the use of threatened species lists. *Trends in Ecology and Evolution* 17, 503–507.

Primack, R.B. (1993) *Essentials of Conservation Biology*. Sinauer Associates, Sunderland, MA.

Reed, D.H., O'Grady, J.J., Brook, B.W., Ballou, J.D. & Frankham, R. (2003) Estimates of minimum viable population sizes for vertebrates and factors influencing those estimates. *Biological Conservation* 113, 23–34.

Risebrough, R. (2004) Fatal medicine for vultures. *Nature* 427, 596–598.

Robinson, R.A. & Sutherland, W.R. (1999) The winter distribution of seed eating birds: habitat destruction, seed depletion and seasonal depletion. *Ecography* 22, 447–454.

Rodrigues, A.S.L., Pilgrim, J.D., Lamoreux, J.F., Hoffmann, M. & Brooks, T.M. (2006) The value of the IUCN Red List for conservation. *Trends in Ecology and Evolution* 21, 71–76.

Roelke, M.E., Martenson, J.S. & O'Brien, S.J. (1993) The consequences of demographic reduction and genetic depletion in the endangered Florida panther. *Current Biology* 3, 340–350.

Schultz, C.B. & Hammond, P.C. (2003) Using population viability analysis to develop recovery criteria for endangered insects: case study of the Fender's blue butterfly. *Conservation Biology* 17, 1372–1385.

Stowe, T.J., Newton, A.V., Green, R.E. & Mayes, E. (1993) The decline of the corncrake *Crex crex* in Britain and Ireland in relation to habitat. *Journal of Applied Ecology* 30, 53–62.

Swinnerton, K.J., Groombridge, J.J., Jones, C.G., Burn, R.W. & Mungroo, Y. (2004) Inbreeding depression and founder diversity among captive and free-living populations of the endangered pink pigeon *Columba mayeri*. *Animal Conservation* 7, 353–364.

Thomas, C.D. (1990) What do real population dynamics tell us about minimum viable population sizes? *Conservation Biology* 4, 324–327.

Townsend, C.R., Begon, M. & Harper, J.L. (2003) *Essentials of Ecology*, 2nd edn. Blackwell Science, Oxford.

Watkinson, A.R., Freckleton, R.P., Robinson, R.A. & Sutherland, W.J. (2000) Predictions of biodiversity response to genetically modified herbicide-tolerant crops. *Science* 289, 1554–1557.

Wayne, R.K. & Morin, P.A. (2004) Conservation genetics in the new molecular age. *Frontiers in Ecology* 2, 89–97.

6 Pest management

Mosquitoes are pests because they carry diseases or because their bites itch, weeds because they reduce farm productivity or spoil our gardens, rats because they feast on stored food or frighten the kids. Pests can carry economic, health, environmental and even aesthetic costs. People want rid of them all.

Chapter contents

6.1 Introduction	140
6.1.1 One person's pest, another person's pet	140
6.1.2 Eradication or control?	141
6.2 Chemical pesticides	146
6.2.1 Natural arms factories	146
6.2.2 Take no prisoners	147
6.2.3 From blunderbuss to surgical strike	147
6.2.4 Cut off the enemy's reinforcements	150
6.2.5 Changing pest behavior – a propaganda war	150
6.2.6 When pesticides go wrong – target pest resurgence and secondary pests	151
6.2.7 Widespread effects of pesticides on nontarget organisms, including people	153
6.3 Biological control	154
6.3.1 Importation biological control – a question of scale	155
6.3.2 Conservation biological control – get natural enemies to do the work	156
6.3.3 Inoculation biological control – effective in glasshouses but rarely in field crops	158
6.3.4 Inundation biological control – using fungi, viruses, bacteria and nematodes	159
6.3.5 When biological control goes wrong	160
6.4 Evolution of resistance and its management	162
6.5 Integrated pest management (IPM)	164
6.5.1 IPM against potato tuber moths in New Zealand	165
6.5.2 IPM against an invasive weed in Australia	166

Key concepts

In this chapter you will

recognize that a pest is simply a species that (some) humans consider undesirable – one person's pest is another person's pet

note that the aim of pest control is often not eradication but maintenance at a level below which further control is unjustified

appreciate that the management of pests, as of endangered species, depends on a thorough understanding of population dynamics theory understand the pros and cons of chemical pesticides, their diversity and varied modes of action

recognize that biological control agents (predators, parasites and pathogens) can be used to deal with pests, but sometimes with adverse consequences for other species

realize that, like all natural populations, pests can evolve – there are many cases of evolution of resistance, particularly to pesticides but also to other agents

conclude that effective pest control usually requires a range of measures to be used in concert to maximize economic benefit but minimize adverse effects – this is Integrated Pest Management (IPM)

6.1 Introduction

'I don't think we did anything wrong – and I'd do the same thing again.' Jenny is adamant. She's a farmer in New Zealand and, like generations before her, she hates rabbits. 'You have to wonder why on earth they were shipped to New Zealand in the first place – something to do with making European settlements more like home.' Like many exotic species, the rabbits thrived and built up extraordinarily dense populations, competing with sheep for fodder and in some areas creating a wasteland. 'Our main weapons against rabbits were shooting and poisoning but the expense involved was never-ending. It is hardly surprising that the farming community was keen on a 'biological' solution – the introduction of a disease that would bring down numbers and keep them down.' A proposal in 1993 to introduce myxomatosis, a viral disease specific to rabbits that causes evident pain and a lingering death, was abandoned after 100,000 people signed a petition. Then in 1995 Australia and New Zealand got together to assess rabbit calicivirus disease as a suitable control agent. Experimentation was meant to be confined to a small island off the Australian coast, but the virus escaped and spread like wildfire across the mainland, killing a large proportion of rabbits as it moved through the landscape. The New Zealand farming community asked for the disease to be imported to New Zealand but the government declined. Then in 1997, farmers took things into their own hands. 'I had my ten minutes of fame when the TV news showed me with a virus-rich concoction of minced rabbit organs in my fridge!' Her farming friends smiled at that, but you can imagine she received hate mail too. 'Environmentalists said that a disease only known in European rabbits might conceivably spread to kiwis – can you believe that?'

Wherever in the world rabbit calicivirus has been deliberately introduced, or arrived in rabbit meat, it has reduced populations by 50–90%, being particularly effective in dry, warm areas. Baby rabbits develop immunity, and mortality may decline with time or the population may follow a cycle of low and high densities. However, there seems no doubt that agricultural productivity improved in parts of New Zealand, at least in the short term.

6.1.1 One person's pest, another person's pet

Those worried about the introduction of rabbit calicivirus disease included pet lovers, because domestic rabbits are derived from the European rabbit (*Oryctolagus cuniculus*) and are therefore susceptible to the disease. In fact, it is by no means unusual for one person's pest to be another person's 'pet'. There are fishermen who would like to cull burgeoning seal populations to protect their fish stocks, but others feel just as strongly that seals should be protected at all costs. There are rabbiters in the UK who cherish the domestic ferrets (*Mustela furo*) they use to chase rabbits from their burrows into waiting nets. But ferrets were imported to New Zealand in an early attempt to control rabbits – their descendants are now responsible for the extinction or endangerment of many of New Zealand's native birds and unique insects. Cat lovers abound, but predation by domestic and feral cats is a leading cause of bird and lizard extinction around the world. Then there are the 'pet' plants beloved by gardeners that escape into the wild and become pests, displacing native species (Section 3.1).

So what exactly do we mean by the term 'pest'? The simple answer is *a species that (some) humans consider undesirable*. This definition covers a multitude of sins. Mosquitoes are pests because they carry diseases or because their bites itch; *Allium* spp. are pest plants (weeds) because when harvested with wheat they make bread taste of onions; rats and mice are pests because they feast on stored food; garden weeds are pests for aesthetic reasons. Pests carry an economic cost, either because of direct damage to health or economic activity, or simply because of a willingness to spend money to counter their nuisance value. People want rid of them all.

And where do pests come from? Many are exotic imports – such as rabbits, zebra mussels and yellow star thistles (Section 1.2.5). Others are native but become an economic problem when a new crop is introduced to a region. For example, when European grasses appeared in New Zealand pasture, the native grass grub (larva of the beetle *Costelytra zealandica*) switched from native tussock grass to the new pasture species and became a widespread pest. Other changes in agricultural practice have also turned native species into economic problems. The tendency to create monocultures, something that went hand in hand with increased mechanization, reduced habitat heterogeneity in a way that was deleterious to the natural enemies of some crop-feeding herbivores, which increased to pest proportions as a result. We will also see later how the application of pesticides can turn previously innocuous species into pests.

Then there are species that my grandparents would not have considered pests, but which my children certainly do. It used to be quite normal for people to eat blemished apples, but our standards are now so exacting that the insects and fungi responsible for blemishes have acquired pest status.

And finally there are pests that seem to arise from nowhere. For example, a significant decline in rice productivity in Colombia was first attributed to soil compaction (so they increased ploughing), then to aphid damage and finally to nematode infestation (prompting very high pesticide use) before the true cause was determined – a newly discovered virus (rice stripe necrosis virus). Such 'emerging infectious diseases' seem to be a modern affliction of plants in much the same way that HIV and mad cow disease have affected human and animal populations (Anderson et al., 2004).

6.1.2 *Eradication or control?*

In Chapter 5 I considered how to apply population dynamics theory to the conservation of endangered species. The trick there was to work out what was necessary to increase density and sustain the population into the future. My aim in this chapter is to do the exact opposite – to drive a pest population extinct or keep density so low that its nuisance value is negligible. Despite the diametrically opposed aims of Chapters 5 and 6, they both rest on the same comprehensive knowledge of population dynamics, genetics and evolution. The population concepts outlined in Boxes 5.1 and 5.2 are extended in Box 6.1 to underpin pest management. These boxes should be considered together.

You might imagine that the general aim of pest controllers would be to eradicate the population concerned. However, this is rarely attempted and even more rarely achieved. It can actually be very difficult to eradicate a pest. This might sound paradoxical when in the last chapter we bemoaned the fact that people are so good at driving endangered species extinct. But, as already noted, the species most vulnerable to extinction tend to be those with life-history traits that are linked to low reproductive rates (*K*-selected: Section 3.4). The majority of pests, on the other hand, possess the opposite traits and have very high reproductive rates (*r*-selected: Box 3.1). Moreover, the success of pests is often related to our provision of abundant and high quality food – pests of rice love it for the same reason we do! Finally, exotic pests often arrive in their adopted countries without many of their natural enemies. Fast reproduction, abundant food, few enemies – a recipe for success if ever there was one.

So eradication may not be feasible, or at least would cost so much in time, effort and money that it is usually not attempted. But there is a general exception, and this concerns invaders that have only just appeared on the scene. The arrival of an exotic species with a high likelihood of becoming a significant pest should be a matter for urgent action, because this is the stage at which eradication is both feasible and easy to justify economically (Simberloff, 2003).

Such campaigns sometimes rely on fundamental knowledge of population ecology. An example is the eradication of the South African polychaete worm *Terebrasabella heterouncinata*, a parasite of abalone and other gastropods that became established near an abalone aquaculture facility in California (Culver & Kuris, 2000). Its population biology was understood sufficiently to know that the parasitic worm was specific to gastropods, that two species in the snail genus *Tegula* were its principal hosts in the area and that large snails were most susceptible to the parasite. Volunteers removed 1.6 million of the larger individuals, thereby reducing the density of susceptible hosts below that needed for parasite transmission, which became extinct. This is an example of 'physical pest control' (the physical removal of pests).

However, in the words of Simberloff (2003), rapid responses to recent invaders will often 'resemble a blunderbuss attack rather than a surgical strike'. Another marine invader provides a graphic illustration. Within 9 days of the discovery of the Caribbean black-striped mussel (*Mytilopsis sallei*) in a small bay near Darwin in Australia, the area was quarantined and treated with 160,000 liters of liquid bleach and 6000 tonnes of copper sulfate. The mussel population was eradicated (Bax et al., 2001), but so was everything else. A blunderbuss approach indeed, but the secret to success was the knowledge that native species would recolonize from nearby source populations while the invader could not. Early action using brute-force methods was also successful for a string of successful eliminations of small populations of weeds such as pampas grass (*Cortaderia selloana*) and ragwort (*Senecio jacobaea*) on New Zealand's offshore islands (Timmins & Braithwaite, 2002). Eradication of a recently established species known to be a problematic invader elsewhere cannot (and should not) wait for new population studies to be performed.

Occasionally, long-established pests have also been successfully eliminated. The global effort to exterminate the smallpox virus provides an outstanding example, and a similar effort to drive the poliomyelitis virus to extinction also looks set to succeed. We can point to more mundane examples too. For example, an eradication campaign in the eastern USA aimed at a root parasite, African witchweed (*Striga asiatica*), is based on strict quarantine of the infested area, the use of herbicide against the plants and soil fumigation to destroy the long-lived seed-bank (Eplee, 2001). But a much more notable success involved the screwworm fly (*Cochliomyia hominovorax*) in a large area of the USA and Central America, and later when it invaded North Africa – it was eradicated by a bit of biological trickery. The fly cements its eggs near a wound on an animal such as a sheep or cow, and occasionally on people (*hominovorax* actually means man eater). The maggots enter and feed in the wound, producing a foul-smelling, pus-discharging sore. The surprising trick was to breed more screwworm flies, but only males – a large factory was established to breed millions of them. By swamping the population with sterilized males (pupae exposed to gamma radiation), the females breed but produce no eggs – the birth rate drops to zero and the population disappears. It really works.

Most often, however, once invaders have established and spread through a new area and are classified as pests, they become just one more species at which the pest manager's armory must be directed. Eradication is not usually the objective. Rather, the aim is to reduce the pest population to a level at which it does not pay to achieve yet more control (the Economic Injury Level). The underlying theory is outlined in Box 6.1.

We can attempt to combat pests by physical means (e.g. simply keeping invaders from arriving, keeping pests away from target crops by planting preferred 'trap' crops in their path or picking them off by hand when they arrive), by 'cultural control' (e.g. rotating crops planted in a field so pests cannot build up their numbers over several years), or using chemical or biological agents.

My emphasis will be on the latter two classes. You will learn how chemical pesticides actually work and, by exploring a range of examples, understand how to avoid unwanted outcomes such as pest resurgence and secondary pest outbreaks (Section 6.2). The basic theory behind these topics is presented in Box 6.1. An alternative to pesticide application is biological control – the introduction or augmentation of natural enemies of the pest (Section 6.3). When this works, the benefits can be huge. But, as with pesticides, there may also be a downside when nontarget species are adversely affected. One of the biggest problems facing pest managers is the tendency of pests to evolve resistance to control agents, whether chemical, biological or cultural – I discuss the implications of evolution of resistance in Section 6.4. The optimal pest control strategy, taking into account the pros and cons of each approach, turns out to be a judicious selection from the complete pest control armory – this is termed integrated pest management (Section 6.5).

Box 6.1 Population dynamics theory 2

> ### Economic Injury Level (EIL)
> Consider a pest of a farm crop. Since every control measure has an associated cost (e.g. payment for pesticide and the equipment and labor to administer it), pest control is only worthwhile if the economic gain from increased crop productivity outweighs the cost of the control measure. The EIL then is the pest population size at which it does not pay to achieve yet more control. Above the EIL, pesticide application is worthwhile because there is a net benefit to the farmer; below the EIL it does not pay.
>
> Every herbivorous insect that feeds on a crop might conceivably achieve pest proportions, but whether it will do so depends on the EIL for that particular species, the densities its population typically achieves and the degree to which its density fluctuates. You saw earlier how a variety of

influential environmental factors determine a species' equilibrium density (or carrying capacity, Box 5.1). This is shown as EP (Equilibrium Population) in Figure 6.1a. In this case, the population stays at a low and relatively stable density, far below its EIL. It is not a pest, but could become one if circumstances changed and its birth or immigration rate increased, or its death or emigration rate decreased.

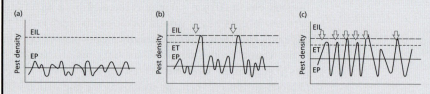

Fig. 6.1 Population fluctuations of hypothetical pests about their equilibrium population (EP) density – set by physicochemical environmental conditions and the pest's interactions with its food and enemies. (a) The EP of this species is well below the economic injury level (EIL) at which it would be worthwhile to apply control measures. (b) Population fluctuations of this pest break through the EIL in some years. Control measures should be applied (arrows) when density reaches the economic threshold (ET), so that the measures take effect before the EIL is reached. (c) This population fluctuates through the ET in most years.

The Economic Threshold (ET)

The hypothetical species shown in Figure 6.1b, on the other hand, has a density that fluctuates markedly and occasionally breaks through the EIL. When a pest population has reached a density at which it is causing economic injury, however, it is generally too late to start controlling it. So let's add another idea into the mix. ET is the density at which action should be taken against a pest to prevent it reaching the EIL. The EP of the pest illustrated in Figure 6.1c is closer to its EIL (than those in Figure 6.1a or b), and its wide population fluctuations take it above the ET most years.

EILs and ETs are predictions based on detailed studies of pest populations and past outbreaks, and usually require the pest controller to monitor the pest, weather conditions and even densities of the pest's natural enemies, because these may help keep pest density down. For example, the prescription for the spotted alfalfa aphid (*Therioaphis trifolii*) on hay alfalfa in California is to apply pesticide in spring or winter when aphids have reached densities of 40 or 50–70 per stem, respectively. In summer and autumn, however, the prescription is more complex: apply when the population reaches 20 aphids per stem, but do not treat the first three cuttings of hay if the ratio of ladybirds (beetle predators of the aphids) to aphids is at least one adult per 5–10 aphids or three ladybird larvae per 40 aphids (Flint & van den Bosch, 1981). The most useful estimates of EIL and ET involve detailed cost–benefit economic analyses (Ramirez & Saunders, 1999) that take into account the current value of the crop – because when the value of the crop at the farm gate is high, it pays to control pests at a lower EIL.

Target pest resurgence

In the history of pesticide use there have been many examples where pest density increases rapidly again soon after application. There are two explanations.

Note first that if the pest population was originally subject to density-dependent processes (lower rates of birth or immigration, or higher rates of death or emigration, at high than low density), reduction of its density by pesticide application will cause density-dependent effects to be relaxed. Thus, the population may quickly return to its original equilibrium level (Figure 6.2a). The speed at which this happens will depend on the pest's intrinsic rate of natural increase (r) (Box 5.1). Pests usually have very high values for r (that's one reason they become pests), so numbers can build up very rapidly and, as a result, several pesticide applications may be needed each year. Recovery of the pest population will be delayed if the pesticide persists in active form in the environment for months or years.

There have also been many records of a pest population recovering after pesticide application to densities far above the original level. This is known as 'target pest resurgence'. A pesticide gets a bad name if, as is often the case, it kills more species than just the one at which it is aimed. In the context of agricultural productivity, the bad name is especially justified if the pesticide kills the

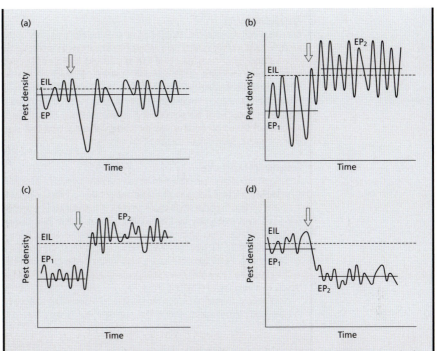

Fig. 6.2 (a) A pest population whose equilibrium population is close to the EIL is knocked down to a low density by pesticide application (arrow), but quickly regains its original density because of the relaxation of density-dependent processes. (b) Target pest resurgence occurs when the pesticide has a stronger or longer-lasting effect on the natural enemies of the pest than on the pest population itself. This allows the pest to rise to a new equilibrium population (EP_2) that is maintained until the natural enemies re-establish themselves. (c) In this case, an innocuous herbivore is fluctuating around an EP that is well below the EIL. After a pesticide has been applied to control another species, the natural enemies of the innocuous herbivore are greatly reduced, allowing it to increase to a much higher equilibrium population (EP_2) where it achieves pest status – a secondary pest has been created. (d) This population trajectory illustrates what can happen when a biological control agent is introduced (arrow) – the extra mortality caused by the introduced enemy shifts the target pest from an equilibrium population where the EIL is frequently breached to a new equilibrium well below the EIL. (Economic thresholds are omitted from these figures for the sake of clarity.)

pests' natural enemies and so contributes to undoing what it was employed to do. In this case, target pest resurgence occurs when treatment kills both large numbers of the pest *and* large numbers of its natural enemies. Pest individuals that survive the pesticide or that migrate into the area find themselves with a plentiful food resource and few, if any, natural enemies. The pest population may then explode to fluctuate around a higher equilibrium population density (Figure 6.2b) until enemy populations re-establish.

Secondary pest outbreaks
The after-effects of a pesticide may involve even more subtle reactions. After pesticide application it may not only be the target pest whose density surges. Alongside the target we may expect a number of innocuous herbivores that have been kept in check by their natural enemies (take Figure 6.1a as an example). If the pesticide has a stronger effect on natural enemies than on the innocuous herbivore itself, this 'potential' pest becomes a real one, and is called a secondary pest (Figure 6.2c).

Biological control

Biological control involves introducing or augmenting an enemy of the target pest. This may be a predator (e.g. a ladybird that eats aphids), a parasitoid (an insect that lays its eggs in a pest insect, which is consumed from the inside out) or a disease organism. The control agent is often imported from the area where the pest originated. Figure 6.2d, which is really a mirror image of Figure 6.2c with the addition rather than subtraction of an enemy, illustrates how the pest declines from an equilibrium population around the EIL to a new, much lower equilibrium, and one that may be maintained because of the persistence, at low densities, of both target and control agent. When successful, biological control has the particular advantage that it only needs to happen once, whereas pesticides need to be used time and again.

Evolution of resistance

At best, chemical pesticides have to be used repeatedly and forever. However, most pesticides have a limited life because pests evolve resistance to them. This is simply natural selection in action. It is almost certain to occur when vast numbers of individuals in a genetically variable population are killed in a systematic way by the pesticide. One or a few individuals may be unusually resistant (perhaps because they possess an enzyme that can detoxify the pesticide). If the pesticide is applied repeatedly, each successive generation of the pest will contain a larger proportion of resistant individuals. Pests typically have a high intrinsic rate of reproduction, and so a few individuals in one generation may give rise to hundreds or thousands in the next, and resistance spreads very rapidly in a population (Figure 6.3).

It has been extremely common for resistance to pesticides to evolve. The evolution of resistance to biological control agents seems less common, but is not unheard of.

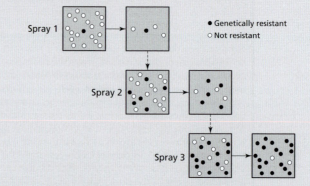

Fig. 6.3 Evolution in a pest population of resistance to a control agent. When the pesticide is first applied (Spray 1) the majority of susceptible individuals are killed while a few are naturally resistant. The latter contribute more offspring to the next generation, passing on the genes that confer resistance, and resistant individuals are better represented when Spray 2 is applied. Again, most of the susceptible individuals are killed, but now there are even more resistant individuals to contribute resistant offspring to the next generation. Within a few generations, most individuals are resistant and the pest controller has to look for another pesticide.

6.2 Chemical pesticides

Chemical pesticides are a key part of the armory of pest controllers but they have to be used cautiously because population theory (Box 6.1) predicts undesirable responses to careless attacks. You have to know your enemy's weaknesses, develop chemical weapons to exploit them, and be ready for the pests to fight back (by evolving resistance). It's an arms race.

6.2.1 Natural arms factories

Plants have been in the defence game for millennia. Relentless attack from herbivores has ensured that individual plants with features that reduce the impact of grazing pass on more than their fair share of genes to future generations (essentially the same story as illustrated in Figure 6.3). The physical defences that have evolved are obvious to every one – just think of stinging nettles and acacia thorns. The chemical defences, on the other hand, are invisible but ubiquitous and amazingly diverse. And people have known for centuries that certain insects can be fought using naturally occurring substances, collectively known as *botanicals*, such as nicotine from tobacco plants (*Nicotiana rustica*) and pyrethrum from flowers in the *Chrysanthemum* genus. Both are nerve agents, disrupting nerve impulses and killing insects. Many natural insecticides are unstable on exposure to light and air, but chemists have overcome this by producing their own versions, such as the synthetic neonicotinoids and pyrethroids that are now among the most widely used of pesticides.

6.2.2 *Take no prisoners*

The use of the so-called *inorganics* goes back to the dawn of pest control and, along with the botanicals (Section 6.2.1), they were the chemical weapons of the expanding army of insect pest managers of the nineteenth and early twentieth century. The inorganic insecticides, mostly metallic compounds or salts of copper, sulfur, arsenic or lead, are primarily stomach poisons effective only against invertebrates with chewing mouthparts. A range of inorganic herbicides was also commonly used against weeds. However, the persistence in the environment of toxic residues has led pest controllers to abandon inorganics, with rare exceptions. Borates (absorbed by plant roots and translocated to above-ground parts) are still sometimes used where no vegetation of any sort is wanted. And you heard in Section 6.1.2 how an invasion of the Caribbean black-striped mussel was greeted by the application of a massive amount of copper sulfate. The fearful inorganics still have some uses but only when the decision is to take no prisoners.

6.2.3 *From blunderbuss to surgical strike*

With advancing knowledge of cellular physiology, a chemical warfare catalog has gradually been put together. Like pyrethrum, *chlorinated hydrocarbons* disrupt nerve-impulse transmission. They are contact poisons (no need to eat them) that are insoluble in water but show a high affinity for fats, thus tending to become concentrated in animal fatty tissue, and to pass along food chains. The most notorious is DDT: a Nobel Prize was awarded for the discovery of its pest control properties in 1948, but it was suspended from all but emergency uses in the USA in 1973. It is still used in South America, Asia and Africa in an attempt to control the mosquitoes that spread malaria. Arguably, in these areas the human costs of malaria outweigh the environmental damage caused by the use of DDT.

One of the biggest problems of DDT and related compounds, such as toxaphene, dieldrin and chlordane, is their toxicity to just about everything, coupled with a very long life in the environment. They can cause particularly severe problems because of *biomagnification* – this happens when a pesticide is present in an organism that becomes the prey of another and the predator fails to excrete the pesticide. It then accumulates in the body of the predator, and so on up the food chain. Top predators (including people) in aquatic and terrestrial food chains, which of course were never intended as targets, can accumulate extraordinarily high doses (Figure 6.4). And this was the cause of a famous decline of peregrine falcons (Section 5.6.2) when calcium metabolism was upset and egg shells became dangerously thin.

The application of chlorinated hydrocarbons might be characterized as a blunderbuss approach, because it is not just the target that gets hurt. But a better analogy would be with the dreadful land mines left behind after human conflict: the weapon

Fig. 6.4 Chlordane, applied as a pesticide on land, is transported to the Arctic via rivers and oceanic and atmospheric circulation. A study in the Barents Sea shows how chlordane is biomagnified during its passage through the marine food chain. Concentrations in seawater are very low. Herbivorous copepods (that feed on phytoplankton) have higher concentrations (measured in nanograms per gram of lipid in the organisms), and predatory amphipods higher concentrations still. The polar cod (*Boreogadus saida*), which feeds on the invertebrates, and cod (*Gadus morhua*), which also includes polar cod in its diet, show further evidence of biomagnification. However, it is the higher steps in the food chain where biomagnification is most marked, because the seabirds that feed on the fish (black guillemots – *Cepphus grylle*) or on fish and other seabirds (glaucous gull – *Larus hyperboreus*) have much less ability to eliminate the chemicals than fish or invertebrates. (After Townsend et al., 2003, based on data in Borga et al., 2001.)

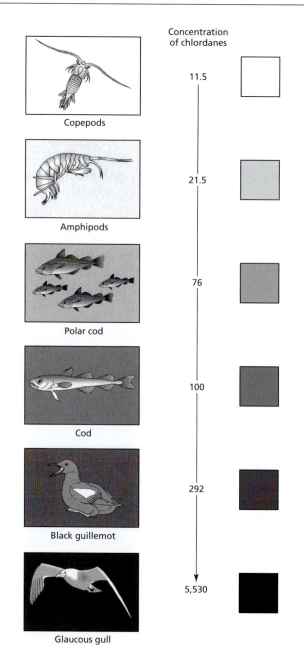

Concentration of chlordanes

Copepods — 11.5

Amphipods — 21.5

Polar cod — 76

Cod — 100

Black guillemot — 292

Glaucous gull — 5,530

continues to kill nontarget organisms for an extended period. Other chlorinated hydrocarbons have been developed with somewhat greater specificity and shorter persistence times (e.g. lindane; Table 6.1).

Organophosphates (including parathion, malathion, bidrin and azodrin) are also nerve poisons, chemically related to the ones developed during the Second World War. Some are very toxic to both insects and mammals but they do not persist so long in the environment. Thus unwanted residues in crops are less of a problem

Table 6.1 Toxicity to nontarget organisms and persistence of selected insecticides. Possible ratings range from 1 (includes zero toxicity) to a maximum of 5. Most unintended damage is done by insecticides that combine persistence with acute toxicity to a range of nontarget organisms, such as the chlorinated hydrocarbons (Section 6.2.3). Organophosphates are also broad spectrum in their effects but are less persistent in the environment, while carbamates are both more specific and shorter-lived (Section 6.2.3). The insect growth regulators (diflubenzuron and methoprene) are the most specific of all, but vary in their persistence time (Section 6.2.4). (Based on Horn, 1988.)

	Toxicity				
	Rat	Fish	Bird	Honeybee	Persistence
DDT (chlorinated hydrocarbon)	3	4	2	2	5
Lindane (chlorinated hydrocarbon)	3	3	2	4	4
Parathion (organophosphate)	5	2	5	5	2
Malathion (organophosphate)	2	2	1	4	1
Carbaryl (carbamate)	2	1	1	4	1
Diflubenzuron (chitin-synthesis inhibitor)	1	1	1	1	4
Methoprene (juvenile hormone analogue)	1	1	1	1	1

than for chlorinated hydrocarbons. And because they are not stored in animal tissue, biomagnification up the food chain has not been a problem either.

Carbamates such as carbaryl have a mode of action similar to the organophosphates, but they are rapidly detoxified and excreted and so are less toxic to nontarget mammals. However, most are extremely toxic to bees (and so can adversely affect crop pollination) and to parasitoid wasps (important natural enemies of insect pests).

Some carbamates are used against weeds, such as asulam that kills by stopping cell division. These contrast with *phenol derivatives*, such as 2-methyl-4,6-dinitrophenol, with a broad spectrum toxicity extending beyond plants to fungi, insects and mammals. They act by uncoupling one of the primary biochemical pathways in the cell – oxidative phosphorylation. The *bipyridyliums* (including two important herbicides: diquat and paraquat) also have wide toxicity. They are powerful and very fast-acting contact chemicals that destroy cell membranes. Then there are the *substituted ureas* (e.g. monuron) and the *heterocyclic nitrogen* herbicides (e.g. metribuzin): somewhat less general in their toxicity, these work by blocking electron transport in the cells.

Final mention should be given to glyphosate (a *glycine derivative*), a nonpersistent herbicide that can be applied by spraying or brushing onto the foliage of any plant; it interferes with essential protein synthesis. The specificity with which this can be applied has been further enhanced by genetic modification of certain crops to make them resistant to the pesticide, in contrast to the luckless weeds in their midst. This might sound like a win–win situation but, as discussed in the previous chapter (Section 5.6.1), the use of genetically modified crops might impact indirectly on endangered animal species that rely on the weeds. (A further risk associated with the use of genetic modification to incorporate pest resistance into a crop will be discussed in Section 6.3.4.)

The historical development of pesticides has been one where general progress has been made from the blunderbuss approach, with nontarget species being just as badly affected as the pest, to something (a bit) closer to a surgical strike – the target is taken out, collateral damage is avoided, and the control agent does not persist for long, travel far or become concentrated up food chains.

6.2.4 *Cut off the enemy's reinforcements*

In the previous section I considered pesticides that disrupt cellular functioning. Another group of pesticides upsets growth and development, thereby reducing the chance of juvenile stages surviving to become reproducing adults.

Insect growth regulators mimic natural insect hormones and enzymes, and hence interfere with normal insect growth and development. As such they are generally harmless to vertebrates and plants, but they may be as effective against a pest's natural insect enemies as against the pest itself. The two main types are the chitin-synthesis inhibitors, such as diflubenzuron, which prevent the formation of a proper exoskeleton when the insect molts, and juvenile hormone analogues, such as methoprene, which prevent insects from moulting into their adult stage, and hence reduce population size in the next generation.

The equivalent herbicides are the *phenoxy* or *hormone* weed killers. These are also relatively selective. For instance, 2,4,5-dichlorophenoxyacetic acid (2,4-D) is highly selective against broad-leaved weeds, whilst 2,4,5-trichlorophenoxyethanoic acid (2,4,5-T) is used mainly to control woody perennials. Put these together and you have the notorious Agent Orange, used to defoliate vast areas in the Vietnam War and with significant health consequences for the combatants. The hormone weed killers appear to act by inhibiting the production of enzymes needed for coordinated plant growth. *Substituted amides* (e.g. diphenamid) and *nitroanilines* (e.g. trifluralin) are largely effective against seedlings rather than adult plants, and are applied to the soil around established plants as a 'pre-emergence' herbicide, preventing the subsequent appearance of weeds and reducing future problems by cutting off reinforcements.

While pest wars have been primarily waged against invertebrate pests and weeds, there are plenty of vertebrate pests too. Sea lampreys (*Petromyzon marinus*) invaded the Great Lakes of North America from the Atlantic Ocean through shipping canals in the early 1900s, contributing significantly to declines in several highly valued fish species. The lampricide TFM (3-trifluoromethyl-4-nitrophenol), when carefully applied for a few days to produce just sufficiently high concentrations in the small streams where larval lampreys develop, kills them without detriment to other fish (although invertebrates and amphibians may be locally affected). The populations of adult lampreys in the Great Lakes, starved of reinforcements, have been very substantially reduced.

6.2.5 *Changing pest behavior – a propaganda war*

In human warfare, propaganda pamphlets dropped behind enemy lines are intended to influence behavior and disrupt the enemy. There is an equivalent here too, developed more recently than the inorganic and organic toxins and thus sometimes called 'third-generation insecticides' (Forrester, 1993). *Semiochemicals* (literally 'chemical signs') are not toxins but chemicals that elicit a change in the behavior of the pest. They are all based on naturally occurring substances, but in a number of cases it has been possible to synthesize them in the laboratory. Sex-attractant pheromones have been used, for example, to control pest moth populations by interfering with mating or to attract moths to cages holding a virus that is deadly to the moth's caterpillars. In a different context, an aphid alarm pheromone has been used to increase the effectiveness of a fungal pathogen against pest aphids in glasshouses by increasing the mobility of the aphids, and hence, their rate of contact with fungal spores (Hockland *et al.*, 1986).

One of the oldest of pest control tools also aims to reduce pests, this time vertebrates, by modifying their behavior. Thus scarecrows are intended to reduce feeding by birds on valued seeds and crops. In similar vein, model (or real) hawks may be placed around airports to reduce bird densities and the risk of bird strikes.

6.2.6 *When pesticides go wrong – target pest resurgence and secondary pests*

Mexicans in the mountain village of Atascaderos noticed a build up in the rat population when these pests appeared in numbers in their food stores in 2003. Their response was to apply poison, but this backfired when cats and other rat-eaters succumbed to the pesticide. Target pest resurgence (discussed in Box 6.1) was dramatic: as a result of the unintended loss of their enemies, each household in the village became infested by as many as 200 rats! (The villagers' response was equally dramatic – they placed advertisements in Chihuahuan newspapers to recruit a replacement army of 700 cats.)

Similar, if less repulsive, examples of target pest resurgence include surges in density of damaging cyclamen mites (*Steneotarsonemus pallidus*) on strawberries after the application of parathion dust, and much higher densities of pacific mites (*Tetranychus pacificus*) attacking vineyards after spraying with carbaryl as compared to unsprayed areas (Debach & Rosen, 1991). In both cases, the natural enemies of the pests were more strongly affected by pesticide application than the pests themselves.

A classic example of pesticides gone wrong concerns the insect pests of cotton in the southern part of the USA. When mass use of organic insecticides began in 1950 there were two primary pests of cotton, the Alabama leafworm (*Alabama argillacea*) and the boll weevil (*Anthonomus grandis*), the latter an invader from Mexico. Smith (1998) estimates that the boll weevil alone has cost cotton producers more than $15 billion since it arrived a century ago. Chlorinated hydrocarbons and organophosphate insecticides were applied fewer than five times a year and initially had spectacular results – cotton yields soared. By 1955, however, three secondary pests had emerged as a result of reductions in natural enemy populations: the cotton bollworm (*Heliothis zea*), the cotton aphid (*Aphis gossypii*) and the false pink bollworm (*Helicoverpa armigera*). The pesticide applications rose to 8–10 per year. This reduced the problem of the aphid and the false pink bollworm, but led to the emergence of five further secondary pests. By the 1960s, the original two pest species had become eight and there were, on average, an unsustainable 28 applications of insecticide per year.

A study in the San Joaquin Valley, California, provides examples of both target pest resurgence and secondary pest outbreaks. Figure 6.5a shows how cotton bollworm, the target pest, resurged after application of azodrin, as a result of pesticide-generated reductions in its natural enemies. Figure 6.5b and c, on the other hand, show how cabbage loopers and beet army worms became secondary pests after natural enemies declined when target pests – lygus bugs in this case – were sprayed with bidrin or a mixture of DDT and toxaphene.

A surge in pest numbers happens because the pest population recovers more quickly than its enemies. Predators will often have a lower intrinsic rate of population increase than their prey and this partly explains the pattern. But the predators may also be slower than the pests to recolonize after the pesticide loses its potency.

Fig. 6.5 Pesticide problems amongst cotton pests in the San Joaquin Valley, California. (a) Target pest resurgence: cotton bollworms (*Heliothis zea*) resurged because the abundance of their natural predators was reduced – the number of damaged bolls was higher. An increase (b) in cabbage loopers (*Trichoplusia ni*) and (c) in beet army worms (*Spodoptera exigua*) were seen when plots were sprayed against the target lygus bugs (*Lygus hesperus*) – both are examples of secondary pest outbreaks. (After van den Bosch et al., 1971.)

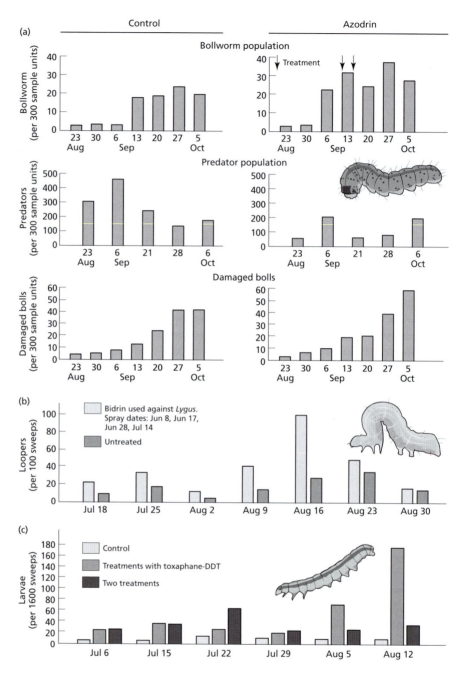

Thus, in their model of the role of dispersal rates of pests and their predators on target pest resurgence, Trumper and Holt (1998) found that resurgent effects were greatest when predator dispersal rate was low and pest dispersal rate was high (Figure 6.6). In a related context, recolonization by natural enemies can be speeded up by providing adjacent refuges for predators from the pesticide. Lee et al. (2001) planted grassland strips as natural enemy refuges adjacent to a corn crop (*Zea mays*)

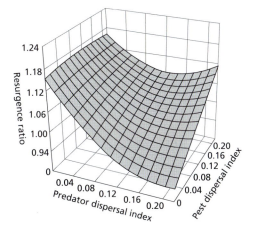

Fig. 6.6 Results of a mathematical model of pest resurgence. A resurgence ratio greater than 1.0 indicates that resurgence occurs after pesticide application. (R is the ratio of pest density after application to density in the absence of application). Predator and pest dispersal indexes are the probabilities that an individual predator or prey, respectively, will disperse into the sprayed field from an adjacent untreated field. Higher dispersal rates of predators reduce the likelihood of resurgence, unless the pests also have high dispersal potential. (After Trumper & Holt, 1998.)

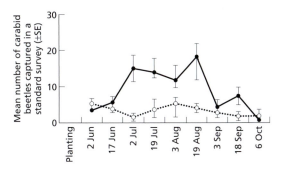

Fig. 6.7 Consequences of the provision of grassland strips (solid circles) as refuges for carabid beetles (predators of pest insects) in fields of corn. Open circles are results where no refuge strip was provided (corn occurred where the strip would have been). A soil pesticide (the organo-phosphate terbufos) was applied at the time of corn planting (except in the refuge and control strips). Carabid beetles achieved much higher densities in the corn-growing season if grassland refuges were present. (After Lee et al., 2001.)

treated with a soil insecticide to protect the crop from corn rootworm larvae (*Diabrotica* spp.). The density of carabid beetles, which are potent natural enemies with good powers of dispersal, recovers to much higher levels than is the case where no refuge strips were provided (Figure 6.7).

6.2.7 Widespread effects of pesticides on nontarget organisms, including people

Sometimes the unintended effects of pesticides have been a whole lot less subtle than target pest or secondary pest resurgence. We have already bemoaned the fact that the general use of herbicides, coupled with more intensive farming methods, has eradicated springtime displays of meadow flowers in Europe (Section 3.1). These are now restricted to areas of peasant farming. And when massive doses of dieldrin were applied to large areas of Illinois in the 1950s to 'eradicate' a grassland pest, the Japanese beetle (*Popillia japonica*), cattle and sheep were poisoned, 90% of cats and a number of dogs were killed, and 12 wild mammals and 19 bird species suffered losses (Luckman & Decker, 1960).

Not surprisingly, given that pesticides kill a range of living organisms, many also pose risks to human health. Indeed, repeated exposure to some pesticides can cause cancer, birth defects, blood disorders, brain damage and kidney problems. And because the nervous systems of insects and humans have much in common, even a single dose of one of the nerve poisons (e.g. organophosphates or carbamates) can lead to dizziness, twitching, seizures or death. For all these reasons, many old pesticides are no longer used, while a very rigorous testing regime is applied to new pesticides and instructions for their safe use are prominently displayed on packaging.

If chemical pesticides brought nothing but problems, and if their use was intrinsically unsustainable, they would already have fallen out of widespread use. But instead their rate of production has continued to increase rapidly because the ratio of cost to benefit for the individual user has generally remained in favor of pesticide use. However, the many examples of undesirable outcomes argue for a precautionary approach in any pest management exercise. Coupled with much improved knowledge about toxicity and the development of more specific and less persistent pesticides, such disasters should generally be a thing of the past. But note that in many poorer countries, the prospect of imminent mass starvation, or of an epidemic disease, are so frightening that the environmental and health costs of using pesticides may sometimes have to be ignored.

6.3 Biological control

Outbreaks of pests occur repeatedly and so does the need to apply pesticides. But pest controllers can sometimes replace chemicals with another tool that can be more effective and, over the long term, often costs a great deal less – biological control. Biological control, or the manipulation of natural enemies of the pest, involves the application of theory about interactions between species and their natural enemies (Box 6.1) to limit the population density of specific pest species. There are three main approaches.

The first is the *importation* of a natural enemy from another geographical area – often the area where the pest originated (Section 6.3.1). The objective is for the control agent to persist and thus maintain the pest below its economic threshold for the foreseeable future. This is a case of a desirable invasion of an exotic species and is often called *classical biological control*.

By contrast, *conservation biological control* involves manipulations to increase the equilibrium density of natural enemies that are already native to the region where the pest occurs naturally or is an invader (Section 6.3.2).

Augmentation is similar to *importation*, but its aim is to temporarily supplement an existing population without the expectation of a long-term increase in the enemy population. Augmentation has to be carried out repeatedly, typically to head off periods of rapid growth of the pest population, and provides control for one or a few pest generations. Augmentation has two classes. *Inoculation* is where the released natural enemies (field-collected or laboratory-reared) serve to inoculate a new crop but control is provided later by their offspring (Section 6.3.3). *Inundation*, on the other hand, is where all control is provided by the released natural enemies

Table 6.2 The record of insects as biological control agents against pests and weeds. (After Waage & Greathead, 1988.)

	Insect pests	Weeds
Control agent species	563	126
Pest species	292	70
Countries	168	55
Cases where agent has become established	1063	367
Substantial successes	421	113
Successes as a percentage of establishments	40	31

themselves (Section 6.3.4). Here the aim is to kill the pests present at the time of the release and, by analogy with the use of chemicals, such agents are sometimes known as *biological pesticides*.

Insects have been important agents of biological control against both insect pests (particularly parasitoids that lay their eggs in or on the bodies of their host) and weeds. Table 6.2 summarizes the extent to which they have been used and the proportion of cases where the establishment of an agent has greatly reduced or eliminated the need for other control measures. In the next sections I present some success stories. Note, however, that most biological control attempts have not been successful, and some have undesirable effects. I will turn to these in Section 6.3.5.

6.3.1 *Importation biological control – a question of scale*

The most classic example of 'classical' biological control concerns the cottony cushion scale insect (*Icerya purchasi*), discovered as a pest of Californian citrus orchards in 1868. By 1886 it had brought the citrus industry to its knees. Species that colonize a new area may become pests because they have escaped the control of their natural enemies. Importation of some of these natural enemies is then, in essence, restoration of the status quo. Ecologists corresponded with colleagues around the world to try to discover the origin of the scale insect and the identity of its natural enemies. This requires a high level of taxonomic skill and is by no means an easy task, particularly if in its native habitat the 'pest' is kept rare by enemies that are also at low density. The search eventually led to the importation to California of two candidate species. The first was a parasitoid, a two-winged fly (*Cryptochaetum* sp.) that laid its eggs on the scale insect, giving rise to a larva that consumed the pest. Twelve thousand of these were despatched from Australia. The other was a predatory ladybird beetle (*Rodolia cardinalis*), of which 500 arrived from Australia and New Zealand. Initially, the parasitoids seemed to have disappeared, but the predatory beetles underwent such a population explosion that, amazingly, all scale insect infestations in California were controlled by the end of 1890. Although the beetles have usually taken the credit, the long-term outcome has been that the beetles keep the scale insects in check inland, but *Cryptochaetum* is the main control near the coast (Flint & van den Bosch, 1981). The economic return on investment in biological control was very high in California and the ladybird beetles have subsequently been transferred to 50 other countries.

Another invasive scale insect was killing off the national tree of the small South Atlantic island of St Helena (the last home of another famous invader – Napoleon Bonaparte). Only 2500 St Helena gumwoods (*Commidendrum robustum*) were left when, in 1991, the South American scale insect *Orthezia insignis* was found to be mounting its attack, killing more than 100 of the trees by 1993. Fowler (2004)

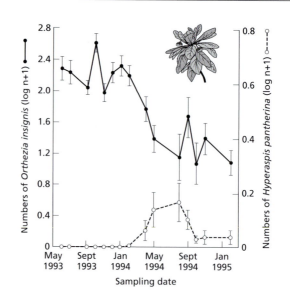

Fig. 6.8 Mean numbers (±SE, log scale) (on continuously monitored 20-cm-branchlets of 30 randomly selected gumwood trees) of the pest scale insect *Orthezia insignis* and its biological control agent, the ladybird *Hyperaspis pantherina*. Mean scale insect numbers dropped from more than 400 adults and nymphs (in September 1993) to fewer than 15 (in February 1995) when sampling ceased. Mean ladybird numbers increased from January to August 1994, coinciding with an obvious decline in scale insects, before ladybird numbers declined again. The highest recorded numbers of ladybirds were 1.3 adults and 3.4 larvae per 20-cm branchlet. (After Fowler, 2004.)

estimated that all remaining individuals of this rare tree would be dead by 1995. Again a ladybird beetle was the savior. *Hyperaspis pantherina* was cultured and released on St Helena in 1993 and as its numbers increased there was a corresponding 30-fold decrease in scale insect numbers (Figure 6.8). Since 1995 no scale outbreaks have been reported and culturing of the ladybirds has been discontinued because the population is maintaining itself at low density in the wild, as good importation biocontrol agents should.

A quite different kind of biological control agent proved successful in Hawaii against the South American climbing vine – banana poka (*Passiflora tarminiana*) – considered the most serious threat to Hawaii's unique high-elevation forests. In 1991 the fungus *Septoria passiflorae* was found on banana poka seedlings in its home range in Colombia. After containment laboratory studies in Hawaii showed the fungus to be specific to banana poka, spore suspensions were sprayed onto leaf surfaces during 1996–97 and produced major epidemics that severely defoliated the vines. The biomass of the weed was reduced by 40–60% within 1 year and by more than 95% 5 years later (Figure 6.9). By 2003, banana poka had been eliminated from most forests (Trujillo, 2005).

It should not be supposed that biological control campaigns always involve just a single agent. For example, in response to the invasion of Australia by the aggressive weed *Mimosa pigra* from tropical America, which forms impenetrable 6-m-high thickets beside rivers and floodplains, so far 11 species of insect and two pathogenic fungi have been released without much evidence of bringing the weed under control (discussed further in Section 6.5.2).

6.3.2 *Conservation biological control – get natural enemies to do the work*

Many pests have a diversity of natural enemies that already occur in their habitat. In the case of the aphid pests of wheat (e.g. *Sitobion avenae*), predators that specialize on aphids include ladybirds and other beetles, heteropteran bugs, lacewings (Chrysopidae), fly larvae (Syrphidae) and spiders. Many of these natural enemies spend the winter in grassy boundaries at the edge of wheat fields, from where they disperse

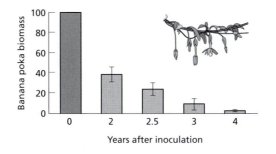

Fig. 6.9 Reduction in banana poka biomass (mean ± SE) after application of the fungus *Septoria passiflorae* at Laupahoe-hoe in the Hilo Forest Reserve, Hawaii. Biomass before application (at time '0') is taken to be 100%, and values from subsequent surveys are shown in relation to this. (After Trujillo et al., 2001.)

to reduce aphid populations around field edges. Farmers can protect grass habitat around their fields and even plant grassy strips in the interior to enhance these natural populations and the scale of their impact on the pests. This example of conservation biological control parallels our earlier discussion of how grassy strips can provide a refuge for natural enemies against pesticide use in cornfields (Section 6.2.6).

Urban settings also have their pest control problems and conservation biological control sometimes works here too. Bagworms, caterpillars of the moth *Thyridopteryx ephemeraeformis*, are among the most important pests of ornamental trees and shrubs in the eastern USA. The caterpillar constructs a bag-shaped shelter of silk and leaf fragments that it never leaves. Adult males (which have no wings) mate by inserting their genitalia into bags of the wingless adult females, whose eggs hatch to release first instar larvae that disperse by 'ballooning' on the wind. Natural enemies include a diversity of wasp parasitoids, such as *Itoplectis conquisitor*, whose larvae consume their caterpillar hosts. The adult parasitoids, on the other hand, often visit flowers to eat pollen and nectar, resources that can increase longevity and reproduction. Ellis et al. (2005) tested the hypothesis that the provision of flowerbeds (planted with flowers in the aster family) in the vicinity of shrubs (*Thuja occidentalis*) would increase the effects of natural enemies and reduce the bagworm population on the shrubs. Figure 6.10 shows how rate of parasitism of bagworms increased with proximity to a flowerbed, emphasizing the potential of this conservation biological control strategy.

Finally, you should note that vertebrates are occasionally used as biological control agents, and great tits (*Parus major*) in apple orchards (*Malus domestica*) provide a

Fig. 6.10 Percentage parasitism (±SE) (by *Itoplectis conquisitor* and other parasitoid wasps) of bagworm pests on shrubs, in relation to distance to the nearest flowerbed (which provides food resources for the adult parasitoids). Bagworms were collected from the wild and 20 were placed on each experimental shrub at the beginning of the 3-month experiment. (After Ellis et al., 2005.)

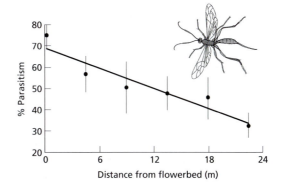

case in point. These birds like to collect caterpillars to feed their young at a time when caterpillars of winter moths *Operophtera brumata* and other species are causing damage in the orchards. Mols and Visser (2002) experimentally manipulated access of great tits to apple trees using netting. They found that foraging by great tits caused a small reduction in caterpillar damage (down from 13.8% to 11.2%) but, in terms of actual fruit yield, the benefit was much bigger (up from 4.7 to 7.8 kg of apples per tree). The only cost to the producer was the erection of nest boxes (2 per hectare) to encourage the birds to breed in their orchards. Conservation biological control clearly has wide application across urban, cropping, forestry and fruit production settings.

6.3.3 *Inoculation biological control – effective in glasshouses but rarely in field crops*

Inoculation as a means of biological control is often used to control invertebrate pests in glasshouses. In this situation, the crop, its pests and their natural enemies are all removed at the end of the growing season, so there is no opportunity to establish a pest–enemy equilibrium. Two particularly important natural enemies used for inoculation are *Phytoseiulus persimilis*, a mite that preys on the spider mite *Tetranychus urticae*, a pest of roses, cucumbers and other vegetables, and *Encarsia formosa*, a chalcid parasitoid wasp of the whitefly *Trialeurodes vaporariorum*, a pest in particular of tomatoes and cucumbers. Adequate distribution of the natural enemies is crucial, and this can sometimes be assisted by means of fans or the construction of 'bridges'. Thus, when *P. persimilis* is introduced to control *T. urticae* on cut roses, the provision of bridges of plastic tape to enhance predator dispersal leads to a 50% reduction in leaves infested with the pest (Casey & Parrella, 2005). This is a modern version of the first recorded use of natural enemies, in China in 324BCE, when artificial bridges between citrus tree branches provided ready access to caterpillar pests for their natural enemies – ants!

Inoculation may also be used under field conditions, particularly where natural enemies fail to colonize crops, or arrive too late in the season to provide effective pest control. However, in their review of studies concerning inoculations of predatory insects and parasitoids into field crops, Collier and Van Steenwyk (2004) found that inoculation failed to reduce pest populations to target levels in 64% of cases. This was for a variety of reasons, including unfavorable environmental conditions for the control agent, dispersal of the agent away from the target area and predation on the agent. These are issues that do not arise under the controlled and enclosed circumstances of the greenhouse. Moreover, in economic terms, inoculation in field crops was usually more expensive than pesticide application, and in a few cases actually cost more than all the other crop production costs combined. On the other hand, augmentation sometimes works well. Thus, when the parasitoid wasp *Aphytis melinus* was inoculated at the rate of 50,000 per hectare into US lemon orchards to control the Californian red scale insect *Aonidiella aurantii*, the pest was reduced to target levels at a cost of only US$102 per hectare, a very small percentage of overall crop production costs (more than $25,000 per hectare; Moreno & Luck, 1992).

Most often insects for inoculation are purchased from a commercial producer, but plant material infested with dormant pests and natural enemies may provide a local source of control agents. Kehrli et al. (2005) collected horsechestnut (*Aesculus hippocastanum*) leaf litter with its associated pests (the invasive horsechestnut leaf-mining moth, *Cameraria ohridella*) and natural enemies (parasitoids) and placed all the material in 'mass-emergence' devices, consisting of large tubs with openings covered by fine mesh that provided easy passage for the small adult parasitoids to emerge, but not the larger moths. Tubs containing 10 kg of leaf litter, and hung in the crown of a horsechestnut tree, increased percent parasitism of *C. ohridella* there.

6.3.4 *Inundation biological control – using fungi, viruses, bacteria and nematodes*

Inundation often involves the use of insect pathogens (fungi, viruses, bacteria) to control insect pests. Each of these is generally present in the natural environment but only at low densities. Their use as 'biological pesticides' requires commercial production and application to swamp the target insect populations. Fungal agents include the insect pathogenic fungus *Beauveria bassiana*, which is used in a commercial spray against locusts and grasshoppers. The fungal spores germinate when they contact the insect's cuticle, penetrating its body cavity where fungal hyphae fill the body. The insects appear mummified and fuzzy because of the fungal growth. Viruses are even more specific in their action, but the expense of production and application means they are not used extensively. One example is '*Lymantria dispar* nuclear polyhedrosis virus', which acts only against the gypsy moth (*L. dispar*), a significant forest pest.

But by far the most significant insect pathogenic agent is the bacterium *Bacillus thuringiensis* (Bt), which can easily be produced on artificial media. After being ingested by an insect, gut juices release powerful toxins and death occurs 30 minutes to 3 days later. Significantly, there is a range of strains of Bt, some specific to caterpillars (many agricultural pests), others to two-winged flies, especially mosquitoes and blackflies (vectors of malaria and onchocerciasis), and yet others to beetles and their grubs (many agricultural and stored product pests). The advantages of Bt are its powerful toxicity against target insects but a lack of toxicity against organisms outside this narrow group (including most of the pest's natural enemies).

The control action of Bt has been extended by genetically modifying crops, including potato, corn and cotton, to express the Bt toxin (crystal protein Cry1Ac); the survivorship of pink bollworm larvae (*Pectinaphora gossypiella*) on genetically modified cotton was 46–100% lower than on nonmodified cotton (Lui et al., 2001). However, concern has arisen about the widespread insertion of Bt into commercial genetically modified crops. This upscaling of the killing power of Bt poses such a constant and high selection pressure that the probability of evolution of resistance is enhanced for what is currently one of the most effective 'natural' insecticides available (Section 6.4).

Finally, the so-called entomopathogenic nematode worms deserve mention. These hold promise as biological control agents of insect pests by infecting, rapidly multiplying and quickly killing their hosts. Worms in the families Steinernematidae and Heterorhabditidae infect insects as juveniles, gaining access through natural openings (spiracles, mouth, anus). The infective juveniles carry in their guts a symbiotic bacterium that is released, multiplies and kills the host, leaving the body contents as a soup used by three or four generations of nematodes. When the soup runs out, many thousands of infective juveniles are released into the environment. Fenton et al. (2000) explored the population dynamics of the nematodes and concluded that they are probably incapable of regulating a host population to a persistent stable equilibrium, but because they are easy to mass-culture they are well equipped for short-term control by inundation. Nematodes are available commercially and can be applied to crops such as mushrooms to control fungus gnats (*Sciara* spp.).

6.3.5 *When biological control goes wrong*

Successful importation biological control can provide a permanent solution because, unlike pesticides, the agent reproduces and maintains itself into the future. But this also has a downside – if the control agent turns out to pose an environmental threat, it may last forever.

As with pesticides, unwanted outcomes of biological control occur and some are spectacular in their extent. Take for example the introduction in the 1970s and 1980s to the Society Islands (French Polynesia) from Florida of the predatory snail *Euglandina rosea* as a biological control agent for a previously introduced crop pest, the giant African snail *Achatina fulica*. These Polynesian islands were once famous for the dramatic diversity of their tree snails (family Partulidae). But the failed biological control agent *Euglandina rosea* has instead driven all but five of the Society Islands' 61 species of tree snails extinct in the wild (Coote & Loève, 2003). (Fifteen species still exist in captive collections in Europe and North America.) The loss of *Partula varia* and *Partula rosea* on the island of Huahine had significant economic and social costs too. These species were used for making shell jewellery, and with their demise many women lost their livelihoods and the artisan's association was closed down.

Euglandina rosea should never have been contemplated as a control agent, because of its generalist diet and the precious biodiversity that it threatened. However, examples have also come to light where even carefully chosen and apparently successful introductions of biological control agents have impacted on nontarget species. Thus biocontrol by the myxoma virus of European rabbits in the UK unexpectedly resulted in extinction of the large blue butterfly *Maculina arion*. This happened because the butterfly uses nests of the ant *Myrmica sabuleti* to rear caterpillars. The ants, in turn, depended on rabbit grazing to maintain the open habitat where they nest. Another example concerns a seed-feeding weevil (*Rhinocyllus conicus*). Introduced to North America to control exotic *Carduus* thistles, the weevil also attacks more than 30% of native thistles (of which there are more than 90 species), reducing thistle densities (by 90% in the case of the Platte thistle *Cirsium canescens*), with consequent adverse impacts on the native picture-winged fly (*Paracantha culta*) which feeds on thistle seeds (Louda et al., 2003a).

Sometimes such indirect food-web effects may even compromise human health. In the early 1970s two species of gallfly (*Urophora* spp.) were introduced to control knapweeds (*Centaurea* spp.) in western North America. The gallflies established and remained host-specific, but they did not control the target plants and became superabundant. Now a rich food source in areas invaded by knapweed, the gallflies make up 85% of the diet of the deer mouse (*Peromyscus maniculatus*), whose population has increased two- or three-fold. So how might this affect human health? Deer mice are also the primary vector of the deadly Sin Nombre hantavirus (see Pearson & Callaway, 2003). More mundanely, the increased mouse populations might, through seed and insect predation, alter biodiversity, compete with other small mammals and, as prey, affect the populations of larger predators (Figure 6.11).

Louda et al. (2003b) reviewed ten biological control projects that included the unusual but worthwhile step of monitoring nontarget effects and concluded that relatives of the target species were most likely to be attacked, with rare native species

Fig. 6.11 (a) Documented (solid lines) and postulated (dotted lines) direct and indirect effects associated with gallflies introduced to control knapweed (b). The flies have not controlled the weeds, and have become superabundant on their specialized hosts, with a positive effect on their deer mice predators (c) and a possible increase in the risk to humans of Hanta virus. The knapweeds compete with native plants (negative effect) and provide food for the deer mice (positive effect). The increased deer mice population may have negative consequences for native plants (seed predation), native insects (predation) and other small mammals (competition). They may also have a positive effect as food for larger predators. (From Pearson & Callaway, 2003.)

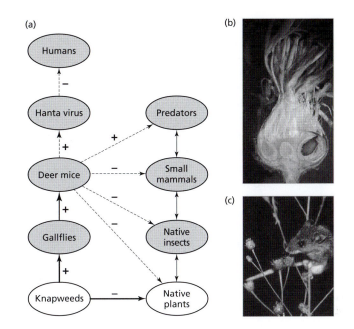

being particularly susceptible. Their recommendations for management included the avoidance of generalist control agents, an expansion of host-specificity testing and the need to incorporate more ecological information when evaluating potential biological control agents. Bourchier (2003) showed how this could be done in an analysis of the risk to 153 nontarget butterfly species in Canada of the release of the parasitoid *Trichogramma minutum*. This parasitoid, which attacks eggs, is used inundatively to control spruce budworm (caterpillar of the moth *Choristoneura fumiferana*), a conifer forest pest. The ecological criteria used to assess risk to each butterfly species included geographical distribution (rare species more vulnerable than widespread ones; butterflies most vulnerable whose range coincides with likely release sites for the parasitoid), timing of egg laying (butterflies most vulnerable that overlap with egg laying of spruce budworm), type of host plant (butterflies associated with conifers more vulnerable than those on deciduous trees, which in turn are more vulnerable than those on shrubs and herbs), egg type (egg masses more vulnerable to parasitoid attack than single eggs) and egg exposure (exposed on leaves more vulnerable than, for example, hidden under bark). Using this process to build an ecological vulnerability list, Bourchier identified several species at particular risk (e.g. *Lycaeides idas* and *Speyeria cybele*) and recommended that prior to any releases of the parasitoid in their areas, careful host-specificity testing of the particular parasitoid strain should be carried out to determine whether the butterfly's potential vulnerability would be realized.

It is not difficult to see that the perfect biological control agent would be specific only to the target, have niche requirements that overlap and a life cycle that is synchronized with the target, and have a greater population growth rate and greater mobility than the pest.

6.4 Evolution of resistance and its management

In the war on pests, it is not unusual for the pest to fight back – chemical pesticides lose their role if the pests evolve resistance (Box 6.1). This problem was often overlooked in the past, even though the first case of DDT resistance was reported as early as 1946 (houseflies, *Musca domestica*, in Sweden). Now, the scale of the problem is illustrated in Figure 6.12, which shows the exponential increase in the number of arthropod invertebrates reported to have evolved resistance and in the number of pesticides against which resistance has evolved. Two thirds of species are resistant to more than one pesticide, and the total number of cases of resistance (species × pesticides) is about five times the number of resistant species. Resistance has been recorded in every family of arthropod pest (including dipterans such as mosquitoes and house flies, as well as beetles, moths, wasps, fleas, lice and mites) as well as in weeds and plant pathogens. Take the Alabama leafworm (Section 6.2.6), a moth pest of cotton, as an example. It has developed resistance in one or more regions of the world to aldrin, DDT, dieldrin, endrin, lindane and toxaphene.

In the face of evolution of resistance, sustainable pest control depends either on the continual development of new pesticides that keep at least one step ahead of the pests or, alternatively, of management strategies that expose the pests to a given pesticide for only a limited time. I discuss this next.

The evolution of pesticide resistance can be slowed by changing from one pesticide to another, in a repeated sequence that is rapid enough that resistance does not have time to emerge. River blindness (onchocerciasis) is a devastating disease that affected millions of Africans. The international community seemed oblivious to its public health and socioeconomic consequences until the 1970s when a United Nations mission to West Africa, and a visit by the president of the World Bank, culminated in the start of a concerted international effort to combat the disease (Benton et al., 2002). Remarkably, as a result of a blend of pest control and a public health campaign, the disease was eliminated by early in the twenty first century. The disease organism, a parasitic worm (*Onchocerca volvulus*), is transmitted by the biting blackfly *Simulium damnosum*, whose larvae live in rivers. By 1999, the blackfly larva control program involved 50,000 km of river and an area of 1235,000 km^2,

Fig. 6.12 Global increases in the number of arthropod pest species reported to have evolved pesticide resistance (dotted line) and in the number of pesticide compounds against which resistance has developed (dashed line). Each pest, on average, has evolved resistance to more than one pesticide, so there are now more than 2500 cases of evolution of resistance (pests × compounds) (solid line). (From Michigan State University's Database of Arthropods Resistant to Pesticides (http://www.pesticideresistance.org/DB/); © Patrick Bills, David Mota-Sanchez & Mark Whalon.)

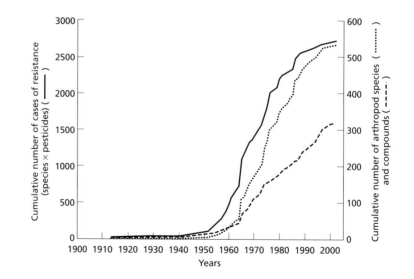

Table 6.3 History of pesticide use against the aquatic larvae of blackflies, the vectors of river blindness in Africa. After early concentration on temephos and chlorphoxim, to which the insects became resistant, pesticides were used on a rotational basis to prevent the evolution of resistance. Note that the bacterium *Bacillus thuringiensis* is a biological agent (Section 6.3.4). (After Davies, 1994.)

Name of pesticide	Class of chemical	History of use
Temephos	Organophosphate	1975 onwards
Chlorphoxim	Organophosphate	1980–90
Bacillus thuringiensis	Biological pesticide	1980 onwards
Permethrin	Pyrethroid	1985 onwards
Carbosulfan	Carbamate	1985 onwards
Pyraclofos	Organic phosphate	1991 onwards
Phoxim	Organophosphate	1991 onwards
Etofenprox	Pyrethroid	1994 onwards

covering seven African countries. A massive helicopter pesticide spraying effort began by using temephos, but resistance in blackfly populations appeared within 5 years. Temephos was then replaced by another organophosphate, chlorphoxim, but resistance rapidly evolved to this too (Yaméogo et al., 2001). The strategy of rotating the use of a range of pesticides, which work in different ways (Table 6.3), has prevented further evolution of resistance, and by 1994 there were few populations that were still resistant to temephos (Davies, 1994).

Even vertebrates have evolved resistance to some pesticides. A naturally occurring substance in certain plants, fluoroacetate, is the basis for the pesticide known as 1080, widely used against vertebrate pests in Australia and New Zealand. Laboratory testing of rabbit populations soon after 1080 was first introduced, and 25 years later, revealed a doubling of the dose needed to kill the rabbits (Twigg et al., 2002). This translated into a reduction from more than a 75% kill rate in a standard pest control operation where rabbits had not previously been exposed, to as low as 50% where they had developed resistance. One approach to counter this problem might be to increase dose rates and the efficiency of 1080 delivery (on food in bait stations), because evolution cannot occur if all individuals are killed. However, this would also increase the likelihood of nontarget mortality. Twigg and his team suggest instead that, as in the blackfly case, a range of measures should be used in concert – including rabbit calicivirus disease (Section 6.1).

Although heritability studies were not performed on the rabbits, it is almost certain that evolved resistance to 1080 had occurred, as shown definitively for Japanese quail (*Coturnix coturnix japonica*) to DDT and for the house mouse (*Mus musculus*) to an anti-blood clotting poison – bromadiolone. The evolution of house mouse resistance involved genetic selection in a biochemical pathway for a single enzyme that is less sensitive to the pesticide (Misenheimer et al., 1994).

I have already pointed out some of the pros (and cons) of biological control as compared to pesticide application. It is worth adding that evolution of resistance to an introduced natural enemy (whether a predator or parasite) has rarely been documented. Evolved resistance to a novel pesticide might involve only a single gene. By contrast, we can expect that 'resistance' to an introduced predator would require simultaneous changes to a number of the target pest's genes (affecting its behavior, morphology and/or physiology). Resistance to an internal parasite might be somewhat easier to achieve, and a famous example concerns the introduction to Australia in the 1950s of the myxoma virus in an early attempt to control rabbits there. Fenner and Ratcliffe (1965) had the foresight to establish baseline strains of both rabbits and virus. They found that rabbit mortality declined from more than 90% to less than 30% within 8 years. Individuals that happened to be least susceptible to the virus were more likely to pass on their genes to future generations – the population did indeed evolve resistance (Figure 6.3 applies to this case just as well as to pesticide resistance). Less expected was the fact that the virus itself evolved to become less virulent. This was because when virulence was very high, rabbits died before the virus could be passed to other rabbits (by mosquitoes feeding on live rabbits). Viruses with an intermediate level of virulence proved to be fittest, contributing most to subsequent generations of the virus.

Another kind of biocontrol agent is the endotoxin-producing bacterium *Bacillus thuringiensis* (Bt – mentioned in Section 6.3.4) but, for example, mosquitoes have evolved resistance to Bt in areas where it has been used intensively. Two strains of the bacterium effective against mosquitoes are Bt subspecies *israelensis* and *sphaericus*; the first produces two classes of endotoxin called 'Bin' and 'Cry' while the second produces another called 'Cyt'. A fundamental principle in the management of the evolution of resistance is that the more complex and potent a toxin mixture, the slower will be the development of resistance. Thus, Park et al. (2005) have genetically engineered forms of the bacterium to produce novel combinations of endotoxins. Those engineered to produce only Cry and Cyt were between 9 and 15 times less potent at killing mosquitoes (*Culex quinquefasciatus*) than others that produced Cry, Cyt and Bin.

Some forms of pest control do not involve pesticides or biological control agents, but resistance can still evolve. Take, for example, the northern corn rootworm (*Diabrotica barberi*), a beetle pest of corn crops in North America. The most effective method of control (a case of so-called 'cultural' control) was simply to rotate the corn crop with another crop every second year; this worked because rootworm eggs could not survive through more than one winter. Then corn producers in some areas began to notice rootworm damage even in rotated crops. This occurred because some individuals now lay eggs that can survive for more than one winter (Levine et al., 1992), an apparent case of evolved resistance to a cultural control method. Judicious insecticide application will be needed in future to supplement the effect of crop rotation.

6.5 Integrated pest management (IPM)

Having considered chemical and biological approaches to pest control separately, the pros and cons of the different approaches should now be clear. However, it is vital to turn to a broader perspective and consider how all the different tools at the pest controller's disposal can be deployed most effectively, both to maximize the economic benefit of reduced pest density and to minimize adverse health and environmental consequences (including the evolution of resistance).

This is what integrated pest management (IPM) is intended to achieve, by combining physical, cultural, chemical and biological control, and the use of resistant crop varieties. IPM accepts that crops, forests and any other managed biological resources are part of a functioning ecosystem, and that management decisions must take into account the way that control actions will affect not only the pest but also other species in the web of community interactions. IPM also involves accepting that the mere presence of a pest is not necessarily a problem; before a potentially disruptive control action is taken, we need to know that the benefit in reducing pest numbers will be justified by the cost of control (Box 6.1). IPM was, in part, a reaction to the

Fig. 6.13 Mean (±SE) percentage of diamond-back moth caterpillars parasitized by the parasitoid *Cotesia plutellae* in a glass-house. Each bar represents eight replicates with 16 plants per replicate. The caterpillars were feeding on control cabbage plants, or on plants treated with a botanical pesticide extracted from the syringa tree (S) (*Melia azedarach*) or the neem tree (N) (*Azadirachta indica*). (After Charleston et al., 2005.)

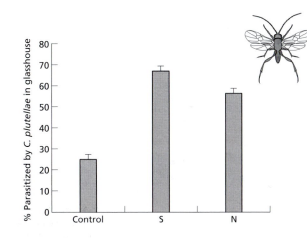

unthinking application of pesticides in the 1940s and 1950s. The IPM philosophy thus considers all possible control actions before one is adopted and, in most cases, integrates several control measures in a compatible manner.

The key word here is 'compatible'. You have already seen how the natural control of potential pests can be disrupted by chemical attack aimed at the pest by dispro-portionately affecting their natural enemies. This can also apply to an imported biological control agent. Thus, the first use of DDT in Californian citrus orchards in 1946–47 led to an outbreak of the (by then) rarely seen cottony cushion scale insect, when the DDT almost eliminated ladybirds (Section 6.3.1) that had success-fully controlled this previously devastating pest. However, careful study can reveal combinations of biological control agents and particular pesticides that are compat-ible or even work better together than alone. This is the case for diamondback moths, a pest of cabbage crops, attacked by a biological control agent, the parasitoid *Cotesia plutellae*, in concert with application of a botanical pesticide (Figure 6.13).

IPM relies heavily on natural mortality factors, such as weather and enemies, and seeks to disrupt the latter as little as possible. The aim is to control pests below the economic injury level, and IPM invariably depends on monitoring the abundance of pests and their natural enemies so that sensible decisions can be taken about whether and when to apply control. Broad-spectrum pesticides are used very spar-ingly or not at all, and the use of any chemical is minimized. The essence of the IPM approach is to make the control measures fit the pest problem, and no two problems are the same—even in adjacent fields. Thus, IPM often involves the devel-opment of computer-based expert systems that can be used by farmers to diagnose pest problems and suggest appropriate responses.

To illustrate the way that managers design and implement integrated pest manage-ment I present two cases – the first against an insect pest (Section 6.5.1) and the second against an invasive weed (Section 6.5.2).

6.5.1 *IPM against potato tuber moths in New Zealand*

The caterpillar of the potato tuber moth (*Phthorimaea operculella*) is a pest of potato crops in New Zealand. An invader from a warm temperate subtropical country, it is most devastating when conditions are warm and dry (i.e. when the environment coincides closely with its optimal niche requirements – Chapter 2). There can be as many as eight generations per year and different generations mine leaves, stems and tubers. The caterpillars are protected both from natural enemies (parasitoids) and insecticides when in the potato tuber, so control must be applied to leaf-mining generations. The IPM strategy for potato tuber moth (Herman, 2000) involves monitoring (female pheromone traps, set weekly from mid summer, are used to attract males, which are counted), cultural methods (the soil is cultivated to prevent soil cracking, soil ridges are molded up more than once, and soil moisture is maintained), and the use of insecticides, but only when absolutely necessary (most commonly the organophosphate, methamidophos). Farmers follow the decision tree shown in Figure 6.14.

6.5.2 IPM against an invasive weed in Australia

A more quantitatively based attempt to define an IPM system concerns the thicket-forming weed *Mimosa pigra* in Australia. You learnt in Section 6.3.1 that several biological control agents have been released, but so far without much success. Buckley et al. (2004) considered biological control, along with all other conceivable weed management approaches, and assessed their likely value by means of a simulation model. (This has much in common with the population viability analyses discussed in the context of species conservation in Chapter 5.) Realistic parameters were included in the model (defining seed production, seed-bank properties, seed dispersal, germination, seedling survival, weed longevity, time to reproduction, etc.) and simulations were run for the following scenarios:

1 *Disturbance alone* – note that native vegetation can outcompete the weed in the absence of disturbance by fire or by water buffalo *Bubalus bubalis*.

2 *Physical and chemical control alone* – note that small isolated infestations can be eradicated by hand-pulling, cutting or herbicides but this approach may need to be prolonged because of regrowth from the base of the plant.

Fig. 6.14 Decision flow chart for the integrated pest management of potato tuber moths (PTM) in New Zealand. Boxed phrases are questions (e.g. 'what is the growth stage of the crop?'), words in arrows are the farmer's answers to the questions (e.g. 'before the tuber has formed') and the recommended action is shown in the vertical box (e.g. 'don't spray the crop'). Note that February is late summer in New Zealand. (From Begon et al., 2006, based on Herman, 2000.)

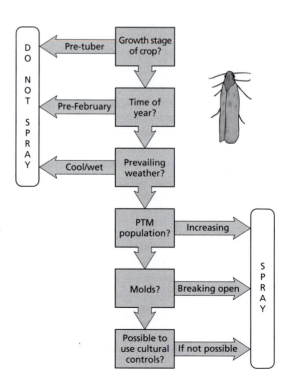

3 *Biological control alone* – involving the two currently most effective species, the stem-mining moths *Neurostrota gunniella* and *Carmenta mimosa*, which can stunt weed growth and reduce reproduction and recolonization.

4 *Integrated weed management* – 25 different single and combination treatments were considered involving biocontrol, herbicide application, mechanical control and fire.

Of all the simulations, the most successful strategy was *integrated weed management* involving biocontrol together with application of herbicide in year 1, mechanical control and fire in year 2, and herbicide again in year 3. This is indeed an integrated pest management strategy.

Summary

Pest eradication or control?

Pests are species that people consider undesirable. They carry an economic cost, either because of direct damage to health or economic activity, or simply because of a willingness to spend money to counter their nuisance value. Many pests are exotic imports, but natives may achieve pest status after a change to agricultural practices – by introducing new crops, creating monocultures or applying pesticides.

The aim of pest control is to drive a pest population extinct or keep density so low that its nuisance value is negligible. It can actually be very difficult to eradicate a pest because most pests have very high reproductive rates. Eradication may be attempted soon after an invader has arrived, when it is most vulnerable, or where the consequences of the pest are particularly devastating – such as the smallpox virus. More often, however, the aim is to reduce the pest population to a level at which it does not pay to achieve yet more control (the 'economic injury level').

Chemical control of pests

A very wide range of chemical pesticides has been developed. Some, known for centuries, are naturally occurring defensive chemicals from plants (botanicals) – although more effective forms are now synthesized in the laboratory. Other ancient pesticides are the simple inorganics such as salts of copper and arsenic, but these are little used today. Increasing knowledge of cellular physiology has seen the development of much more sophisticated killers, including nerve poisons and disrupters of natural cell function. The ideal pesticide would only affect the target species and quickly break down in the environment. But there are many notorious cases of pesticides that affect a wide range of species, persist for years, and pass along food chains, accumulating in top predators.

Target pest resurgence and secondary pest outbreaks

Pest density may recover after pesticide application to densities far above the original level. This 'target pest resurgence' occurs when treatment kills both large numbers of the pest *and* large numbers of its natural enemies. Pest individuals that survive the pesticide or that migrate into the area find themselves with a plentiful food resource and few, if any, natural enemies. And it may not only be the target pest whose density surges. A number of innocuous herbivores, previously kept in check by natural enemies, may achieve the status of 'secondary pest' because a pesticide reduces numbers of their natural enemies.

Biological control

Biological control involves introducing or augmenting an enemy (predator, parasite or pathogen). The control agent is often imported from the area where the pest originated. When successful, the pest declines to a much lower equilibrium density, and one that may be maintained because of the persistence, at low densities, of both target and control agent. Biological control only needs to happen once, whereas pesticides need to be used time and again. But as with pesticides, unwanted outcomes of biological control occur when the control agent affects nontarget native species. And if a biological control agent turns out to pose an environmental threat, this may last forever.

Evolution of resistance

In fact, most pesticides have a limited life because pests evolve resistance to them. Not all individuals in a genetically variable population will necessarily be susceptible to the pesticide. These contribute more genes to future generations and, if the pesticide is applied repeatedly, each successive generation of the pest will contain a larger proportion of resistant individuals. Resistance does not usually take long to appear and it can spread very rapidly in a population. In the face of evolution of resistance, sustainable pest control depends either on the continual development of new pesticides or management strategies that expose the pests to a given pesticide for only a limited time. It has been extremely common for resistance to pesticides to evolve, but it is also seen occasionally in response to biological control agents.

Integrated pest management (IPM)

IPM considers how all the different tools at the pest controller's disposal can be deployed most effectively to maximize economic benefit and to minimize adverse health and environmental consequences. IPM views crops and plantations as part of functioning ecosystems, and management decisions take into account the effects of control actions not only of pest but also other species in the web of community interactions. IPM arose, in part, in reaction to the unthinking application of pesticides in the 1940s and 1950s. The philosophy considers all possible control actions (physical, cultural, chemical, biological and resistant crop varieties) before one is adopted and, in most cases, integrates several control measures in a compatible manner.

The final word

Jenny would so love to eradicate the rabbits on her farm (Section 6.1). *'After I released rabbit calicivirus disease, rabbit numbers certainly came crashing down, sheep production improved and my annual poisoning costs disappeared. But now, a few years later, the population seems to be on the rise. We are not using anything like the quantity of poison we used to, but this year the helicopter will have to go up for a poison carrot drop on some parts of the farm.'*

Do you think that the reduction in potency of rabbit calicivirus disease (RCD) is a result of evolution of resistance? How would you check this? The presence of RCD in New Zealand has now prompted the government to legalize its use. Devise, in outline, an IPM strategy for rabbit control that includes RCD. What other measures would you include?

References

Anderson, P.K., Cunningham, A.A., Patel, N.G., Morales, F.J., Epstein, P.R. & Daszak, P. (2004) Emerging infectious diseases of plants: pathogen pollution, climate change and agrotechnology drivers. *Trends in Ecology and Evolution* 19, 535–544.

Bax, N., Carlton, J., Mathews-Amos, A. et al. (2001) Conserving marine diversity through the control of biological invasions. *Conservation Biology* 451, 145–176.

Begon, M., Townsend, C.R. & Harper, J.L. (2006) *Ecology: from individuals to ecosystems*, 4th edn. Blackwell Publishing, Oxford.

Benton, B., Bump, J., Seketeli, A. & Liese, B. (2002) Partnership and promise: evolution of the African river-blindness campaigns. *Annals of Tropical Medicine and Parasitology* 96, 5–14.

Borga, K., Gabrielsen, G.W. & Skaare, J.U. (2001) Biomagnification of organochlorines along a Barents Sea food chain. *Environmental Pollution* 113, 187–198.

Bourchier, R.S. (2003) Receptor characterization of nontarget butterflies for risk assessment of biological control with the egg parasitoid *Trichogramma minutum* (Hymenoptera: Trichogrammatidae). *Canadian Entomologist* 135, 449–466.

Buckley, Y.M., Rees, M., Paynter, Q. & Lonsdale, M. (2004) Modelling integrated weed management of an invasive shrub in tropical Australia. *Journal of Applied Ecology* 41, 547–560.

Casey, C.A. & Parrella, M.P. (2005) Evaluation of a mechanical dispenser and interplant bridges on the dispersal and efficacy of the predator, *Phytoseiulus persimilis* (Acari: Phytoseiidae) in greenhouse cut roses. *Biological Control* 32, 130–136.

Charleston, D.S., Kfir, R., Dicke, M. & Vet, L.E.M. (2005) Impact of botanical pesticides derived from *Melia azedarach* and *Azadirachta indica* on the biology of two parasitoid species of the diamondback moth. *Biological Control* 33, 131–142.

Collier, T. & Van Steenwyk, R. (2004) A critical evaluation of augmentative biological control. *Biological Control* 31, 245–256.

Coote, T. & Loève, E. (2003) From 61 species to five: endemic tree snails of the Society Islands fall prey to an ill-judged biological control programme. *Oryx* 37, 91–96.

Culver, C.S. & Kuris, A.M. (2000) The apparent eradication of a locally established introduced marine pest. *Biological Invasions* 2, 245–253.

Davies, J.B. (1994) Sixty years of onchocerciasis vector control – a chronological summary with comments on eradication, reinvasion, and insecticide resistance. *Annual Review of Entomology* 39, 23–45.

Debach, P. & Rosen, D. (1991) *Biological Control by Natural Enemies*, 2nd edn. Cambridge University Press, Cambridge.

Ellis, J.A., Walter, A.D., Tooker, J.F. et al. (2005) Conservation biological control in urban landscapes: manipulating parasitoids of bagworm (Lepidoptera: Psychidae) with flowering forbs. *Biological Control* 34, 99–107.

Eplee, R.E. (2001) *Striga asiatica* (L.) O. Kuntz (Witchweed) eradication from the USA. In: *Invasive Alien Species: a toolkit of best prevention and management practices* (R. Wittenberg & M.J.W. Cock, eds), p. 36. CAB International, Wallingford, Oxon, UK.

Fenner, F. & Ratcliffe, R.N. (1965) *Myxomatosis*. Cambridge University Press, London.

Fenton, A., Norman, R., Fairbarn, J.P. & Hudson, P.J. (2000) Modelling the efficacy of entomopathogenic nematodes in the regulation of invertebrate pests in glasshouse crops. *Journal of Applied Ecology* 37, 309–320.

Flint, M.L. & van den Bosch, R. (1981) *Introduction to Integrated Pest Management*. Plenum Press, New York.

Forrester, N.W. (1993) Well known and some not so well known insecticides: their biochemical targets and role in IPM and IRM programmes. In: *Pest Control and Sustainable Agriculture* (S. Corey, D. Dall & W. Milne, eds), pp. 28–34. CSIRO, East Melbourne.

Fowler, S.V. (2004) Biological control of an exotic scale, *Orthezia insignis* Browne (Homoptera: Orthexiidae), saves the endemic gumwood tree, *Commidendrum robustum* (Roxb.) DC. (Asteraceae) on the island of St Helena. *Biological Control* 29, 367–374.

Herman, T.J.B. (2000) Developing IPM for potato tuber moth. *Commercial Grower* 55, 26–28.

Hockland, S.H., Dawson, G.W., Griffiths, D.C., Maples, B., Pickett, J.A. & Woodcock, C.M. (1986) The use of aphid alarm pheromone (*E*-β-farnesene) to increase effectiveness of the entomophilic fungus *Verticillium lecanii* in controlling aphids on chrysanthemums under glass. In: *Fundamental and Applied Aspects of Invertebrate Pathology* (R.A. Sampson, J.M. Vlak & D. Peters, eds), p. 252. Foundation of the Fourth International Colloquium of Invertebrate Pathology, Wageningen.

Horn, D.S. (1988) *Ecological Approach to Pest Management*. Elsevier, London.

Kehrli, P., Lehmann, M. & Bacher, S. (2005) Mass-emergence devices: a biocontrol technique for conservation and augmentation of parasitoids. *Biological Control* 32, 191–199.

Lee, J.C., Menalled, F.D., & Landis, D.A. (2001) Refuge habitats modify impact of insecticide disturbance on carabid beetle communities. *Journal of Applied Ecology* 38, 472–483.

Levine, E., Oloumisadeghi, H. & Fisher, J.R. (1992) Discovery of a multiyear diapause in Illinois and South-Dakota northern corn rootworm (Coleoptera, Chrysomelidae) eggs and incidence of the prolonged diapause trait in Illinois. *Journal of Economic Entomology* 85, 262–267.

Louda, S.M., Arnett, A.E., Rand, T.A. & Russell, F.L. (2003a) Invasiveness of some biological control insects and adequacy of their ecological risk assessment and regulation. *Conservation Biology* 17, 73–82.

Louda, S.M., Pemberton, R.W., Johnson, M.T. & Follett, P.A. (2003b) Non-target effects – the Achilles' heel of biological control? Retrospective analyses to reduce risk associated with biocontrol introductions. *Annual Review of Entomology* 48, 365–396.

Luckmann, W.H. & Decker, G.C. (1960) A 5-year report on observations in the Japanese beetle control area of Sheldon, Illinois. *Journal of Economic Entomology* 53, 821–827.

Lui, Y.B., Tabashnik, B.E., Dennehy, T.J. et al. (2001) Effects of Bt cotton and Cry1Ac toxin on survival and development of pink bollworm (Lepidoptera: Gelechiidae). *Journal of Economic Entomology* 94, 1237–1242.

Misenheimer, T.M., Lund, M., Baker, A.E.M. & Suttie, J.W. (1994) Biochemical basis of warfarin and bromadiolone resistance in the house mouse, *Mus musculus domesticus*. *Biochemical Pharmacology* 47, 673–678.

Mols, C.M.M. & Visser, M.E. (2002) Great tits can reduce caterpillar damage in apple orchards. *Journal of Applied Ecology* 39, 888–899.

Moreno, D.S. & Luck, R.F. (1992) Augmentative releases of *Aphytis melinus* (Hymenoptera: Aphelinidae) to suppress California redscale (Homoptera: Diaspididae) in southern California lemon orchards. *Journal of Economic Entomology* 85, 1112–1119.

Park, H-W., Bideshi, D.K. & Federici, B.A. (2005) Synthesis of additional endotoxins in *Bacillus thuringiensis* subsp. *morrisoni* PG-14 and *Bacillus thuringiensis* subsp. *jegathesan* significantly improves their mosquitocidal efficiency. *Journal of Medical Entomology* 42, 337–341.

Pearson, D.E. & Callaway, R.M. (2003) Indirect effects of host-specific biological control agents. *Trends in Ecology and Evolution* 18, 456–461.

Ramirez, O.A. & Saunders, J.L. (1999) Estimating economic thresholds for pest control: an alternative procedure. *Journal of Economic Entomology* 92, 391–401.

Simberloff, D. (2003) How much information on population biology is needed to manage introduced species. *Conservation Biology* 17, 83–92.

Smith, J.W. (1998) Boll weevil eradication: area-wide pest management. *Annals of the Entomological Society of America* 91, 239–247.

Timmins, S.M. & Braithwaite, H. (2002) Early detection of new invasive weeds on islands. In: *Turning the Tide: eradication of invasive species* (D. Veitch & M. Clout, eds), pp. 311–318. Invasive Species Specialist Group of the World Conservation Union (IUCN), Auckland, New Zealand.

Townsend, C.R., Begon, M. & Harper, J.L. (2003) *Essentials of Ecology*, 2nd edn. Blackwell Science, Oxford.

Trujillo, E.E. (2005) History and success of plant pathogens for biological control of introduced weeds in Hawaii. *Biological Control* 33, 113–122.

Trujillo, E.E., Kadooka, C., Tanimoto, V., Bergfeld, S., Shishido, G. & Kawakami, G. (2001) Effective biomass reduction of the invasive weed species banana poka by *Septoria* leaf spot. *Plant Diseases* 85, 358–361.

Trumper, E.V. & Holt, J. (1998) Modelling pest population resurgence due to recolonization of fields following an insecticide application. *Journal of Applied Ecology* 35, 273–285.

Twigg, L.E., Martin, G.R. & Lowe, T.J. (2002) Evidence of pesticide resistance in medium-sized mammalian pests: a case study with 1080 poison and Australian rabbits. *Journal of Applied Ecology* 39, 549–560.

van den Bosch, R., Leigh, T.F., Falcon, L.A., Stern, V.M., Gonzales, D. & Hagen, K.S. (1971) The developing program of integrated control of cotton pests in California. In: *Biological Control* (C.B. Huffaker, ed.), pp. 377–394. Plenum Press, New York.

Waage, J.K. & Greathead, D.J. (1988) Biological control: challenges and opportunities. *Philosophical Transactions of the Royal Society of London, Series B* 318, 111–128.

Yaméogo, L., Crosa, G., Samman, J. et al. (2001) Long-term assessment of insecticides treatments in West Africa: aquatic entomofauna. *Chemosphere* 44, 1759–1773.

7 Harvest management

Harvesting always involves reducing a population below its carrying capacity. By doing this, density-dependent factors are relaxed and the population responds by raising its birth rate and dropping its death rate – producing a surplus that can be harvested without further reducing the population. Harvest managers aim to take the maximum sustainable yield (the highest the population can support indefinitely) or the optimum economic yield (producing the highest profit per organism harvested). But this is easier said than done.

Chapter contents

7.1 Introduction   173
     7.1.1   Avoiding the tragedy of the commons   173
     7.1.2   Killing just enough – not too few, not too many   174
7.2 Harvest management in practice – maximum sustainable yield (MSY) approaches   178
     7.2.1   Management by fixed quota – of fish and moose   178
     7.2.2   Management by fixed effort – of fish and antelopes   181
     7.2.3   Management by constant escapement – in time   182
     7.2.4   Management by constant escapement – in space   183
     7.2.5   Evaluation of the MSY approach – the role of climate   184
     7.2.6   Species that are especially vulnerable when rare   185
     7.2.7   Ecologist's role in the assessment of MSY   186
7.3 Harvest models that recognize population structure   186
     7.3.1   'Dynamic pool models' in fisheries management – looking after the big mothers   187
     7.3.2   Forestry – axeman, spare which tree?   190
     7.3.3   A forest bird of cultural importance   191
7.4 Evolution of harvested populations – of fish and bighorn rams   191
7.5 A broader view of harvest management – adding economics to ecology   193
7.6 Adding a sociopolitical dimension to ecology and economics   195
     7.6.1   Factoring in human behavior   195
     7.6.2   Confronting political realities   197

Key concepts

In this chapter you will

recognize that sustainable harvesting is a primary aim of resource management, permitting future generations to enjoy the same opportunities as we do

note that sustainable harvesting is most challenging when the organisms are small and hidden from view, with birth and death rates that respond sensitively to climatic fluctuation, and where multiple exploiters are involved (ocean fish are more difficult to manage than forest trees)

---

appreciate that the maximum sustainable yield may be harvested by fixed quota, fixed effort or constant escapement strategies, with varying levels of administrative expense and risk of 'getting it wrong'

understand that the most realistic harvest models recognize that different age classes have different mortality rates and contribute differentially to reproduction

appreciate that harvesters exert a profound selection pressure that can cause evolution to smaller size-at-maturity, with implications for sustainable yields

recognize that management recommendations differ when the objective is expressed in terms of profit rather than size of harvest

appreciate that sociopolitical factors often come into play when governments make decisions about harvest management

7.1 Introduction

Elena, a Guatemalan, remembers when forest harvesting was a free for all. *'Before 1950 we tapped latex from chico trees – that's what they make chewing gum from. Later some foreign companies were allowed to cut down our mahogany trees – they made big money but most of the forest is now gone. We were heartbroken about the loss of the mahogany. But many of us were also unhappy when some international conservation organizations got involved and imposed their rules on us.'* That was about 25 years ago. There was a lot of resentment and some engaged in illegal logging at that time. But the government listened to local opinion and now, together with 75 other families, Elena has what they call a community forestry concession. *'And, guess what? We have been granted Forest Stewardship Council Certification so that people know that what we sell comes from a forest that is cared for. The area we manage is more than 10,000 hectares but we harvest from just 400 hectares a year. We take only some of the bigger trees (wider than 55 cm) including mahogany, but never more than two from each hectare. It's great to be working for ourselves, in our own forest'.* Elena is proud of what her community is achieving.

7.1.1 Avoiding the tragedy of the commons

We all depend on the harvesting of living resources, whether these are domesticated (cabbages, plantation trees, beef cattle) or wild (natural forest trees, bushmeat, marine fish). Sustainable harvesting is a primary aim of resource management, so future generations can enjoy the same opportunities as we do. And achieving sustainability does not always pose a difficulty. A producer of cabbages, for example, has simply to plant an appropriate number of seeds, take care of the growing plants and harvest the entire crop to take to market for the highest economic returns. Much the same applies to foresters when they 'clear-cut' a plantation of trees. And the same can be said for a fishing club that 'seeds' a pond with trout reared in captivity for the enjoyment of anglers during a brief fishing season, or a game hunting company that releases captive reared animals into the gun sights of its clients. As long as the economic returns justify the costs, these activities are sustainable – because although all the plants and animals die in the process, sufficient seeds, fish and game can be reared and reintroduced into the exploited ecosystem.

Things become trickier when the plants or animals live and reproduce in the wild. Thus, foresters who harvest natural forests have a more complex situation on their hands because sufficient trees of various ages need to remain untouched so that harvests for future generations are protected. In their favor, the foresters know the details of the resource they are managing – trees are big and don't move – and a forestry company may not have the challenge of managing a resource used by other exploiters too.

Sometimes, though, as you heard from Elena, forests are available for exploitation by a community of exploiters – the resource then is a *good* that is *held in common*. This can lead to what Hardin (1968) called the 'tragedy of the commons' because when individuals use a public good, they do not necessarily bear the entire cost of their actions – which may in the long term be borne by others. The story goes that the best short-term strategy is for individuals to selfishly try to exploit more than their share, gaining in the short term but leading inevitably to overexploitation and fewer resources for later generations. That is what happened in earlier times in Elena's forest. Given human nature, the selfish approach is often to be expected in an unregulated situation, but the tragedy can be avoided if neighbors agree on appropriate behaviors and rules (lore) or if competing exploiters are regulated by government (law) so that they take only their share of a sustainable harvest.

The difficulties of managing resources are most acute when, in contrast to trees, the exploited organisms are hidden from view, relatively short lived, and have birth and death rates that are highly responsive to the vagaries of climate. Determining what constitutes a sustainable harvest is then fraught with difficulty. For these reasons, and because they are also *wild* and *held in common*, marine fisheries are truly difficult to manage. Their importance as a human food supply has meant that the most sophisticated of management regimes have been developed for these denizens of the deep but, even so, the strategies don't always work, as you will see in this chapter.

7.1.2 *Killing just enough – not too few, not too many*

I noted in Chapters 5 and 6 that the aims of conservation and pest control are diametrically opposed – conserve endangered species by reducing mortality but get rid of pests by killing as many as possible. Harvest management is based on the same population dynamics theory, but its objective lies somewhere in between. In this case exploiters aim to kill as many as possible, so that human benefit from a food resource or forest timber is maximized, but not too many – which would lead to the exploited population shrinking to a size that is economically insignificant or even biologically extinct. Simultaneously avoiding both overexploitation and underexploitation is not as easy as it sounds. Box 7.1 presents the essential population theory that underpins both how to work out the maximum sustainable yield (MSY) and how to devise ways of achieving it. This extends further my treatment of population dynamics theory and can be viewed in combination with the boxes in Chapters 5 and 6.

The MSY concept is central to harvest management, and it brings into stark relief some of the core concepts that managers need to bear in mind. However, there are a number of critical limitations to the simple application of the MSY approach – I will highlight these in Section 7.2 by considering examples of harvest management in practice.

A major simplification in the MSY approach is to treat the population as a number of identical individuals, ignoring all aspects of population structure such as size or age classes with their different rates of growth, survival and reproduction. Alternative approaches that incorporate population structure will be considered in Section 7.3.

You have seen in earlier chapters about applications of population theory (managing endangered populations and pest control) that a full understanding had to include an evolutionary dimension. Harvest management is no different. Harvesters exert a very strong selection pressure on the populations they exploit, and inevitably this leads to evolutionary changes in life-history features, and particularly to a reduction in body size of the harvested animals. The implications of evolution for harvest management will be discussed in Section 7.4.

Finally, and as with other topics in this book, harvesting needs to be viewed in a broad context. Thus, harvest managers must combine ecological considerations (the well-being of the exploited population) with economic (the profits being made from the operation) and social considerations (local levels of employment and the maintenance of traditional lifestyles). The economic and sociopolitical dimensions of harvesting will be discussed in Sections 7.5 and 7.6.

Box 7.1 Population dynamics theory 3

This is the last of three boxes that set out population dynamics theory. Here I extend ideas presented in Box 5.1 about population growth and carrying capacity to show, in simple terms, how it is possible to identify a sustainable harvest.

Logistic growth and the underlying patterns in births and deaths
You saw in Box 5.1 that a combination of density-dependent and density-independent factors combine to determine a population's carrying capacity (K), the population size about which the population fluctuates. Every population has the potential to increase exponentially, because parents have the capacity to produce more offspring than are needed just to replace themselves. However, real populations do not increase indefinitely; instead they increase until density-dependent factors (e.g. too little food per individual) bring birth rate and death rate roughly into balance. Mathematicians call this pattern of population growth *logistic* (Figure 7.1a; the equation that describes logistic growth is given in Box 5.1).

It is instructive to consider for different segments of the logistic growth curve what is happening in terms of births and deaths (Figure 7.1b). The numbers of both births and deaths progressively increase because as we move to the right in Figure 7.1b there are ever more individuals available to give birth or to die. But, in addition, density-dependent forces mean that the per capita birth rate can be expected to fall at higher densities (e.g. because of food limitation) and the per capita death rate to rise (e.g. because of increased probabilities of starvation or predation or disease). In the exponential phase of the curve (when the population is still at low density), resource availability is high, birth rate is high and death rate low. At higher densities, when density-dependent factors start taking effect, the environmental brakes are applied and the rate of population growth declines, because of a decrease in birth rate and/or an increase in death rate. Finally, when the population reaches the carrying capacity, birth rate is matched by death rate and the population stabilizes. A population with exactly equal birth and death rates must, by definition, stay the same.

When is a harvest sustainable?
How do these considerations of birth and death rates help to identify a sustainable harvest? Note that the net recruitment at any particular density in Figure 7.1c is a surplus of births over deaths that can be removed without causing the population to decline from that density. Thus, as long as we know the density of the population to be harvested and understand the relationship between net recruitment (surplus) and density, we can harvest the surplus without reducing the population size. This is a sustainable harvest (or yield). It is low when the population is very small and low also, at the other end of the scale, when competition for resources is intense. And it is crucial to note that when the population is at carrying capacity there is no net recruitment, so there can be no sustainable harvest. In other words, you cannot both harvest and maintain a population at carrying capacity. Harvesting always involves reducing a population below its carrying capacity. By so doing, density-dependent factors are relaxed and the population responds by raising its birth

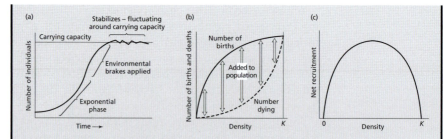

Fig. 7.1 (a) Logistic population growth in which population size increases exponentially at first but as numbers increase with time the rate of increase first slows down and then stabilizes around the carrying capacity. The environmental brakes are supplied by density-dependent factors, such as competition for food or risk of predation, which, at high density, reduce the per capita birth rate and/or increase the per capita death rate. (b) When time is replaced by population density on the horizontal axis, the patterns in numbers of births and deaths at each density can be highlighted. The difference between the birth and death curves gives us the number of individuals actually added to the population in a unit of time when the population is at a given density. (If there are more births than deaths a net surplus is produced.) Largest numbers are added to the population at some intermediate density, when there are plenty of individuals to produce offspring and density-dependent factors are not yet fully in operation. As density increases further, the number added per unit time declines until finally the birth and death curves cross. At this point, the number of births is exactly matched by the number of deaths and the population stabilizes (at K the carrying capacity). (c) This figure shows net recruitment (surplus of births over deaths) per unit time at different densities. Net recruitment is greatest at some intermediate density (see how the longest arrows in (b) are at intermediate densities) and declines to zero at K.

rate and dropping its death rate – producing a surplus that can be harvested without further reducing the population. It is in this sense that the harvest is sustainable.

The maximum sustainable yield (MSY) – putting theory into practice
The ideal harvesting operation takes neither too little nor too much. Consider Figure 7.1c and decide what would constitute the *maximum sustainable* harvest. Yes, the maximum is obtained by harvesting at the density where net recruitment is itself maximal – the highest point on the curve. For a perfect 'logistic' curve this occurs when density is 50% of carrying capacity K. But in real populations, the shape of the curve is not necessarily symmetrical. In large mammals, for example, the maximum yield occurs at a density only slightly less than K. However, the maximum sustainable yield is always at a density that is less than K.

Let's ignore for now some real difficulties in devising a harvesting strategy, such as how to determine the precise relationships between net recruitment and density for real populations. Assuming you know the MSY, how can you arrange to harvest it without danger to the population? Two related methods are to take a fixed quota each year (i.e. the calculated MSY) or to apply a fixed effort (in terms of numbers of exploiters and time spent hunting or fishing) set so as to achieve the MSY.

Fixed quota
Harvesting at a fixed rate (i.e. a fixed number of individuals removed during a given period of time), otherwise known as 'taking a *fixed quota*', is sustainable if the harvesting rate line crosses the net recruitment curve. When these lines cross, harvesting and recruitment rates are equal and opposite and the number removed per unit time by the harvester equals the number recruited per unit time by the population. In Figure 7.2 three different harvesting rates are illustrated (horizontal dotted lines). The highest of these (H_H) is to be avoided at all costs. When harvesting rate is greater than even the maximum recruitment rate, the population is doomed to extinction because more

Plate 1.5 (Fig. 1.5 on page 8) Extinctions in Singapore since the early 1800s – green and blue bars represent recorded and inferred extinctions, respectively. (After Sodhi et al., 2004.)

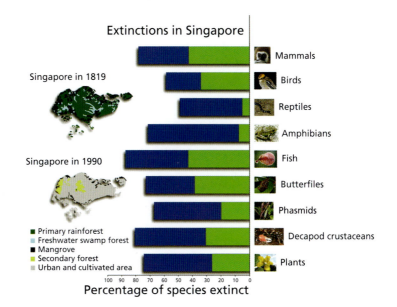

Extinctions in Singapore

Singapore in 1819

Singapore in 1990

- Primary rainforest
- Freshwater swamp forest
- Mangrove
- Secondary forest
- Urban and cultivated area

Mammals
Birds
Reptiles
Amphibians
Fish
Butterfiles
Phasmids
Decapod crustaceans
Plants

Percentage of species extinct
100 90 80 70 60 50 40 30 20 10 0

Plate 10.6 (Fig. 10.6 on page 269) Predicted population growth rate (λ) of wood thrush populations in relation to forest cover (colored cells) in the eastern USA. The extent of their breeding range is shown by dotted lines. Blue areas depict source populations that are currently safe (λ > 1.0), while pink areas show sink populations likely to go extinct (λ < 1.0). Safe populations are generally those in larger forest fragments and/or in landscapes containing only a small proportion of nonforest habitat. (After Lloyd et al., 2005.)

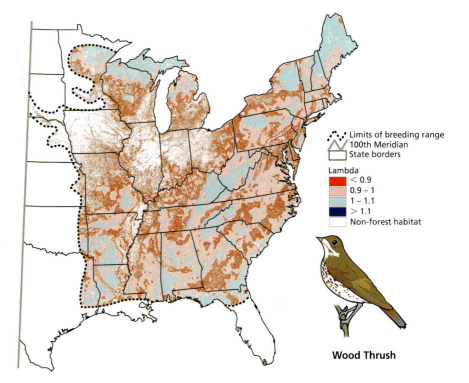

Limits of breeding range
100th Meridian
State borders

Lambda
- < 0.9
- 0.9 – 1
- 1 – 1.1
- > 1.1
- Non-forest habitat

Wood Thrush

Plate 10.7 (Fig. 10.7 on page 271) Simulated spread on ocean currents along the north Pacific coast of Canada of particles introduced at a depth of 2 m to represent fish larvae, from each of ten marine protected areas (outlined in white) over a 90-day period. Particles from each protected area are shown in a unique color (see key). Grey areas represent Vancouver Island and the Canadian mainland. The continental shelf is shown in pink. NK, Naikoon; BISR, Banks Island sponge reef; BY, Byers; MBSR, Middle Bank sponge reef; LS, Laredo Sound; GIB, Goose Island Bank; HK, Hakai; SI, Scott Islands; QCS, Queen Charlotte Sound; BS, Blackfish Sound. (After Robinson et al., 2005.)

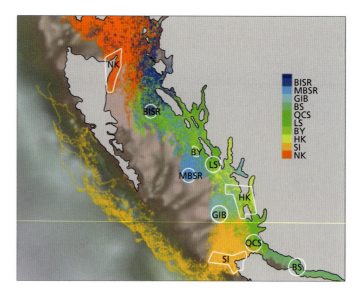

Plate 10.12 (Fig. 10.12 on page 278) Distribution of biodiversity hotspots, showing numbers of species of globally threatened birds plus amphibians mapped on an equal area basis (each grid cell is 3113 km^2). (After Rodrigues et al., 2006.)

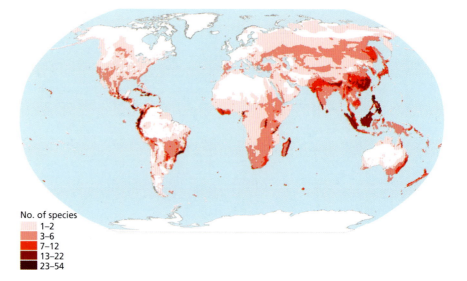

No. of species
1–2
3–6
7–12
13–22
23–54

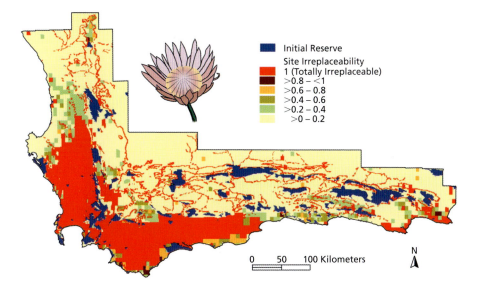

Plate 10.14 (Fig. 10.14 on page 281) Map of South Africa's Cape Floristic Region showing site *irreplaceability* values for achieving a range of conservation targets in the 20-year conservation plan for the region. Irreplaceability is a measure, varying from zero to one, which indicates the relative importance of an area for the achievement of regional conservation targets. Existing reserves are shown in blue. (After Cowling et al., 2003.)

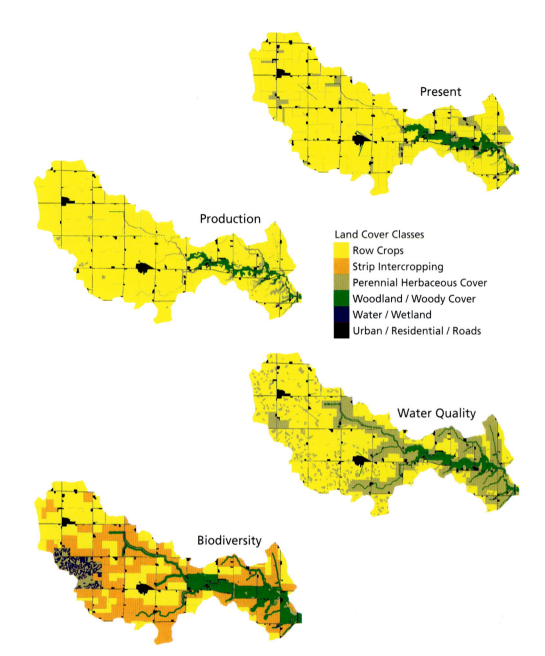

Present

Production

Land Cover Classes

- Row Crops
- Strip Intercropping
- Perennial Herbaceous Cover
- Woodland / Woody Cover
- Water / Wetland
- Urban / Residential / Roads

Water Quality

Biodiversity

Plate 10.16 (Fig. 10.16 on page 285) Present landscape (top right) and alternative future scenarios for the Walnut Creek catchment area in Iowa, USA. In comparison to the present situation, note the increase in row crops at the expense of perennial cover in the Production scenario. In the Water Quality scenario, note in particular the increase in perennial cover (pasture and forage crops) and wider riparian buffers. Finally in the Biodiversity scenario, note the increase in strip intercropping, the wide riparian buffers and the extensive prairie, forest and wetland restoration reserves. (From Santelmann et al., 2004.)

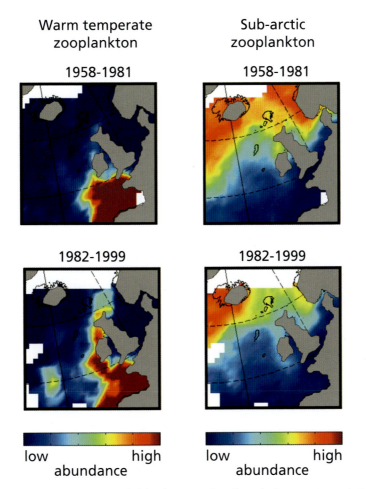

Plate 11.1 (Fig. 11.1 on page 294) The northerly shift by about 1000 km of zooplankton species typical of warm temperate conditions in the northeast Atlantic over the last 40 years. The scale represents the relative abundance and species richness of warm temperate species in samples taken from different geographical locations. By contrast, the colder-water species that are typical of subarctic conditions have contracted their range. (From Hays et al., 2005.)

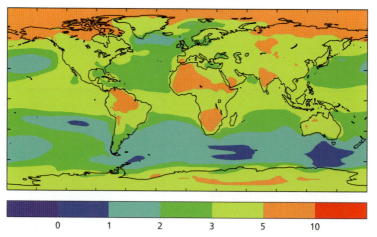

(a) Temperature change

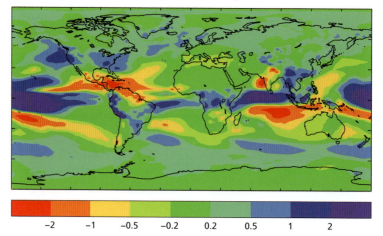

(b) Precipitation change

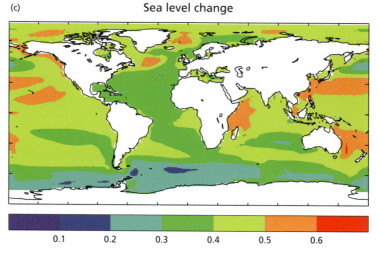

(c) Sea level change

Plate 11.2 (Fig. 11.2 on pages 294–5) Predicted annual changes in (a) average surface air temperature (°C), (b) average precipitation (mm day^{-1}) and (c) average sea level rise (m) from 1960–1990 to 2070–2100, when carbon dioxide concentration in the atmosphere is expected to have doubled. These predictions are based on a coupled atmosphere–ocean general circulation model (HadCM2) of the Hadley Centre for Climate Prediction and Research, which assumes a midrange global average temperature increase of 3.2°C. (© Crown copyright 2005, Published by the Met Office Hadley Centre.)

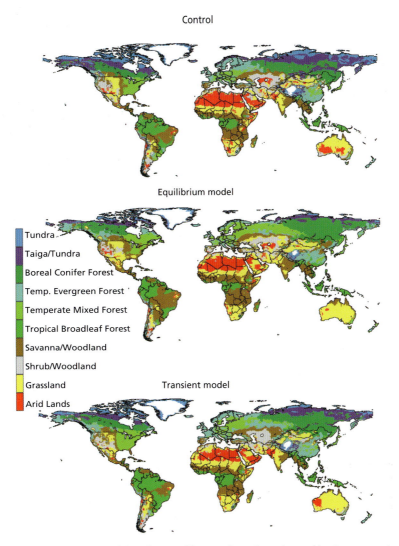

Plate 11.3 (Fig. 11.3 on page 296) Simulated distribution of biomes throughout the world using contrasting climate change models. Biomes are regional ecosystems characterized by distinct types of vegetation fitted to the prevailing physical conditions. (a) The 'control' output represents the current distribution of biomes. (b) Output from an 'equilibrium' model, in which the doubling of CO_2 is assumed to produce its full temperature effect instantaneously. (c) Output for a 'transient model', in which atmospheric CO_2 responds to dynamic feedback between atmosphere and oceans, achieving only about 65% of the eventual predicted temperature change at the time of CO_2 doubling. Under both scenarios, forests shift northwards to currently unforested areas. However, in temperate latitudes the transient model produces much lower increases in temperature and larger increases in precipitation than the equilibrium model, with consequent differences in the distribution of biomes in the two cases. (Tundra is a treeless plain, Boreal (= northern) Conifer Forest consists of needle-leaf trees, while Taiga/Tundra is intermediate between the two, with tundra vegetation and scattered conifer trees. Savanna is grassland with scattered trees.) (From Neilson & Drapek, 1998.)

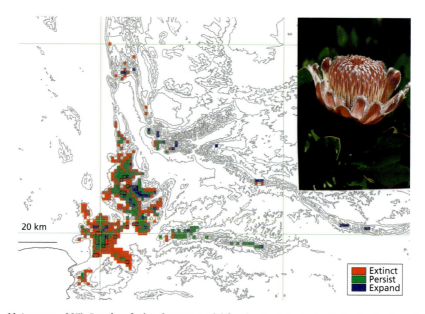

Plate 11.4 (Fig. 11.4 on page 297) Results of a bioclimatic model for the plant *Protea lacticolor* in the Cape Floristic Region of South Africa. Its current distribution is represented by red and green squares on the map. According to the modeled climate change scenario, the species is predicted to persist in the green areas in 2050 but will be extinct in the red areas, because here the future patterns of temperature and water availability will be beyond the plant's climatic envelope. Blue squares represent new areas that could be inhabited by the plant provided their dispersal capabilities (or action by conservation managers) can get them there. (From Hannah et al., 2005.)

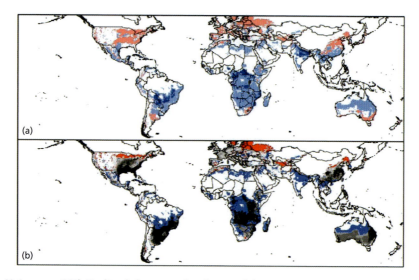

Plate 11.6 (Fig. 11.6 on page 300) Predicted changes to distribution of the Argentine ant between now and 2050. The map represents the average predictions of four global climate change scenarios. Red areas are those predicted to improve for Argentine ants, whereas blue areas are predicted to worsen for the species. (After Roura-Pascual et al., 2004.)

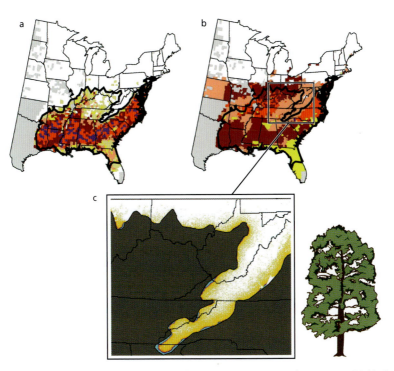

Plate 11.8 (Fig. 11.8 on page 302) (a) Current distribution of sweetgum trees across the eastern USA (darker colors show greater habitat quality – an index of density and size of trees) with the 1971 range outlined in black. (b) Predicted distribution of sweetgum in 2100, assuming that rate of dispersal can keep up with change in habitable area. (c) Actual occupied area in 2100 is predicted to be very much less than the potential area, because of dispersal restrictions. This panel, which focuses in on the current range boundary (indicated by the box in (b)), shows the presently occupied range in gray and the surprisingly small addition by 2100 of newly occupied habitat immediately to the northeast. According to the predicted distribution in (b), all of the area depicted in (c) could, in theory, be occupied by sweetgum if only they could shift fast enough. (From Iverson et al., 2004.)

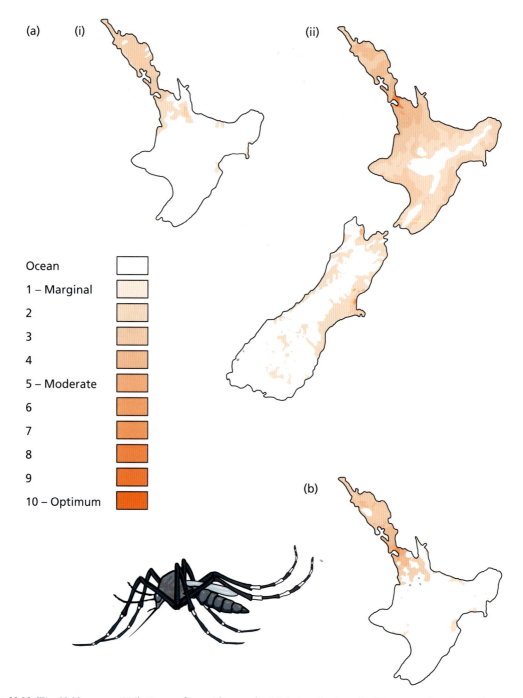

Plate 11.12 (Fig. 11.12 on page 308) Dengue fever risk maps for (a) *Aedes albopictus* for (i) present climatic conditions and (ii) for 2100 under a climate change scenario. (b) The fundamental niche of *Aedes aegypti* means that it cannot survive anywhere in New Zealand at present, but it could inhabit parts of the North Island as shown under the climate change scenario by 2100. (After de Wet et al., 2001.)

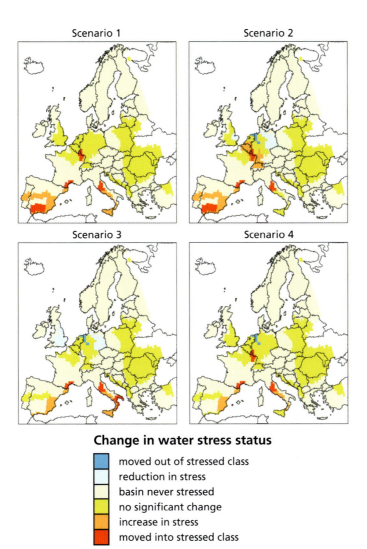

Change in water stress status

- moved out of stressed class
- reduction in stress
- basin never stressed
- no significant change
- increase in stress
- moved into stressed class

Plate 11.13 (Fig. 11.13 on page 309) Stress status of river catchment areas throughout Europe by 2080 according to four different climate change models. In a few areas, water stress is reduced in comparison to no climate change (blue), in many places there is no change (yellow), but in some areas water stress is set to increase (orange, red). (From Schroter et al., 2005.)

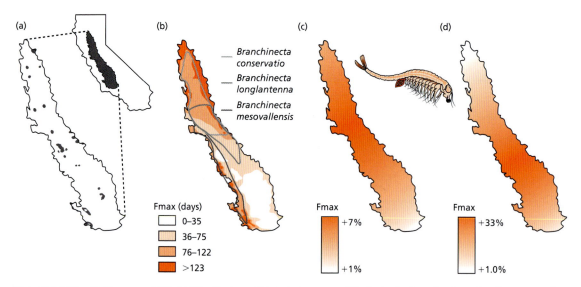

Plate 11.14 (Fig. 11.14 on page 310) (a) Distribution of biological reserves (black shading) in the Central Valley region of California, USA (inset). The current set of protected areas only includes a small proportion of locations where vernal pools are found. Thus, temporary pond dwellers, such as fairy shrimps (*Branchinecta* spp.), are at risk. (b) The current distribution of F_{max} – the annual average number of days of continuous inundation of ponds. F_{max} is a key factor determining the success of fairy shrimps. The current ranges of three fairy shrimp species are shown in outline. (c) Predicted changes to F_{max} under a cooler and drier scenario (–1°C, –10% precipitation). (d) Predicted changes to F_{max} under a warmer and wetter scenario (+3°C, +30% precipitation). The manager's objective will be to maintain shrimp biodiversity by adding new protected areas that incorporate the necessary future range of F_{max} values. (From Pyke, 2005.)

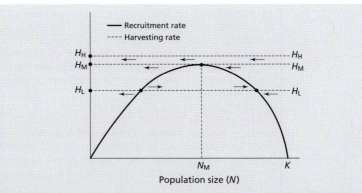

Fig. 7.2 Fixed quota harvesting. The figure shows a single recruitment curve and three fixed quota harvesting curves: high quota (H_H), medium quota (H_M) and low quota (H_L). Arrows in the figure refer to changes to be expected in abundance under the influence of the harvesting rate to which the arrows are closest. Black circles are equilibruim points. At H_H the only 'equilibrium' is when the population is driven to extinction. At H_L there is a stable equilibrium at a relatively high density, and also an unstable breakpoint at a relatively low density. The MSY is obtained at H_M because it just touches the peak of the recruitment curve (at a density N_M): populations greater than N_M are reduced to N_M, but populations smaller than N_M are driven to extinction. (After Begon et al., 2006.)

individuals are consistently being taken than can be replenished by recruitment. In other words, if we overestimate the MSY, even slightly, the population will be driven extinct. The lowest harvesting rate shown (H_L) crosses the recruitment curve at two points – at a relatively low population density and at a relatively high density quite close to carrying capacity. Where the lines cross, the harvest is, by definition, sustainable but it is not as big as it could be. Of particular interest, then, is harvesting rate H_M, which just touches the recruitment rate curve at its peak. This is the highest harvesting rate that the population can match with its own recruitment – in other words, the MSY – the largest harvest that can be removed from the population regularly and indefinitely. The MSY is obtained from the population by depressing it to the density at which the recruitment curve reaches its peak.

There are some profound difficulties in using the MSY as a basis for harvest management, which will be discussed in the light of real harvests in Section 7.2. For now, just one of the problems will be highlighted. The MSY density (N_m) is an equilibrium (gains = losses), but when harvesting is based on the removal of a fixed quota (H_M in Figure 7.2), the equilibrium is very fragile. If, by chance, the actual density exceeds the MSY density, then H_M exceeds the recruitment rate and the population declines towards N_m. This, in itself, is satisfactory. But if the actual density is ever so slightly less than N_m, then H_M will again exceed the recruitment rate so that density will decline even further – indeed, if the fixed quota is maintained the population will decline to extinction (see direction of arrows along the H_M line in Figure 7.2). Things are not so bad at the lower harvesting rate (H_L): in this case there is a stable equilibrium where the lines cross at high density, but an unstable one where they cross at low density (so only if actual density were much lower than assumed would the population drop to extinction). But H_L does not provide an MSY.

In the quest for an MSY, in other words, a fixed quota strategy is reasonable if you have perfect knowledge and the world is wholly predictable. In the real world of fluctuating environments and imperfect data sets, though, this is a very risky approach.

Fixed effort

The risk can be reduced if instead of taking a fixed quota we regulate harvesting effort (e.g. in terms of the number of 'boat-days' in a fishery or 'gun-days' in a hunted waterfowl population). If harvesting effort is constant, the yield will simply be proportional to the population size (e.g. 50

gun-days may yield 100 birds when the population is small, but 200 birds when it is twice as large, and 300 birds when it is three times as large). A plot of harvesting rate for constant effort is thus a straight line, increasing linearly with density and starting at zero (with zero effort). Four fixed effort lines are plotted in Figure 7.3a. There is an 'optimum' effort giving rise to the MSY, E_M, while efforts both greater (E_H; steeper effort line) and smaller than this (E_L) provide smaller yields. The relationship between yield from the harvest and effort is shown in Figure 7.3b: the MSY is gained with effort E_M.

Both a fixed quota (H_M) and a fixed effort (E_M) can be used to provide the MSY. But the really good thing about the fixed effort approach is that the equilibrium (where harvest rate equals recruitment rate) is stable, in contrast to the unstable equilibrium of the fixed quota. Now, in contrast to Figure 7.2, if density drops below N_M (Figure 7.3), recruitment exceeds the harvesting rate and the population recovers. In fact, there needs to be a very considerable overestimate of E_M before the population is overexploited and driven to extinction (E_{OV} in Figure 7.3).

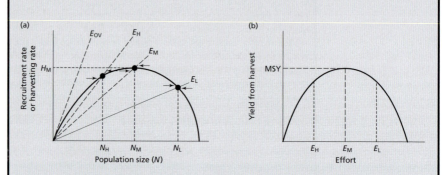

Fig. 7.3 Fixed effort sampling. (a) The maximum sustainable yield (H_M) is obtained with effort E_M, leading to a stable equilibrium at density N_M. At higher (steeper dotted line) or lower efforts, there are also stable equilibria where the effort lines cross the recruitment curve, but the yields are less than the maximum sustainable. The very high effort line (E_{ov}) does not cross the recruitment curve; in other words, harvesting rate is always greater than recruitment rate, the population is overexploited and will be sent extinct. Curves, arrows and dots as in Figure 7.2. (b) The relationship between yield from the harvest and effort applied. The MSY occurs with an effort of E_M.

7.2 Harvest management in practice – maximum sustainable yield (MSY) approaches

Despite all its shortcomings, the MSY concept dominated resource management for many years in fisheries, forestry and wildlife exploitation. Prior to 1980, for example, every one of the 39 management agencies for marine fisheries around the world managed on the basis of an MSY objective (Clark, 1981), and in many harvests the MSY concept is still the guiding principle.

7.2.1 *Management by fixed quota – of fish and moose*

On a specified day in the year, the hunting season (for a given species) or a fishery is opened and the cumulative catch is logged. The permitted quota is set at the estimate of MSY for the exploited species and, when the quota has been taken, the hunt is closed for the rest of the year. The ideal situation would occur where considerable data are available on net recruitment and density – so that a plot like that in Figure 7.2 can be used to estimate the MSY. But such data are rarely available and, even when they are, the pattern does not necessarily conform to the idealized form in Box 7.1. Thus, Figure 7.4a shows estimates for both recruitment (of

Fig. 7.4 (a) Estimates of recruitment of young fish and the size of the spawning stock of Pacific whiting from 1977 to 2001. (b) When recruitment is plotted against biomass (a measure of population size), there is no sign of the idealized domed shape pattern drawn in Figure 7.2. (After Ishimura et al., 2005.)

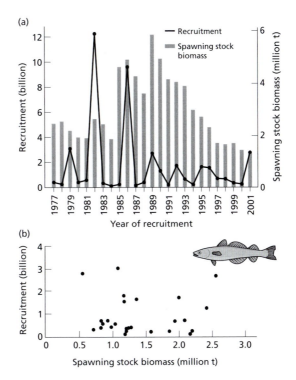

2-year-old fish – the youngest stage that can be reliably sampled) and spawning stock biomass for Pacific whiting (*Meluccius productus*). When recruitment each year (the data for 2-year-old recruits are shifted back 2 years) is plotted against spawning stock biomass (a measure of population size N), the domed curve we were expecting is not evident at all (Figure 7.4b). The considerable yearly variation in biomass and, more particularly, recruitment no doubt reflect variation in environmental conditions, obscuring any underlying dome-shaped relationship in Figure 7.4b.

Another data-intensive approach to estimate MSY involves the use of historical records of annual yield against hunting effort because, as noted in Figure 7.3b, MSY is the maximum yield in a plot of yield against effort. However, even this approach can be flawed, as shown in Figure 7.5. In 1975, the International Commission for the Conservation of Atlantic Tunas (ICCAT) used the available data (1964–73) to plot the yield–effort relationship for yellowfin tuna (*Thunnus albacares*) in the eastern Atlantic. It looked like they had reached the top of the curve – with a sustainable yield of around 50,000 tons. With time, however, effort rose further and it became clear that the top of the curve had not been reached. A reanalysis in 1985 suggested a sustainable yield of around 110,000 tons. It seems that the top of the curve has now been achieved, and the MSY is 110,000 tons.

A famous example of the use of fixed quotas is provided by the Peruvian anchovy (*Engraulis ringens*), which from 1960 to 1972 was the world's largest fishery. Fisheries ecologists advised that the MSY was around 10 million tonnes annually, and regulations were put in place to limit catches accordingly. But the regulations were not properly enforced, the fishing capacity of the fleet expanded and the annual catch went above 10 million tonnes (Figure 7.6). In 1972 the catch crashed, at least partly

Fig. 7.5 Estimated yield–effort relationships for the eastern Atlantic yellowfin tuna (*Thunnus albacares*) on the basis of data for 1964–73 (International Commission for the Conservation of Atlantic Tuna (ICCAT), 1975) and 1964–83 (ICCAT, 1985). (After Hunter et al., 1986; Hilborn & Walters, 1992.)

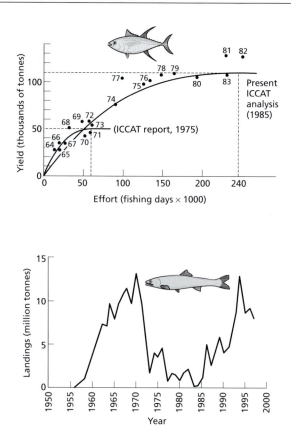

Fig. 7.6 Landings of the Peruvian anchovy since 1950. The catch crashed in 1972 because of overfishing, compounded by climatic variation, and its recovery was slow because a moratorium was not immediately instituted. (After Jennings et al., 2001.)

because of overfishing, although this was compounded by profound climatic variation. A moratorium on fishing would have been the sensible next step, but this was not politically feasible because so many people were dependent on the anchovy industry for employment. So fishing continued, albeit with smaller harvests, and the stock took more than 20 years to recover.

A simpler way to set an MSY begins with an estimate of the carrying capacity of a population: in bird and mammal harvests this may be an expert's opinion of density attainable, but in the case of a fishery it is what was present before exploitation. Of course, these things are only known imperfectly, but reasonable 'guestimates' can be made. Recall from Box 7.1 that with perfectly 'logistic' growth, the MSY occurs at half of carrying capacity ($0.5K$) or half of unexploited biomass ($0.5B_0$) Thus, the MSY quota can be set equal to the net recruitment expected when the population size is $0.5K$ or $0.5B_0$. This kind of rule of thumb is often used for fisheries, but the more conservative rule of $0.4K$ or $0.4B_0$ is usually adopted.

When the exploited population is easier to observe, other approaches are available to set a realistic quota. In the case of moose (*Alces alces*) in Canada, for example, aerial surveys reveal the numbers of bulls, cows and calves within a particular management area. Not only does this provide information on population size, but also recruitment rate. The quota of licences to shoot moose is then set to match the current recruitment rate (after allowing for other natural mortality that will occur). This is a sustainable yield at its most basic. On the other hand, the wildlife agency may want to see an overall reduction in moose, perhaps because they are overexploiting their own food resources, and so sets a higher quota (or vice versa if the population has declined due to adverse weather). Similar approaches are used around the world for harvests of game birds, waterfowl, mammals, and even reptiles such as crocodiles, with quotas of animals permitted to be taken from each hunting area (often quite small in extent) set according to local circumstances.

7.2.2 *Management by fixed effort – of fish and antelopes*

Theory tells us there is less risk of overexploitation with fixed effort than fixed quota strategies (Box 7.1); as a result many harvests are managed by legislative regulation of effort. And this is despite the difficulty of measuring and controlling effort in a precise manner. For example, an agency may issue a fixed number of gun licenses with the expectation that this will produce constant hunting effort from year to year, but the skill of the hunters is uncontrolled. Similarly, regulating the size and composition of a fishing fleet, as a means of controlling effort, leaves the weather to chance.

Consider again the Pacific whiting (Figure 7.4). The fishery for this species is shared between Canadian and US fishing fleets. But in the 1990s the two national agencies could not agree how to share the catch and, as a result, the *total allowable catch* (TAC) was exceeded, probably contributing to reductions in the size of the stock during these years (Figure 7.4a). In 2003, a treaty was signed to improve management of Pacific whiting, with 73.88% of the TAC going to the USA and 26.12% to Canada. The method of determining the TAC is of the fixed effort variety, using what is known as the '40–10 harvest strategy'. This rule automatically imposes the constraint that fishing effort should not exceed that required to produce the MSY – taken as equal to recruitment rate when the population is 40% of unexploited biomass (i.e. $0.4B_0$) – and fishing ceases if the stock biomass drops to 10% of unexploited biomass ($0.1B_0$). Between $0.4B_0$ and $0.1B_0$ there is a straight-line effort curve (as for E_M in Figure 7.3a), but with the added safeguard that at low population density (10% or less of B_0) no catch is taken at all. Ishimura et al. (2005) carried out simulations of a population model (like that used for population viability analysis in Section 5.4) to test the effectiveness of the 40–10 rule for Pacific whiting. They concluded that the rule worked well – producing acceptably large annual catches, with small variation in catch from year to year, and only a small number of years when the fishery would have to be closed.

The Saiga antelope (*Saiga tatarica*) has been hunted for centuries in the deserts of Central Asia for its meat, hide and horns. Until 1990 the population was intensively managed by the Soviet state, but management broke down and poaching increased after the break-up of the Soviet Union. Milner-Gulland (1994) produced a simulation model, based on extensive information about birth rates and death rates and environmental variation, to test whether fixed quota or fixed effort management regimes would be best for the antelope's future management. Males are worth more than females because their horns, used for medicine, are much more valuable than meat. Thus, the model considered the consequences of including different proportions of males and females in the catch. Allowing a constant number of individuals to be taken is administratively more straightforward, but this is a poor strategy for the long-term survival of the population (Figure 7.7a). To avoid the well-known risk of overexploitation in a fixed quota harvest, Milner-Gulland suggests that hunting would have to be kept to a very low level (about 2% of the population per year, or

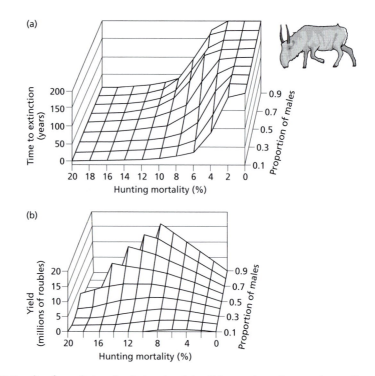

Fig. 7.7 Results of population simulations involving Saiga antelopes harvested according to fixed quota or fixed effort. (a) A fixed quota of animals is harvested each year (expressed as the proportion of the population that the fixed number represents). The risk of overexploitation using a fixed quota strategy is graphically illustrated by the rapid increase in the probability of extinction when the number taken is more than around 2% of the population. Note that harvests containing more males are somewhat less risky (because males are not the ones producing young). (b) When a fixed proportion of the population is taken each year (equivalent to fixed effort), the maximum yield occurs at values of around 10% or more. Yield is expressed here in terms of millions of Russian roubles, with highest values for harvests that include more males with their valuable horns. (After Milner-Gulland, 1994.)

about 20,000 animals). Note that the quota can be somewhat larger (equivalent to 3 or 4% of the population) if the catch is made up predominantly of males. This is because females are the ones that produce offspring and most males are surplus to requirements in this respect. Using a fixed effort approach, on the other hand, about 10% of the population can be safely harvested. Figure 7.7b is like the yield vs effort curve in Figure 7.3b, but note that the yield is shown in terms of the value of the harvest (in Russian roubles) rather than individuals harvested. The harvest is much more valuable when a large proportion of harvested individuals are males, because of the value of their horns.

7.2.3 Management by constant escapement – in time

Another approach, based around the MSY idea, does not prescribe the number of animals to be included in the harvest but focuses instead on the number that are to be left behind. The idea, then, is to leave a minimum fixed number of breeding individuals at the end of each hunting season, a strategy called *constant escapement*. This is a particularly safe option because it rules out the accidental removal of all

Fig. 7.8 Monthly squid catches by licensed vessels in the Falkland Islands where a constant escapement management strategy is used. Note that there are two peaks in catches each year – this is because there are two fishing seasons (February–May and August–October). Dotted lines (1984–86) represent estimated rather than actual catches. The total annual catches have held up well, indicating that the constant escapement strategy is a sustainable approach. (After des Clers, 1998.)

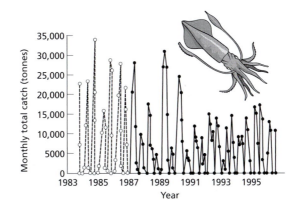

breeding individuals before breeding has occurred. It has a downside, however, and that is the expense of continuous monitoring through the hunting season. Constant escapement is particularly appropriate for annual species that mature together, breed once and then die, because these lack the buffer provided by immature individuals in longer-lived species (Milner-Gulland & Mace, 1998). The Falkland Islands Government uses a constant escapement strategy for the annual *Loligo* squid. Stock sizes are assessed weekly from mid-season onwards and the fishery is closed when the stock falls to $0.3–0.4B_0$. After 10 years of this management regime the squid fishery is holding up well, indicating that the approach is sustainable (Figure 7.8).

Note that the fishery for Pacific whiting (Section 7.2.2), a nonannual species, also has an element of constant escapement in that the fishery is not opened in a year when the stock has fallen below $0.1B_0$. And the approach can be used for mammal hunting too. Stephens et al. (2002) used simulation models to compare the outcomes for a population of alpine marmots (*Marmota marmota*) of fixed quota, fixed effort and constant escapement harvesting. In the latter case, harvesting only occurred during years in which the population exceeded a given threshold and exploitation continued until that threshold was reached. These social mammals are hunted in parts of Europe but the model was based on extensive data available from a nonhunted population. Stephens' team found that constant escapement harvesting provided the highest mean yields and this was coupled with a low risk of extinction.

7.2.4 Management by constant escapement – in space

A different way to approach a constant escapement strategy is to place spatial controls on fishing or hunting, as opposed to the temporal controls of closing a season early or not having a season at all. While nature reserves around the world have usually been established to conserve biodiversity, some could also play a role in harvest management. The protection of 10% of a hunted population in reserves (e.g. shell fish in coastal marine reserves, reef fish in coral reef reserves, Saiga antelope in national parks) is potentially equivalent to providing a fixed escapement of $0.1B_0$. However, to be truly equivalent, the unexploited reserve should then be the source of recruits to the harvest at large. And this depends on whether the reserves are naturally sources or sinks of recruits, as determined by dispersal powers in the face of factors such as ocean currents. These ideas were discussed in the context of dispersal behavior in Section 4.2.4.

It is easy to conceive some potential benefits of marine protected areas for harvest management. First, for species that don't move too much, the individuals in the reserve are likely to be spatially protected from fishing and achieve higher density. Second, elevated density within the reserve may result in net migration to fished areas, either by random dispersal or via density-dependent spillover. And third the unfished population in the reserve can be expected to consist of larger individuals, which are more fecund and thus might contribute disproportionately to recruitment in the fished area (Willis et al., 2003).

When marine protected areas are established, increases in densities and sizes of target species are commonly noted. For example, after 10 years of protection from fishing in a reserve in the French Mediterranean, target fish species were two to three times more abundant inside the reserve than outside, larger species were ten times more abundant, and the mean size of some of the species was about 50% greater (Harmelin-Vivien et al., 1995). Such results are promising, lending weight to the hypothetical benefits of reserves. But, as Sale et al. (2005) point out, while there are now a fair number of cases where density and size increases have been documented, the changes are often very slight, and very rarely has the acid test been applied – to show that beneficial effects are felt outside the reserve. The crucial gaps in our knowledge relate to the distance and direction of dispersal, particularly of larval stages, but also of juveniles and adults. These gaps will need to be filled, and more clearcut demonstrations of benefit to fisheries made, before marine protected areas can be rigorously incorporated into harvest management strategies. Sale and his colleagues anticipate that marine protected areas will become part of the fisheries manager's armory but they bemoan what has often been uncritical advocacy by scientists, which may erode confidence in marine science and marine scientists.

7.2.5 *Evaluation of the MSY approach – the role of climate*

To understand what underlies the concept of a maximum sustainable yield is to understand a lot about how to manage a harvest. Theory tells us when we plot net recruitment against population size (or fishing effort), the curve will be hump-shaped with a maximum that gives harvesters something to aim for. However, putting that understanding into practice, or in other words 'finding the top of the curve', is far from straightforward. Indeed, as Hilborn and Walters (1992) wryly point out, 'You cannot determine the potential yield from a fish stock without over-exploiting it'. After amassing data for a number of years, managers may think they have found the top of the curve, when in fact they have not (Figure 7.5 – yellowfin tuna) or the graph they produce may not have a curve that can be discerned at all – usually the result of year-to-year vagaries in climatic conditions that influence the production of eggs or survival of eggs or juveniles in the population. If climate were actually constant we might see a hump-shaped curve for the Pacific whiting fishery in Figure 7.4, but instead there is no sign of it. In the face of this uncertainty, the message that management by constant effort (or constant escapement) is safer than by constant quota is even more important.

Fisheries collapses often result from straightforward overfishing and human greed, but climate may also be implicated. Without doubt fishing pressure exerts a strain on the ability of a population to recruit sufficiently to counteract losses. But the immediate cause of a collapse, in one year rather than any other, may be unusually unfavorable environmental conditions. For example, prior to the major crash of the Peruvian anchovy fishery in 1972–3, the population had already suffered a small dip in the upward rise in catches in the mid-1960s resulting from an 'El Niño event' (Figure 7.6). Such an event involves the incursion of warm tropical water from the north, reducing the upwelling of nutrient-rich water that normally fuels the anchovy's food chain. Subsequent increases in fishing effort meant that the effects of later El Niño events, particularly in 1973 but also in 1983, were much more severe. One positive aspect of a crash associated with large-scale climate events, is that the population is more likely to recover when favorable conditions have returned than would be the case if the crash were solely the result of overfishing.

7.2.6 *Species that are especially vulnerable when rare*

A further shortcoming of the yield curve concept needs to be highlighted. Even in the absence of climatic variation, some species may not exhibit hump-shaped curves like those in Figure 7.3a and b. This can happen for two reasons. First, recruitment rate may be particularly low at the smallest population sizes: for instance, recruitment of young salmon may be perilously low at low densities because of intense predation by larger fish, while recruitment of baby whales may be low at low densities simply because males and females may never meet. The same outcome – higher mortality than expected at low density – occurs if harvesting efficiency is higher when the population is small. For instance, many sardines, anchovies and herrings are especially prone to capture at low densities because they form a small number of large schools that follow predictable migratory paths and can be intercepted by fishing boats. In both cases, the species are particularly prone to overexploitation because they lack the ability to compensate by rapid population growth when rare.

Dulvy et al. (2003) list 32 species of marine fish, mammals and birds that, as a result of overexploitation, have been driven locally extinct (e.g. common skate *Dipturus batis* and sea otter *Enhydra lutris*) or globally extinct (e.g. the flightless great auk *Alca impennis* and Steller's sea cow *Hydrodamalis gigas*). The syndrome associated with extinction risk is ease of capture, large size and slow reproductive rate – essentially the same pattern as for the terrestrial megafauna (e.g. mammoths and giant sloths) that went extinct 11,000 years ago as a result of human hunting in North America. I have already noted in Section 3.4 how large body size tends to correlate closely with extinction risk.

However, despite the theoretical risk of extinction by overfishing, there is substance to the claim that economic extinction will usually occur before biological extinction. In other words, the fishery goes extinct before the fish. Paradoxically, however, nontarget species (by-catch), taken incidentally in a fishery, are more at risk of extinction. Some of the larger species that went extinct some time ago were the target of unmanaged hunting. Nowadays, it is still large species that are most vulnerable but these are predominantly by-catch species, including turtles (e.g. *Caretta caretta*), porpoises and dolphins (e.g. *Stenella coeruleoabla*), and albatrosses (e.g. *Diomedia exulans*). As by-catch, these large animals are vulnerable to extinction on a second count too. Although target species may be subject to a sustainable harvesting regime, as long as their harvest continues the by-catch species will also be taken, and quite possibly at unsustainable levels. Monitoring of by-catch requires additional management expenditure together with the costs of developing innovative gear – such as turtle excluder devices in trawl nets, bird-scaring lines to keep seabirds from baited hooks and acoustic pingers to alert marine mammals to the presence of gill nets. Most fishing nations have no such program at present (Lewison et al., 2004).

7.2.7 Ecologist's role in the assessment of MSY

The ecologist's role in harvest management is in *stock assessment*: making quantitative predictions about the response of a harvested population to alternative management approaches and addressing questions such as whether a given fishing effort will cause the spawning stock to decline, or whether nets of a smaller mesh size will allow the recruitment rate to recover. In the past, it has often been assumed that stock assessment could be done simply by careful monitoring of total catch and fishing effort. But you have seen how difficult it is to determine the MSY and to devise strategies to capture it in a sustainable manner. Certainly, complex mathematical methods exist for performing the necessary calculations, but these are themselves based on assumptions (not necessarily tested or true) about the underlying dynamics of the population. However, management decisions must be made, and the best possible stock assessments should be the basis. At the same time, we should set about filling our knowledge gaps. The so-called 'adaptive management' approach involves seeking a balance between, on the one hand, probing for information (directed experimentation) and, on the other, minimizing losses in short-term yield while avoiding long-term overfishing (Hilborn & Walters, 1992). The process is 'adaptive' because according to the monitored outcome of some appropriately cautious management regime, the approach can be fine-tuned and tried again. Ecologists are crucial in this process because who can better appreciate the uncertainties and provide appropriately enlightened interpretations?

Finally, and to be brutally realistic, managing most marine fisheries to achieve maximum sustainable yields will be very difficult to achieve. There are generally too few researchers to do the work and, in many parts of the world, no researchers at all. The term *data-less management* has been applied to situations where local villagers follow simple prescriptions to make sustainability more likely – for example locals on the Pacific island of Vanuatu were provided with some simple principles of management for their trochus (*Tectus niloticus*) shellfishery (stocks should be harvested every 3 years and left unfished in between) with an apparently successful outcome (Johannes, 1998).

7.3 Harvest models that recognize population structure

The models of harvesting described in Box 7.1 and Section 7.2 are known as 'surplus yield' models. They are both useful and used by managers, but they have a variety of limitations, as you have seen. Perhaps their biggest drawback is that they ignore population structure, and this is bad for two reasons. The first is that 'recruitment' is not the simple process envisaged in MSY curves, but a complex outcome of adult survival and growth, adult fecundity, and juvenile survival and growth, each of which may respond independently to changes in density and harvesting. The second drawback is that most harvests focus on a portion of the population (e.g. fish that are large enough to sell, mature trees for buildings or furniture, large deer with antlers worth displaying). The approach that takes these complications into account involves the construction of models like those already described for population viability analysis (Section 5.4). These usually take the form of matrix models encapsulating survivorships and reproductive rates in an age-structured population. In fact, I have already mentioned the results of such models in this chapter – recall the Saiga antelope example (Figure 7.7), where a harvest including more males could be sustained at a higher level because many males are surplus to requirements when it comes to producing offspring.

7.3.1 'Dynamic pool models' in fisheries management – looking after the big mothers

In fisheries management, age-structured models are known as *dynamic pool* models. Their general structure is illustrated in Figure 7.9. The submodels (recruitment rate, growth rate, natural mortality rate and fishing rate) combine to determine the size of the fished population, the proportion that can be exploited and how this translates into a yield to the fishing community. In contrast to the MSY approach, this yield (as a biomass) depends not only on the number of individuals caught but also on their size (past growth). Moreover, the catchable biomass depends not on a basic estimate of surplus recruitment but on a more complex combination of natural mortality, harvesting mortality, individual growth and recruitment into the age classes that are targeted. The crucial point is that in the case of the dynamic pool approach, a harvesting strategy can include not only a harvesting effort, but also the partitioning of that effort amongst the various age classes present.

A classic example of a dynamic pool model in action concerns the Arcto-Norwegian cod fishery (*Gadus morhua*) (Garrod & Jones, 1974). The age-class structure of the cod population in the late 1960s was used to predict, for a period of 25 years, the effects on yield of different fishing efforts and different mesh sizes in the trawl nets. The significance of varying the size of mesh is that the bigger it is, the larger

Fig. 7.9 The dynamic pool approach to fishery harvest management, illustrated as a flow diagram. There are four main submodels: growth rate of individuals, recruitment rate into the population, the natural mortality rate (other than from fishing) and fishing mortality rate. Each submodel can itself be broken down into more complex and realistic systems. Solid lines and arrows refer to changes in biomass under the influence of these submodels. Dotted lines refer to external influences and effects of one submodel on another. Harvest yields are estimated under various management regimes characterized by particular values inserted into the submodels. (From Begon et al., 2006, after Pitcher & Hart, 1982.)

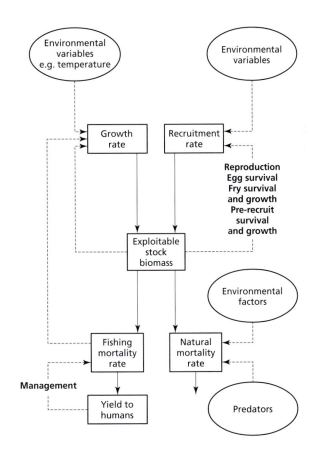

the fish that will escape. We know that recruitment varies from year to year and it was particularly good in 1969. As this large year class works its way through the population, it can be seen as the temporary peak in years 6–8 of this hypothetical management regime (Figure 7.10), but this is only incidental to the point to be made here. The simulations show that overall catches achieved greatest levels in the top graph, where fishing effort was relatively low (26%). By contrast, the bottom graph shows how a high fishing effort (45%) produces lower, and declining catches. Just as important in determining catches was mesh size: the largest mesh size (160 mm) consistently produced the greatest yield (for any fishing effort). Thus, by letting more and larger fish escape, future catches are increased. Together, low effort and large mesh size give the fish more opportunity to grow (and reproduce) before they are caught, which is important because yield is measured in biomass, not simply in numbers. Efforts of 45%, regardless of mesh size, or of 33% coupled with a small mesh size (130 mm) all resulted in consistent declines in catches as a result of overexploitation.

Sadly, Garrod and Jones' recommendations were ignored. Mesh sizes were eventually increased but not for 10 years and then only from 120 to 125 mm. Fishing effort never dropped below 45% and catches of 900,000 tonnes were being taken in the

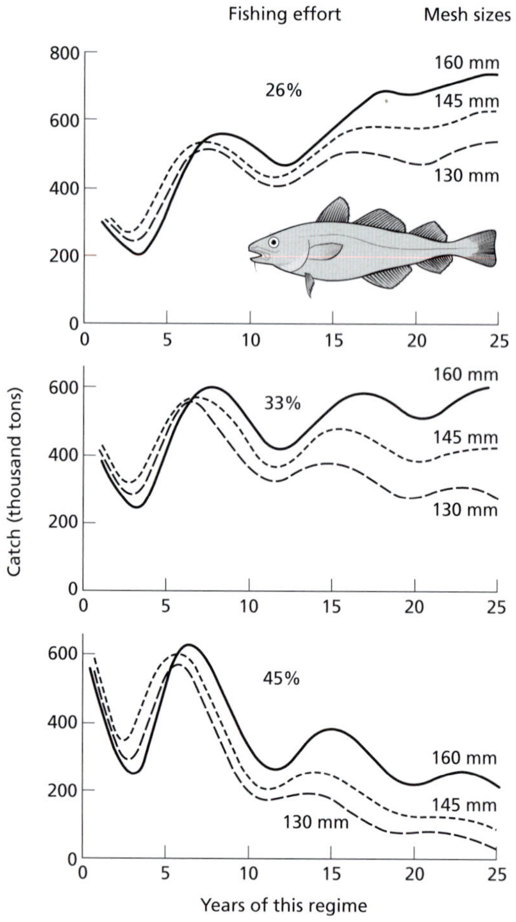

Fig. 7.10 Predictions of a dynamic pool model for a cod fishery under three levels of fishing effort and with three different mesh sizes. The largest effort (45%, bottom panel) is clearly unsustainable, regardless of the mesh size used. Largest sustainable catches are achieved with a low fishing effort (26%, upper panel) and a large mesh size. (After Pitcher & Hart, 1982).

late 1970s. It is not surprising that later surveys showed that these and other North Atlantic cod stocks were very seriously depleted as a result of overfishing. This species reaches sexual maturity around the age of 4 years, but the cod have been so heavily exploited that now some 1 year olds are harvested and almost all 2 year olds are taken each year, leaving only 4% of 1 year olds to survive to age 4 (Cook et al., 1997).

Indigenous harvesters have long had their own 'regulations' to reduce the chance of overexploitation. In their harvest of moi (*Polydactylus sexfilis*), a fish associated with sandy beds near the shore, Hawaiian customary fishers take only intermediate-sized fishes, leaving both juveniles and large females. Thus, they went a stage further than the approach of increasing the mesh-size of nets, which, while reducing the numbers of larger individuals taken, nevertheless invariably capture the largest individuals in the population. The good sense of the Hawaiian strategy has recently been reinforced by the discovery that large females of some fish not only produce exponentially more offspring (something known for a long time) but also each of their eggs is more likely to achieve adulthood. The black rockfish (*Sebastes melanops*), off the coast of Oregon, USA, is a long-lived fish that produces live young. Bobko and Berkeley (2004) noted that, as usual, bigger fish produce more eggs to be fertilized. However, the proportion of these that are in fact fertilized is itself greater in larger females and, in addition, larvae produced by older females grow more than three times as fast and survive starvation more than twice as long as do larvae produced by younger females (Figure 7.11).

We do not know how widespread is this effect of fish size on the quality of offspring, but where it occurs how could the knowledge be incorporated into fisheries management? Scuba divers and spear fishers could take it on board, because these recreational fishers can select the individuals they target. The same would be true for crabs and lobsters caught in pots, since large individuals can be freed unharmed.

Fig. 7.11 Relationship between maternal age and (a) growth rate of larval black rockfish in laboratory experiments and (b) time by which 50% of larvae have died due to starvation. Larvae from older mothers grow substantially faster and survive much longer. (After Berkeley et al., 2004.)

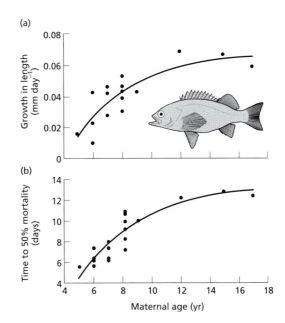

Table 7.1 Values used in a matrix model of a mixed beech forest in the South Island of New Zealand. Trees are classified into 11 size classes according to diameter at breast height (DBH, a standard forestry measure). Growth is expressed as increases (mm yr^{-1}) in DBH. Recruitment and mortality rates are per tree per year. Smaller trees have zero or small recruitment rates and higher mortality rates than their larger counterparts. (From Efford, 1999.)

Size class (mm)	Trees (per hectare)	Growth (mm yr^{-1})	Recruitment rate	Mortality rate
0–100	742	2.56	0	0.0748
100–200	61	2.55	0.029	0.0236
200–300	29	2.85	0.079	0.0075
300–400	25	3.11	0.155	0.0075
400–500	21	3.19	0.257	0.0075
500–600	22	2.88	0.383	0.0075
600–700	15	2.62	0.536	0.0075
700–800	11	2.32	0.713	0.0075
800–900	7	2.04	0.916	0.0075
900–1000	4	1.75	1.144	0.0075
1000+	6	1.47	1.398	0.0075

Net- and line-fishing methods, on the other hand, catch fish before their size is known and usually damage them so that subsequent release is not feasible. Marine protected areas, as long as they have the appropriate properties to contribute to the harvest (Section 7.2.4), may have the additional benefit of protecting the really big mothers.

By way of summarizing the management measures available for fisheries, you can envisage these as either *input* or *output controls*. Input controls include limiting the number of fishing vessels permitted or restricting the size and fishing power of their gear. *Output controls*, on the other hand, include restrictions on the size composition of the catch, the total allowable catch of fish to be taken by the fishing fleet, or of individual quotas for each vessel. Economic checks on catch size, such as government taxes on fish landed, are also sometimes used as output controls.

7.3.2 *Forestry – axeman, spare which tree?*

Models used to manage harvests in native forests follow a similar procedure to the dynamic pool fisheries models, but in many respects the problem is more tractable – the trees can be counted and measured and do not move around, and often only a single company is responsible for harvesting them. In the case of a proposal to sustainably harvest a mixed native beech forest (*Nothofagus* spp. and *Dacrydium cupressinum*) in New Zealand, a matrix model was used of the type described in Section 5.4, in which tree size classes (measured according to diameter at breast height, DBH) were used instead of age classes. The number of trees in each size class (0–100 mm, 100–200 mm, etc.) changes from year to year as some die or are felled, and others grow across the boundary to the next size class. Saplings are considered to be recruited to the smallest tree size class once they attain a height of 1.4 m. Known rates of growth, reproduction (number of sapling recruits per individual per year) and mortality were fed into the model (Table 7.1). Finally, a sustainable yield was estimated by finding the mortality rate in each size class that will maintain the current size composition, and treating that mortality as a harvestable surplus. To allow for uncertainty, the proposal was to only take 50% of this surplus, at least in the first phase of an adaptive management approach.

As usual with any model of a complex system, one can argue about the most appropriate model structure and parameter values. In particular, as Efford (1999) pointed out, we do not know for certain that harvest mortality substitutes for natural mortality rather than adding to it. Incidentally, this is a problem with fisheries models too. These generally accept that there is density-dependence in natural mortality so that the decline in population density brought about by harvesting reduces natural mortality and allows the population to rebound. This is the basis of the hump-shaped yield curves, but it is only a 'reasonable' assumption that has rarely been demonstrated. Taken overall, the proposals for sustainable beech forestry, removing a small number of trees (sometimes by helicopter) that will be replenished by natural recruitment, conform to best practice and are used in many forests around the world – including the one described by Elena in Guatemala (Section 7.1). However, the plan was not acted upon. At a time of a general election in New Zealand, public opinion that native forests should be left completely untouched led to government rejection of the proposal.

7.3.3 *A forest bird of cultural importance*

Deep in the New Zealand beech forest are found small populations of one of the largest pigeons in the world, known to the indigenous people as kereru (*Hemiphaga novaeseelandiae*). The species is legally protected because the formerly large populations have shrunk due to habitat loss and, in particular, predation by ferrets and other mustellids introduced to New Zealand in an abortive attempt to control rabbits (Section 6.1.1). The demise of kereru has been particularly painful for Maori people, for whom it was important both for sustenance and cultural reasons: dying female elders (kuia) would be given a kereru. Many Maori hope that with appropriate action some populations can be brought back to their former status, which would permit a small harvest of animals for cultural purposes.

A population viability analysis (much like the age-structured forest model) was performed, using population sizes and carrying capacities, and birth and death rates estimated from real populations. The birds breed first when 1 year old, produce one egg per year (although as few as 40% may breed in a bad year) and may live to at least 14 years. Up to 81% of adults survive each year, but as few as 9–15% of eggs and juveniles survive, mainly as a result of predation of eggs and nestlings by mustellids, as well as rats and other introduced mammals. Simulated populations all went extinct within 25 years, whether modeled for mainland or island populations: only the most optimistic scenario for an island population persisted for 100 years (with 100% of females breeding each year and 15% of young surviving). However, even this modeled population quickly went extinct if an annual harvest was taken. Figure 7.12 shows the probability of kereru extinction on the island for four levels of mammalian predator control. Only with the highest level of control (essentially a 100% reduction in mammalian predators) could a small annual harvest of four or five kereru be taken without increasing the risk of the population going extinct. Against this background, Maori have implemented plans to reduce predation and protect habitat in mainland reserves (mammals trapped or poisoned; predator-proof fence erected). They say they will only lift their rahui (prohibition of any harvest) 'when the pigeons are once again a pest in nearby orchards'.

7.4 Evolution of harvested populations – of fish and bighorn rams

Size-selective harvesting, whether for bighorn sheep with the largest horns or for the biggest fish to sell at market, can be expected to result in rapid evolution, by selecting for animals that mature at a smaller size. The reasoning goes like this. In an unharvested population a few small individuals may mature earlier than the rest. If the population is then harvested in a way that takes mainly large individuals, the

Fig. 7.12 Results of a population viability analysis of an island population of New Zealand woodpigeons. In the absence of predator control, the population is predicted to go extinct within 100 years. Only at the highest levels of predator control (equivalent to removing all mammalian predators) is the population safe enough to take a small annual harvest of the pigeons. (After Keedwell, 1995.)

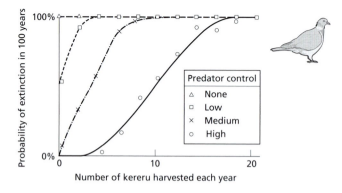

few small but mature animals are likely to provide a disproportionate number of offspring to the next generation and, as long as size-at-maturity is heritable, future generations will be dominated by smaller fish. This is evolution in action and the process is the same as that described for evolution of resistance to pesticides (Box 6.1, Figure 6.3).

Does this pattern of evolution actually happen in harvested populations? Apparently yes. On Ram Mountain in Canada, for example, trophy bighorn rams (*Ovis canadensis*), the ones that hunters most prize, now have smaller bodies and horns (Figure 7.13a,b). Similarly, the size at which cod mature suffered a dramatic decline between 1959 and 1979 in the north Atlantic (the solid circles in Figure 7.13c). In both cases there is strong evidence that the changes were evolutionary responses to size-selective harvest. Thus, for example, actual changes in the cod population are closely paralleled by simulations where size-at-maturity is assumed to be heritable, but not when it is determined simply by random environmental factors (Figure 7.13c).

These evolutionary changes are of more than passing interest to harvest managers because evolution of earlier maturation threatens the size and long-term sustainability of yields. Individuals that mature earlier invest less in growth – so biomass yield declines – and, being smaller, they have lower reproductive outputs overall – so the population is less able to compensate for harvest mortality. We can draw an analogy with foresters and farmers. These would be ridiculed if the seeds they planted each generation were taken from the smallest, least productive parents available. Harvesters of wild animals, when they selectively remove the largest and most productive animals are basically doing what the farmers and foresters avoid – producing successive generations that are smaller and less productive.

What lesson can managers take from these results? In essence, they should follow the practice of indigenous fishers such as those in Hawaii, and throw back both the largest and the smallest individuals. The trouble, as I have already noted, is that net and line fishers catch (and damage) fish before their size is known. In these cases, setting aside nonharvested areas, where large fish are safe, may sometimes be the appropriate option.

Baskett et al. (2005) used population simulations to ask a number of questions about how to avoid evolution and reductions in sustainable yields. They found, for example, that setting up nonharvested areas could protect both size-at-maturity and yield, but only in fisheries that are already heavily exploited. On the other hand, in

Fig. 7.13 Changes over 25 years in (a) mean body weight and (b) horn length of 4-year-old bighorn rams on Ram Mountain, Alberta, Canada. (After Coltman et al., 2003.) (c) Changes in mean size-at-maturity of cod (*Gadus morhua*) in one area of the northwest Atlantic. Solid circles represent mean size-at-maturity during each of 1959–64, 1965–69, 1970–74 and 1975–79. Simulated values for the same time periods are for populations in which body size-at-maturity is assumed to be heritable (diamonds) or where variation is not determined genetically but simply according to random environmental effects (squares). The actual change in size-at-maturity corresponds closely with the model with heritability, indicating that it is an evolutionary response to size-selective fishing. (After Baskett et al., 2005.)

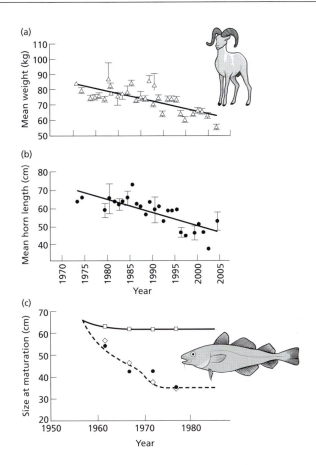

cases where harvest mortality is low, protected areas may provide no benefit, and may even reduce overall yield because part of the fishery is removed from the harvest. When they compared the use of nonharvested areas with more traditional management approaches, Baskett and his team found that increasing the minimum size limit protected to some extent against both evolution to small size-at-maturity and lower yields. But only when a maximum size limit was also imposed (fish above this size must not be caught) were the benefits comparable to those provided by protected areas. Maximum size limits cannot by applied to certain fishing methods but, on the other hand, such an approach would be feasible in terrestrial mammal hunts, if hunters would agree to forego the largest trophies. But this is an unlikely outcome when, for example, bighorn sheep trophy hunters are prepared to pay hundreds of thousands of dollars at auction for a permit to hunt (Coltman et al., 2003). Quite a conundrum.

7.5 A broader view of harvest management – adding economics to ecology

The most obvious shortcoming of a purely ecological approach to harvest management is its failure to recognize that when harvesting is a business enterprise, the monetary value of the harvest must be set against the costs of obtaining it. I noted for pest management that the aim is usually not to go the whole hog, but to reduce the pest population to a level at which it does not pay to achieve yet more control

(Box 6.1). Equally, it makes no sense at all to strive to obtain the last few tonnes of an MSY if the money spent in doing so could be more effectively invested in another means of food production. The basic idea is illustrated in Figure 7.14a. Looked at from an economic point of view the aim should be to maximize not total yield but net value – the difference between the gross value of the harvest and the costs of obtaining it. The costs have two components: fixed costs (interest payments on ships or factories, insurance, etc.) and variable costs that increase with harvesting effort (fuel, crew's expenses, fish processing, etc.). The first thing to note is that the economically optimum yield (EOY) is less than the MSY, and for this reason alone the chance of overfishing is reduced. Note, however, that the difference between the EOY and the MSY is least when most costs are fixed (so that the 'total-cost' line has a shallower slope; Figure 7.14b). And this is most likely to be the case in high-investment, highly technological operations such as deep-sea fisheries. As a result, these are more prone to overfishing even with management aimed at optimizing economic returns.

Spiny lobsters (*Panulirus argus*) provide the most valuable fishery in Cuba. Fixed costs in the fishery are US$24,840 per boat, and variable costs consist of US$171 per boat per day together with processing costs of US$687 per tonne of raw lobster. Most lobsters are sold whole (value about US$12 per kg) but 30% as frozen tails (US$27.87 per kg). Some lobsters are more valuable than others, as was the case with male as opposed to female Saiga antelope (Figure 7.7). Putting together all the economic information for the lobster fishery in a particular year (2002), revenues were US$55.87 million, fixed costs were US$2.91 million, variable costs US$6.56 million, leaving a profit of US$46.41 million.

Previously the fishery has been managed simply according to biomass yield and not in terms of economic profit. In their model, based on extensive economic and biological data, Puga et al. (2005) determined that the long-term maximum profit (US$1.7306 per lobster recruited into the fishery) would be obtained with a fishing effort of 24,016 boat fishing days, slightly lower than needed to obtain the MSY (26,405 days; US$1.7260 per recruit), but higher than what is currently the case (18,315 days in 2000–2; US$1.6950 per recruit). On the other hand, the larger effort

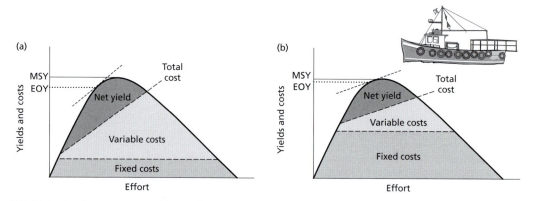

Fig. 7.14 (a) In this diagram, gross yield in dollars is represented by the familiar hump-shaped curve. The economic optimum yield (EOY), that which maximizes profit to the fishing industry, occurs at the point on the effort curve where the difference between gross yield and total costs (fixed plus variable) is greatest. At this point, the gross yield and total cost lines have the same slope. The EOY is obtained at a smaller effort than needed to obtain the MSY (top of the curve). (b) The effort required to obtain the EOY lies closer to the MSY when fixed costs make up a larger proportion of total costs.

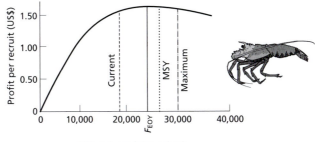

Fig. 7.15 Relationship between profit and fishing effort in the Cuban spiny lobster fishery. This curve is the top segment of Figure 7.14a – the net yield – expressed as profit per lobster recruited into the fishery. The fishing effort needed to optimize the economic yield (F_{EOY}) is shown as a solid vertical line. The effort needed to produce the maximum sustainable yield (MSY; dotted line) is slightly greater. The current effort (dashed line) is smaller than the optimum, whereas the maximum effort (sometimes exerted in the fishery) is higher than both the optimum for economic returns and the maximum sustainable yield – such effort would lead to overexploitation of the fishery. (From data in Puga et al., 2005.)

occasionally seen in the fishery (30,000 days) can be expected to produce lower profits (US$1.7050) and overexploit the spawning stock (Figure 7.15).

7.6 Adding a sociopolitical dimension to ecology and economics

'Social' factors enter the sustainable harvesting equation in two different ways: because of the need to factor human nature into harvest management procedures and because of political realities that sometimes work against taking ecological advice.

7.6.1 Factoring in human behavior

Management plans may fail if they simply assume people will conform with the requirements of achieving an MSY or EOY but take no account of the way harvesters will actually behave in changing circumstances. Harvesting involves a predator–prey interaction: it makes no sense to base plans on the dynamics of the prey alone while simply ignoring the dynamics of the predators – us (Begon et al., 2006). A particularly graphic example of humans behaving as classical predators is shown in Figure 7.16: the anticlockwise predator–prey spiral is precisely what ecologists predict when owls feed on mice, or lions on antelopes – but this is for seal fishermen catching fur seals (*Callorhinus ursinus*) in the North Pacific in the last years of the nineteenth century. What you see is extra vessels entering the fleet when the stock is abundant, but leaving it when seals are rarer. The most important point, however, is the inevitable time lag in this response. Managers must remain wary because whatever they might propose, a perfect, equilibrium match between stock size and effort is never likely to be achieved. Hilborn and Walters (1992) wisely note that 'the hardest thing to do in fisheries management is reduce fishing pressure'. At least the seal fishers were eventually able to switch their effort to another stock waiting to be exploited (halibut, *Hippoglossus stenolepis*). When alternative targets are unavailable, or cannot be exploited with the specialist gear in use, the temptation to continue effort in an overexploited fishery may be irresistible.

Fig. 7.16 The fleet size of the North Pacific fur seal fishery (predators) responded to the size of the seal herd (prey) between 1882 and 1900 by exhibiting an anticlockwise predator–prey spiral. (After Hilborn & Walters, 1992, from data of Wilen, 1976, unpublished observations.)

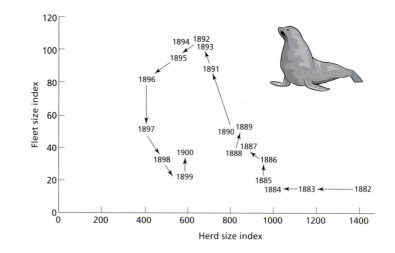

In the strict sense, a harvester who inadvertently fails to follow rules and regulations and illegally kills an animal is a poacher. But the term is usually reserved for the more overtly antisocial behavior of criminals who operate outside the management regime. The Saiga antelope (Section 7.2.2) provides an example of how poaching can be taken into account. Poachers are thought to take 16% of the legal antelope harvest each year, three quarters of which are males with their valuable horns. This sex bias actually makes rather little difference to the size of the Saiga population, because females give birth to almost as many babies despite the loss of a proportion of males. However, the legally taken yield, expressed in Russian roubles, is significantly reduced in the poached population compared to one with no poaching (Figure 7.17) because of the smaller number of valuable males left for the lawful hunters. Whenever poaching is a problem, and it almost invariably is, two steps need to be taken: (i) improve policing and compliance with the rules; (ii) set the total allowable catch at a level that takes poaching into account so that the stock is not overexploited.

Another problem related to human behavior has recently come to light. The harvest of bushmeat in tropical forests (involving about 400 species of mammals) poses a significant threat to the biodiversity of these species-rich forests. But

Fig. 7.17 The effects of poaching on the legal yield (millions of Russian roubles) of Saiga antelopes in relation to hunting effort (expressed as proportion killed) in a constant effort regime. The graph for the population where poaching occurs (squares and solid line) shows a substantial reduction in legal yield compared to the graph for the unpoached population (circles and dashed line). (After Milner-Gulland, 1994.)

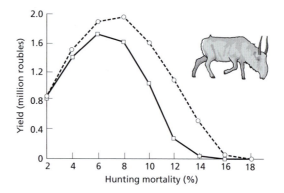

Fig. 7.18 (a) Annual counts of hunters in forest reserves in Ghana are inversely related to supply of fish to the region. Each data point represents one year, and there are 30 years of data. (b) Year-to-year change (expressed as λ, the fundamental net per capita rate of increase – see Box 5.1) in biomass of large mammals in the reserves. When $\lambda = 1$, biomass has not changed; values greater than 1 represent an increase in biomass and values less than 1 represent a decrease. Years with a lower than average supply of fish had higher than average declines in mammal biomass, and vice versa. (After Brashares et al., 2004.)

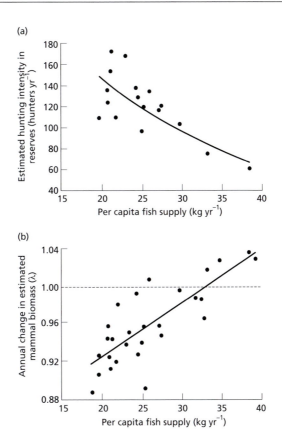

bushmeat provides a vital food resource to local people, especially when their supply of protein from a completely different harvest – fish from the sea – is short. In Ghana, for example, the annual consumption of bushmeat is estimated to be at least 385,000 tonnes, compared to 490,000 tonnes of fish. Both harvests are at great risk of over-exploitation, but the problem is compounded because it turns out that the harvests are inextricably linked (Figure 7.18). Brashares et al. (2004) have shown how bush-meat hunting increases in wildlife reserves across Ghana during periods of low national fish production, accelerating the declines in wildlife abundance during these periods. When few fish are available to eat, the price of fish and the amount of bushmeat sold in rural markets both rise. It seems clear that people treat bush-meat as a substitute for fish. Determining economically optimum yields in such a situation is fraught with difficulty because of the linkage in the markets – coordinated management is called for. Brashares' team suggests two urgent steps. First, access to the fishery of large and technologically advanced foreign fishing fleets should be limited. Second, and given the parlous state of both harvests, regional livestock and agriculture needs to be developed sufficiently to take pressure off the overexploited wild resources.

7.6.2 Confronting political realities

The ecological and economic models of sustainable fisheries are, of course, never perfect. But even if they were, we could still not be confident that ecological recommendations would always be implemented. Political realities, reflecting the political persuasion of the government in power or the political 'clout' of powerful lobbies, often intervene.

While an optimum economic yield (with its biggest dollar return per animal taken) might be achieved using a technological advanced fishing fleet of large ships, practical politics might nevertheless dictate that a large fleet of small, individually inefficient boats be maintained in an area where there are no alternative means of employment. This stand would be more likely from a socialist government (or by any government clinging tenuously to power and desperate for the support of the local community concerned).

Then there is the frequently encountered situation where a strong commercial lobby successfully argues that a catch reduction, indicated by ecologists on sustainability grounds, should not be fully implemented. Such a position is not difficult to argue, and is often won, because of the mathematical uncertainty of precisely what reduction is needed. This was the state of affairs, for example, in the cod fishery described in Section 7.3.1.

Sometimes it is recreational rather than commercial fishers that muster sufficient political clout – to block, for example, the plans of conservationists for marine protected areas in their locality. The fishers' argument is likely to be that they should not be asked to forego fishing opportunities when the potential fishery advantages of a marine reserve have not actually been established, and when sustainability could be attained by other means (perhaps by reducing the catch of commercial fishers).

But it is not always the exploiter lobby that gets the upper hand. Consider whale hunting where, despite the possibility of managing a sustainable catch of certain of the smaller species, countries such as Japan are vilified globally by conservationists and conservation-minded nations for wishing to hunt these large and intelligent vertebrates. Commercial whaling would be more widespread now were it not for restrictions imposed by the International Whaling Commission.

Thus it is that sociopolitical nuances often help to explain why action for sustainability is slowed down, or accelerated, or doesn't take place at all. Thankfully, in recent decades there has been a large-scale shift from the thoughtless push to exploit without restraint, as more and more people recognize the risk of overexploitation and the principles of sustainable harvesting.

Summary

Sustainable harvesting

Sustainable harvesting involves managing the use of living resources, taking neither too little nor too much, so that future generations can enjoy the same opportunities as we do. The task is most difficult when the organisms are small and hidden from view, with birth and death rates that respond sensitively to climatic fluctuation, and where multiple exploiters are involved. Avoiding the 'tragedy of the commons' is a prime aim. When individuals use a public good, they do not necessarily bear the entire cost of their actions – which may in the long term be borne by others. The selfish strategy is for individuals to exploit more than their share, gaining in the short term but leading inevitably to overexploitation and fewer resources for later generations. The tragedy can be avoided if neighbors agree on appropriate rules (lore) or if competing exploiters are regulated by government (law).

Maximum sustainable yield (MSY) approaches to management

Harvesting always involves reducing a population below its carrying capacity. By doing so, density-dependent factors are relaxed and the population responds by raising its birth rate and dropping its death rate – producing a surplus that can be harvested without further reducing the population. The maximum sustainable yield (MSY) is obtained by harvesting at the density where net recruitment is itself maximal.

Two related harvesting strategies are to take a fixed quota each year (i.e. the calculated MSY) or to apply a fixed effort (in terms of numbers of exploiters and time spent hunting or fishing) set so as to achieve the MSY. A 'fixed quota' strategy would be safe if managers had perfect knowledge and the world was wholly predictable. In the real world of fluctuating environments and imperfect data sets, though, this is a very risky approach. The risk can be reduced if instead managers regulate harvesting effort (e.g. in terms of 'gun-days' in a hunted waterfowl population). Another approach, 'constant escapement', does not prescribe the number of animals to be included in the harvest but focuses on the number to be left behind. This is relatively safe because the accidental removal of all breeding individuals is ruled out. A different way to approach a constant escapement strategy is to place spatial controls on fishing or hunting. The protection of 10% of a hunted population in reserves is potentially equivalent to providing a fixed escapement of 10%.

The ecologist's role in harvest management is in 'stock assessment' – making quantitative predictions about the response of a harvested population to alternative management approaches (different fishing efforts, mesh sizes, etc.).

Harvest models that recognize population structure

The MSY approach takes a simplistic approach by assuming the population consists of identical individuals, ignoring the different size or age classes with their different rates of growth, survival and reproduction. The approach that takes these complications into account involves the construction of models like those already described for population viability analysis (Chapter 5) in the form of matrix models encapsulating survivorships and reproductive rates in an age-structured population. The crucial point to note is that with models that incorporate population structure, a harvesting strategy can include not only a harvesting effort, but also the partitioning of that effort amongst the various age classes. Age-structured fisheries models are known as 'dynamic pool' models, but an equivalent approach is used for forestry management.

Evolution of harvested populations

When managers use population theory to guide their actions they need to understand that there is usually an evolutionary dimension. Harvesters exert a strong selection pressure on the populations they exploit, leading to evolutionary changes in life-history features. Size-selective harvesting, whether for bighorn sheep with large horns or for the biggest fish to sell at market, results in rapid evolution by selecting for animals that mature at a smaller size. This has dramatic consequences for sustainable yields because individuals that mature earlier invest less in growth – so biomass yield declines – and, being smaller, they have lower reproductive outputs overall – so the population is less able to compensate for harvest mortality.

A socioeconomic view of harvest management

One shortcoming of a purely ecological approach to harvest management is failure to recognize that when harvesting is a business enterprise, the monetary value of the harvest must be set against the costs of obtaining it. Economic models have been developed to estimate the economically optimum yield (EOY), which is invariably less than the MSY. 'Social' factors also enter the sustainable harvesting equation – because of the need to factor human nature (poaching, etc.) into harvest management procedures and because political realities sometimes work against taking ecological advice.

The final word

Elena, the Guatemalan forester, was frankly amused to hear about the effect of hunting on the size of trophy animals. *'We laughed when I told my friends about how hunting for the big animals made them evolve to be smaller – the hunted ones strike back! My parents said this explained why the fish market seems to sell much smaller fish than they remember from their younger days. Thank goodness the same thing does not happen with our forest trees.'*

Check back over Chapters 5, 6 and 7 to review the importance of understanding evolutionary theory when managing endangered species, pests and harvests. What mistakes will managers make if they do not believe in evolution?

References

Baskett, M.L., Levin, S.A., Gaines, S.D. & Dushoff, J. (2005) Marine reserve design and the evolution of size at maturation in harvested fish. *Ecological Applications* 15, 882–901.

Begon, M., Townsend, C.R. & Harper, J.L. (2006) *Ecology: from individuals to ecosystems*, 4th edn. Blackwell Publishing, Oxford.

Berkeley, S.A. et al. (2004) Maternal age as a determinant of larval growth and survival in a marine fish, *Sebastes melanops*. *Ecology* 85, 1258–1264.

Bobko, S.J. & Berkeley, S.A. (2004) Maturity, ovarian cycle, fecundity, and age-specific parturition of black rockfish (*Sebastes melanops*). *Fisheries Bulletin* 102, 418–429.

Brashares, J.S., Arcese, P., Sam, M.K., Coppolillo, P.B., Sinclair, A.R.E. & Balmford, A. (2004) Bushmeat hunting, wildlife declines, and fish supply in West Africa. *Science* 306, 1180–1183.

Clark, C.W. (1981) Bioeconomics. In: *Theoretical Ecology: principles and applications*, 2nd edn. (R.M. May, ed.), pp. 387–418. Blackwell Scientific Publications, Oxford.

Coltman, D.W., O'Donoghue, P., Jorgenson, J.T., Hogg, J.T., Strobeck, C. & Festa-Bianchet, M. (2003) Undesirable evolutionary consequences of trophy hunting. *Nature* 426, 655–658.

Cook, R.M., Sinclair, A. & Stefansson, G. (1997) Potential collapse of North Sea cod stocks. *Nature* 385, 521–522.

des Clers, S. (1998) Sustainability of the Falkland Islands *Loligo* squid fishery. In: *Conservation of Biological Resources* (E.J. Milner-Gulland & R. Mace, eds), pp. 225–241. Blackwell Science, Oxford.

Dulvy, N.K., Sadovy, Y. & Reynolds, J.D. (2003) Extinction vulnerability in marine populations. *Fish and Fisheries* 4, 25–64.

Efford, M. (1999) Analysis of a model currently used for assessing sustainable yield in indigenous forest. *Journal of the Royal Society of New Zealand* 29, 175–184.

Garrod, D.J. & Jones, B.W. (1974) Stock and recruitment relationships in the N.E. Atlantic cod stock and the implications for management of the stock. *Journal Conseil International pour l'Exploration de la Mer* 173, 128–144.

Hardin, G. (1968) The tragedy of the commons. *Science* 162, 1243–1248.

Harmelin-Vivien, M.L., Harmelin, J.G. & Leboulleux, V. (1995) Microhabitat requirements for settlement of juvenile sparid fishes on Mediterranean rocky shores. *Hydrobiologia* 300/301, 309–320.

Hilborn, R. & Walters, C.J. (1992) *Quantitative Fisheries Stock Assessment*. Chapman & Hall, New York.

Hunter, J.R., Argue, A.W., Bayliff, W.H. et al. (1986) *The Dynamics of Tuna Movement: an evaluation of past and future research*. FAO Fisheries Technical Paper No. 277. Food and Agriculture Organization of the United Nations, Rome.

Ishimura, G., Punt, A.E. & Huppert, D.D. (2005) Management of fluctuating fish stocks: the case of the Pacific whiting. *Fisheries Research* 73, 201–216.

Jennings, S., Kaiser, M.J. & Reynolds, J.D. (2001) *Marine Fisheries Ecology*. Blackwell Science, Oxford.

Johannes, R.E. (1998) Government-supported village-based management of marine resources in Vanuatu. *Ocean Coastal Management* 40, 165–186.

Keedwell, R.J. (1995) A Preliminary Population Viability Analysis of Kereru (*Hemiphaga novaeseelandiae*) Populations and Implications for Customary Use. Wildlife Management Report No. 74. Zoology Department, University of Otago, Dunedin, New Zealand.

Lewison, R.L., Crowder, L.B., Read, A.J. & Freeman, S.A. (2004) Understanding impacts of fisheries by catch on marine megafauna. *Trends in Ecology and Evolution* 19, 598–604.

Milner-Gulland, E.J. (1994) A population model for the management of the Saiga antelope. *Journal of Applied Ecology* 31, 25–39.

Milner-Gulland, E.J. & Mace, R. (1998) *Conservation of Biological Resources*. Blackwell Science, Oxford.

Pitcher, T.J. & Hart, P.J.B. (1982) *Fisheries Ecology*. Croom Helm, London.

Puga, R., Vazquez, S.H., Martinez, J.L. & de Leon, M.E. (2005) Bioeconomic modelling and risk assessment of the Cuban fishery for spiny lobster *Panulirus argus*. *Fisheries Research* 75, 149–163.

Sale, P.F., Cowen, R.K., Danilowicz, B.S. and eight others (2005) Critical science gaps impede use of no-take fishery reserves. *Trends in Ecology and Evolution* 20, 74–79.

Stephens, P.A., Frey-Roos, F., Arnold, W. & Sutherland, W.J. (2002) Sustainable exploitation of social species: a test and comparison of models. *Journal of Applied Ecoloogy* 39, 629–642.

Willis, T.J., Millar, R.B., Babcock, R.C. & Tolimieri, N. (2003) Burdens of evidence and the benefits of marine reserves: putting Descartes before des horse? *Environmental Conservation* 30, 97–103.

8 Succession and management

Succession is the relatively predictable sequence of change in community composition that occurs after a disturbance: it is pointless to plan our harvests, restoration, biosecurity or conservation projects on the assumption of constant conditions when very often the reality is disturbance, succession and constant change.

Chapter contents

8.1 Introduction   203
8.2 Managing succession for restoration   206
    8.2.1 Restoration timetables for plants   206
    8.2.2 Restoration timetable for animals   208
    8.2.3 Invoking the theory of competition–colonization trade-offs   209
    8.2.4 Invoking successional-niche theory   209
    8.2.5 Invoking facilitation theory   210
    8.2.6 Invoking enemy-interaction theory   215
8.3 Managing succession for harvesting   216
    8.3.1 Benzoin 'gardening' in Sumatra   216
    8.3.2 Aboriginal burning enhances harvests   217
8.4 Using succession to control invasions   219
    8.4.1 Grassland   219
    8.4.2 Forest   220
8.5 Managing succession for species conservation   221
    8.5.1 When early succession matters most – a hare-restoring formula for lynx   221
    8.5.2 Enforcing a successional mosaic – first aid for butterflies   222
    8.5.3 When late succession matters most – range finding for tropical birds   223
    8.5.4 Controlling succession in an invader-dominated community   223
    8.5.5 Nursing a valued plant back to cultural health   224

Key concepts
In this chapter you will

recognize that succession may be driven by external physical change (allogenic) or by internal
    species interactions (autogenic) prompted by disturbance
understand that the shift from pioneer to mid- to late-successional plant communities is mediated
    by species traits – notably colonizing ability, competitive status, facilitative power, and vulnera-
    bility to herbivores
realize that a succession that appears to have reached a climax is, however, usually a mosaic of
    early-successional patches within a late-successional matrix

---

appreciate that farmers battle with later-successional species when they maintain a pioneer-like crop – some indigenous harvest practices are more straightforward

note that disturbance by fire is an essential feature of many successions – but humans have sometimes converted small, patchy fire regimes into raging infernos

grasp that restoration of abandoned fields, clear-cut forests, degraded marshes and dunes requires knowledge of the relevant successional processes

see that late-successional stages are often harder to invade than early stages, providing another reason to restore mature communities

understand that many endangered animals and plants are characteristic of particular successional stages – their conservation relies on successional theory

8.1 Introduction

Jack loved the woodland cabin he built in Oklahoma. *'All the hassles of city life just evaporated when I got back to the peace of my forest – you should have seen the spring in my step and the smile on my face!'* But Jack's worry level began to rise after seeing TV broadcasts from California of forest blazes that destroyed homes. He began a campaign to persuade his State authorities to do something. *'I wrote to suggest that the local stream should be dammed to provide a reservoir of water for fire fighting. My next idea was to cut down large strips of forest beyond the area with houses.'* Some of his neighbors agreed, but others didn't – preferring to keep things natural because, after all, that was why they chose to live there in the first place. Jack's house was one of those that burnt down late last summer.

In the natural world very little is constant. If conditions simply stayed the same, ecological processes would be likely to move towards an equilibrium or steady state. But disturbances to the status quo are common in every kind of habitat. Thus, in many parts of the world fires are natural occurrences, destroying much of what is in their path, whether vegetation, animals or houses, but contributing to a natural ecological mosaic of renewal and change. Hurricanes, floods and volcanic eruptions are further examples of disturbances that affect ecological communities. My focus in Chapters 2–4 was the application of knowledge about individual organisms. Then I moved to the population level, exploring how population theory can be applied to the management of endangered species, pest control and harvest management. Now I shift to the community level of organization, and simultaneously consider all the species that occur together in the shifting ecological mosaic.

Ecological communities have properties that are the sum of the properties of the individuals they contain plus the interactions that take place among them (predator–prey, competitive, mutualistic, etc.). The interactions are what make the community more than the sum of its parts. Community ecologists are interested in how groupings of species are distributed, and the ways these groupings can be influenced by abiotic and biotic factors. One of the foundations of community ecology is the concept of *ecological succession* – this is the relatively predictable sequence of change in community composition observed after a disturbance. The theory of ecological succession is summarized in Box 8.1. In this chapter you will learn that applied ecologists cannot afford to ignore the reality of ecological succession. It is pointless to plan our restoration schemes (Section 8.2), harvests (Section 8.3), biosecurity policies (Section 8.4) or conservation projects (Section 8.5) on the assumption of constant conditions when very often the reality is constant change.

Box 8.1 The theory of ecological succession

Allogenic succession

Sometimes community composition shifts because of slow physical change, such as when silt gradually builds up on the bed of a small lake or in a coastal salt marsh – leading to the eventual replacement of the aquatic community by a forest. The changing abiotic conditions are mainly responsible for shifts in community composition as the niche requirements of a sequence of species are met in turn. Such successions are called allogenic (caused by outside influence). The shifting species composition depends on the slowly changing physical environment, while interactions among the species themselves are of secondary importance.

Autogenic succession

In many other cases successions begin as a result of disturbance (an external physical factor) but the sequence in community composition results primarily from interactions among the species. These successions are known as autogenic (caused by internal influence). Gaps, both large and small, are opened up in forests by disturbances caused by high winds, fires, earthquakes, lumberjacks or simply by the death of a tree through disease or old age. Disturbances in grassland are caused by fires, frost, burrowing or grazing animals, or by the ploughs of farmers. On rocky shores or coral reefs, gaps are opened by severe wave action during hurricanes, tidal waves, battering by logs or the fins of careless snorkellers. Predictable sequences in the appearance and disappearance of species during autogenic successions result primarily because different species have different strategies for exploiting resources – early species colonize efficiently and grow fast, whereas later species can tolerate lower resource levels and grow to maturity in the presence of early species, eventually outcompeting them. Sometimes the early species actually enhance the success of the later ones, perhaps by increasing the nutrient status or water-holding capacity of the soil.

Primary and secondary successions

Autogenic successions can be divided into primary and secondary successions. A primary succession occurs when a newly exposed landform has not been influenced by a previous community – primary successions occur on volcanic lava flows (Figure 8.1), in craters left by meteors, on substrate exposed when glaciers retreat and on freshly formed sand dunes.

Secondary successions, on the other hand, occur when the vegetation of an area has been partially or completely removed, but where well-developed soil and seeds remain. The loss of trees through disease, high winds, fire or felling starts a secondary succession, as does the abandonment by a farmer of a previously cultivated field (or of a suburban garden when its gardener loses interest). In secondary successions on abandoned farmland the typical sequence starts with annual weeds, proceeding to herbaceous perennials, shrubs, early-successional trees and finally late-successional trees. In some parts of the world, however, successions on abandoned fields culminate in a grassland community (Figure 8.2).

Successions commonly take decades, centuries or even millennia to run their course from 'pioneer' to 'climax' community compositions. But exactly the same ecological processes are responsible for the succession of seaweeds on a boulder overturned by a wave, and this may be complete within a few years.

Species traits determine the course of succession

Early-successional plants have a series of correlated traits, including high fecundity, small seeds, effective dispersal, rapid growth when resources are abundant, and poor growth and survival when resources are scarce. Late-successional species usually have the opposite traits, including an ability to grow, survive and compete when resources are scarce. Note that early-successional species have the general characteristics of r-species, and late-successional species those of K-species (Box 3.1). In the absence of further disturbance, late-successional species eventually outcompete early species by reducing resources (light and nutrients) beneath the levels required by the early-successional species. Early species persist for two reasons: (i) because their dispersal ability and high fecundity permits them to colonize and establish in recently disturbed sites before late-successional species can arrive; or (ii) because rapid growth in unshaded conditions allows them to temporarily outcompete late-successional species even if these arrive at the same time. The first mechanism is known as a competition–colonization trade-off and the second as a successional niche (conditions suit particular species because of their niche requirements) (Rees

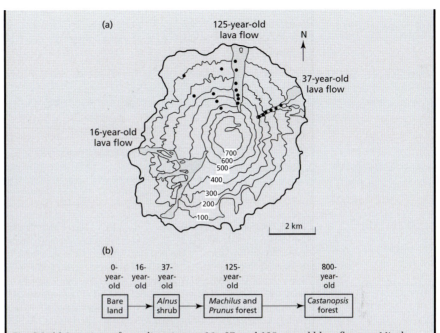

Fig. 8.1 (a) Locations of sampling sites on 16-, 37- and 125-year-old lava flows on Miyake-jima Island, Japan. Sites outside these flows are at least 800 years old. Altitudinal contours are shown in meters. (b) In the earliest stage of succession, soil was sparse and lacking in nitrogen, an essential plant resource: the only vegetation consisted of a few small alder trees (*Alnus sieboldiana*). In the older plots (37–800 years old), 113 species were recorded, including ferns, herbs, lianas and trees. This primary succession consisted of: (i) colonization of the bare lava by the nitrogen-fixing alder; (ii) facilitation (through improved nitrogen availability) of mid-successional *Prunus speciosa* and the late-successional evergreen tree *Machilus thunbergii*; (iii) establishment of a mixed forest in which *Alnus* and *Prunus* were shaded out (i.e. outcompeted); and (iv) competitive replacement of *Machilus* by the longer lived *Castanopsis sieboldii*. (After Kamijo et al., 2002.)

et al., 2001). In addition, some early species change the abiotic environment in ways (e.g. by providing shade or improving soil quality) that make it easier for later species to establish and thrive – this is known as facilitation. A final species interaction may also play an important role in determining the course of succession. Predation, whether of seeds or mature vegetation, can slow or stop a successional sequence in its tracks. Thus, apart from competition–colonization trade-off, successional niche and facilitation, we have to add a fourth mechanism – interactions with enemies – to fully understand ecological succession.

The concept of the successional mosaic and the central role of disturbance

Some disturbances are on a very large scale. Thus, thousands of hectares of forest may be simultaneously destroyed by a fire caused by lightning or by a careless camper. The subsequent successional sequence starts at the same time over this large area and proceeds more or less in synchrony. However, a forest that appears to have reached a stable community structure when studied on a large scale will always be a mosaic of miniature successions. And the same is true for a grassland or rocky shore community. Every time a tree falls, a grass tussock dies or a boulder is turned over, an opening is created in which a new mini-succession starts. The resulting spatio-temporal pattern is called a successional mosaic.

The frequency with which disturbance occurs influences the patchy nature of the successional mosaic. But it also helps determine the total number of species in the area. Where disturbances

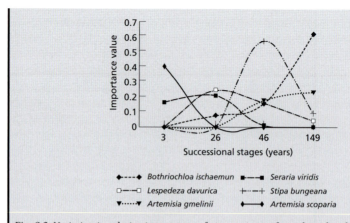

Fig. 8.2 Variation in relative importance of six species on four plots abandoned by farmers for known periods of time (3, 26, 46 and 149 years) during a secondary succession on the Loess Plateau in China. In the first stage of succession *Artemesia scoparia* and *Seraria viridis* were most characteristic; at 26 years *Lespedeza davurica* and *S. viridis* dominated; at 46 years *Stipa bungeana*, *Bothriochloa ischaemun*, *A. gmelinii* and *L. davurica* were most important; finally at 149 years *B. ischaemun* and *A. gmelinii* were dominant. The early-successional species were annuals and biennials with high seed production. By 26 years, the perennial herb *L. davurica* had outcompeted *A. scoparia*. The 46-year-old plot was characterized by the highest species richness. The dominance of the 'climax' grass *B.ischaemun* at 149 years was related to its perennial nature, ability to spread by vegetative means and its high competitive ability. (After Wang, 2002.)

occur very frequently, succession cannot proceed far beyond the pioneer stage and few species are represented. By contrast, if disturbances are very infrequent much of the mosaic will consist of late-successional species that have outcompeted and excluded the pioneers – species richness is again likely to be relatively low. But in situations where disturbances occur at an intermediate frequency, species from all successional stages are represented and the area's species richness can be expected to be at a maximum. This is the essence of the *intermediate disturbance hypothesis* (Connell, 1978).

8.2 Managing succession for restoration

The restoration of natural communities demands a thorough understanding of the theory of succession (Box 8.1). First, managers need to be aware of the timetables of natural recovery of plant and animal communities (Sections 8.2.1, 8.2.2). But just as important is recognition of the underlying successional process, something that varies from case to case: competition–colonization trade-off (Section 8.2.3), successional niche (Section 8.2.4), facilitation (Section 8.2.5), interactions with enemies (Section 8.2.6).

8.2.1 Restoration timetables for plants

In many parts of the world fires are now less frequent but more intense, blazing over larger spatial scales than was the case before human influence. By taking active steps to reduce the likelihood of forest or grassland fires, highly flammable 'fuels' have been allowed to build up (including living biomass and the dead organic matter that accumulates beneath vegetation), so that when a fire is started by lightning or a carelessly discarded cigarette it burns long and hard.

The nature of many plant communities is strongly influenced by the local fire regime. A natural pattern of fire intensity and frequency will have selected for plants able to survive the fires or with seeds keyed to germinate after the fire goes out. The great plains of North America evolved under the influence of relatively small-scale fires. The probability of fire is greatest where a large biomass has accumulated during a period without fire. Recently burnt patches, with their dominant tallgrass vegetation, tend to attract grazing animals (most notably the bison – *Bos bison*) that reduce grass biomass, increase the amount of bare ground and shift the plant community towards nongrass herbaceous species. These changes reduce the likelihood of fire and of further grazing, because the animals now move to more recently burnt areas. The patch then shifts back to a tallgrass successional stage, again increasing the likelihood of fire and of further grazing in future (Figure 8.3). Thus, small-scale fires and grazing together produce a shifting mosaic across the landscape, enhancing heterogeneity and promoting biodiversity.

The traditional management practice when restoring these grasslands has been to minimize patchy disturbance, leading to a reduction in patchiness of the successional mosaic. By contrast, Fuhlendorf and Engle (2004) recommend that management should involve application of spatially discrete fires together with free access of grazing animals to a diversity of patches in a large landscape mosaic.

In contrast to the previous example (a shifting prairie mosaic), the goal of restoration ecology is sometimes a relatively stable and homogeneous successional stage. When the aim is to restore land previously under agriculture, managers need not intervene if they are prepared to wait for natural succession to run its course. Thus, abandoned rice fields in mountainous central Korea proceed from an annual grass stage (*Alopecurus aequalis*), through forbs (*Aneilema keisak*), rushes (*Juncus effusus*) and willows (*Salix koriyanagi*), to reach within 10–50 years a species-rich and stable alder woodland community (*Alnus japonica*). In this case the only active intervention worth considering is the dismantling of artificial rice paddy levees to allow the land to drain, and to accelerate, by a few years, the early stages of succession (Lee et al., 2002).

Fig. 8.3 Diagram of the dynamics of a patch of prairie in a shifting mosaic landscape where each patch experiences similar but out-of-phase dynamics. Ovals represent the key factors of fire and grazing while squares represent the communities within a single patch in relation to time since fire disturbance. Solid arrows indicate positive (+) and negative (–) feedbacks in which plant community structure is influencing the probability of fire and grazing. Forbs are nongrass herbaceous species. (From Fuhlendorf & Engle, 2004.)

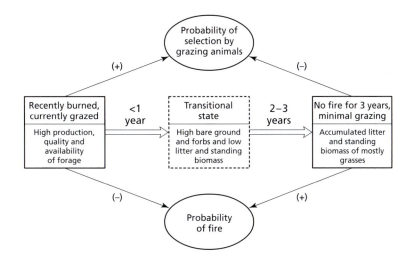

On the other hand, the recovery of intensively grazed English pasture to species-rich meadow is a slow and unreliable process (Pywell et al., 2002). But recovery can be speeded up by sowing a species-rich mixture of seeds of desirable plants adapted to the prevailing conditions (also discussed in Sections 3.2.1 and 4.3).

8.2.2 Restoration timetable for animals

Restoration involves recovery not just of plants but of animals too. After opencast mining of coastal sand dunes in northeastern South Africa has ceased, the sand dune communities slowly recover to the characteristic dune forest vegetation. This can be considered a primary succession (Box 8.1) because mining leaves no trace of the previous community. However, since 1978 regeneration has been accelerated in some locations by actively reshaping dunes to resemble their predisturbance topography, and by covering them with topsoil containing seeds of early-successional plants. In effect, the managers convert the slow primary succession to a more rapid secondary succession (Box 8.1). Millipedes perform an important ecosystem role by feeding on dead organic matter and thus contribute to the decomposition process. They colonize spontaneously regenerating sand dunes, gradually approaching the species composition of unmined dunes, but the process is very much more rapid in the rehabilitated dunes (Figure 8.4).

Tidal salt marshes are now much rarer because many have been drained for agriculture or because tidal cycles have been obstructed by tide gates, culverts and dykes. The restoration of tidal action, and thus of links between the marshes and the wider coastal ecosystem in Connecticut, USA, leads to recovery of salt marsh vegetation (an allogenic succession – Box 8.1). These restored marshes take at least 10–20 years to achieve 50% coverage of their salt marsh plants (principally *Spartina alterniflora*, *S. patens* and *Distichlis spicata*) and characteristic salt marsh animals follow a comparable timetable (Warren et al., 2002). For example, the high marsh snail *Melampus bidentatus* reaches densities comparable to intact marshes after 20 years and the bird community also takes 10–20 years to reach a characteristic marshland community composition. Marsh 'generalists' that forage and breed in upland areas as well as tidal wetlands, such as song sparrows *Melospiza melodia* and red-winged blackbirds *Agelaius phoeniceus*, feature early in the restoration sequence. But these are replaced later by marsh 'specialists' such as marsh wrens *Cistothorus palustris*, snowy egrets *Egretta thula* and spotted sandpipers *Actitis macularia*. Typical fish communities recover more quickly, within 5 years.

Fig. 8.4 A measure of similarity in the composition of millipede communities (Bray–Curtis index) is used to compare regenerating sand dunes and natural undisturbed sand dune forests in South Africa. Triangles represent spontaneously regenerating sites, and squares represent actively rehabilitated dunes. The millipede community achieves 50% similarity in species composition compared to unmined dunes after about 15 years on rehabilitated dunes, but in the case of spontaneously regenerating dunes this takes 50 years or more. Twenty-two species of millipede were identified in the study. Between 12 and 19 are found in undisturbed dune forest, while 12 species occurred after 24 and 54 years in spontaneously regenerating and rehabilitated dunes, respectively. (After Redi et al., 2005.)

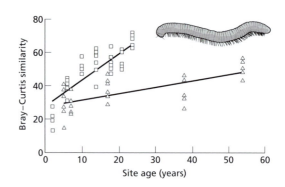

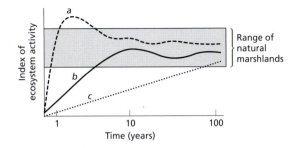

Fig. 8.5 Recovery of different aspects of ecosystem functioning during active marsh restoration. Curve *a* describes processes linked to hydrology (e.g. sedimentation, rates of carbon and nitrogen accumulation in marsh soil) which are quickly restored. Curve *b* describes the slower recovery of biological processes such as rate of production of biomass and decomposition. Curve *c* describes marsh soil development (e.g. total amounts of carbon and nitrogen in the soil, as opposed to their rates of accumulation) which recovers most slowly of all. (After Craft et al., 2003.)

Thus, restoration of a natural tidal regime sets marshes on a path toward full recovery of their ecological functioning, but this generally takes one or more decades. Community recovery can be speeded up by actively planting salt marsh species, but certain ecosystem properties (e.g. the carbon and nitrogen content of marsh soils) do not recover as quickly as others (e.g. deposition of fine sediment) (Figure 8.5).

8.2.3 *Invoking the theory of competition–colonization trade-offs*

The late-successional stage of many successions is delayed because the more competitive late-successional species are much slower to colonize than the pioneer species (competition–colonization trade-off – Box 8.1). This knowledge can be used by managers to speed up certain restoration projects.

An example comes from tropical forest, vast areas of which have been displaced by agriculture. When the land is retired from production the early- and mid-successional species tend to dominate succession for a century or more. Many of the late-successional trees, still represented in distant forest fragments, have large seeds that depend on animal dispersal by birds, bats or primates; they are in short supply in the landscape and of the few that arrive after succession has started, most die as seeds or seedlings. Martinez-Garza and Howe (2003) refer to the 'retarded' community as a pioneer desert. But they note, in Panama and Costa Rica for instance, that when large-seeded late-successional trees such as *Dipteryx panamensis* and *Genipa americana* are planted by hand in pastures (before succession has really got under way) they have a much higher probability of survival. It seems that the competition–colonization trade-off can be circumvented by intervening to get late-successional species off to an unusually good start. In this way up to 70 years of succession can be bypassed.

8.2.4 *Invoking successional-niche theory*

The place of species in a succession does not invariably depend on their probability of colonization. Many species are restricted to a particular moment in the successional timetable because their niche requirements (Chapter 2) are met only at that time.

When sand mining ceases, the surface substrate is deficient not only in seeds but also in soil organic matter and plant nutrients. An additional problem in the subtropical Bongil Peninsula in New South Wales, Australia, has been invasion by the aggressive perennial grass *Imperata cylindrica*. Cummings et al. (2005) devised experiments to identify the barriers limiting restoration of these invader-dominated sand-mined sites. They first tested the hypothesis that the barrier to regeneration was poor native seedling establishment because of competition by *I. cylindrica*, but burning of the grassland and weed control, coupled with native seed planting, did not result in regeneration of native woody cover. So competition was not the most important factor. A second experiment involved the addition to the soil of organic material (mulch) and this significantly improved survival and growth of planted native species. These results support the idea that the absence of appropriate abiotic conditions was inhibiting native succession. In other words, when appropriate niche conditions prevailed, the succession moved forward.

The South African sand mine restoration, discussed in Section 8.2.2, provides a further illustration of the role of changing niche conditions, this time for successional patterns in the dung beetle community. Africa has a rich diversity of dung beetles, whose ecosystem role involves the consumption and burying of dung from a variety of vertebrates. Dung beetles were sampled in eight stands of regenerating vegetation at different stages of succession after cessation of mining, and the communities were compared with those found in unmined, mature sand dune forest sites. Profound abiotic changes occur during this succession, as in other forest successions, with decreasing air temperatures and increasing humidities associated with the shade offered by more mature vegetation (Figure 8.6a, b). Insects vary in their ability to operate at high and low temperatures and in humid or dry conditions, and the changing abiotic conditions are reflected in the representation of different dung beetles during the succession (Figure 8.6c–f).

8.2.5 *Invoking facilitation theory*

Facilitation is a factor that can induce species switches during many successions (Box 8.1). One species may benefit another by enhancing local abiotic conditions, by improving the supply of plant nutrients or by enhancing the probability of reproduction. I consider examples of each of these in restoration projects.

After thousands of years of forest clearance for agriculture, timber production and urbanization, most forests in the Mediterranean area of Europe have disappeared. Restoration projects in such degraded shrub-dominated habitats have generally begun by removing these shrubs, on the assumption that they compete with newly planted tree seedlings. But Gómez-Aparicio et al. (2004) challenged this idea, arguing that because Mediterranean environments are prone to stress from high temperatures and low water availability, the pioneer shrubs might have a positive effect on the establishment of seedlings of desirable mid-successional woody shrubs and of trees. If the pioneer shrubs act as facilitators of succession to woody vegetation they can be referred to as *nurse plants*, which it would be foolish to remove.

Between 1997 and 2001 in the Sierra Nevada area of southeast Spain, more than 18,000 seedlings of 11 woody species were planted in a series of experiments: (i) under a variety of nurse shrubs (16 species); (ii) at high or low altitude; and (iii) on sunny-dry or shady-wet slopes. Overall, in a remarkably high 75% of cases the presence of shrubs increased the probability of seedling survival.

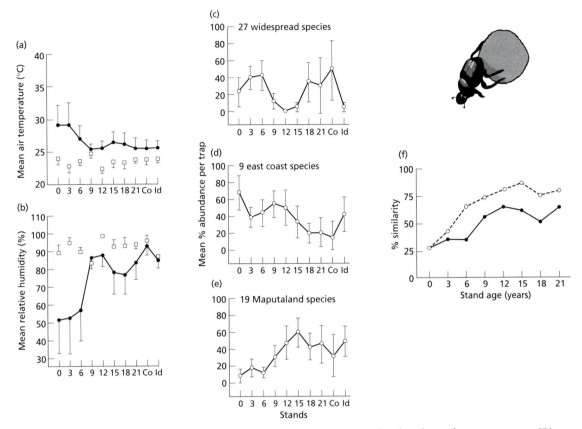

Fig. 8.6 Changes to (a) mean air temperature and (b) mean humidity (daytime, closed circles; night, open squares + SD) during succession to dune forest in South Africa after cessation of sand mining (0, 3, 6, 9, 12, 15, 18 and 21 years later) compared to undisturbed, mature sand dune coastal (Co) and inland (Id) forests. (c) The most widespread of dung beetle species, with their excellent powers of dispersal, show no particular pattern with succession and (d) neither do species characteristic of the wider local area (east coast species), but the most restricted and local Maputaland species (e) are increasingly represented as succession proceeds. (f) The dung beetle communities become increasingly similar to mature dune forest as succession proceeds, and this is particularly the case for shade-specialist species (open circles); closed circles represent all species combined. (After Davis et al., 2002, 2003.)

The average effects of nurse shrubs were calculated for different categories of target woody plants, different categories of nurse plants and different abiotic conditions (Figure 8.7). The benefit provided by nurse shrubs was evident for the survival of seedlings of evergreen and deciduous trees, particularly oak trees (*Quercus* spp.) and *Acer opalus*, reflecting the late-successional status of these species and their requirement for shade. However, mid-successional woody shrubs such as *Rhamnus alternus* and *Crataegus monogyna* also benefited significantly. On the other hand, the shade-intolerant montane pines were not facilitated by the presence of nurse plants (Figure 8.7a). Among the nurse plants, legumes provided the strongest facilitation for woody species. Legumes have root nodules containing bacteria that fix atmospheric nitrogen into a form that plants can use: in this way legumes enhance soil nutrient status in these nitrogen-limited soils (Figure 8.7b). The legumes, and

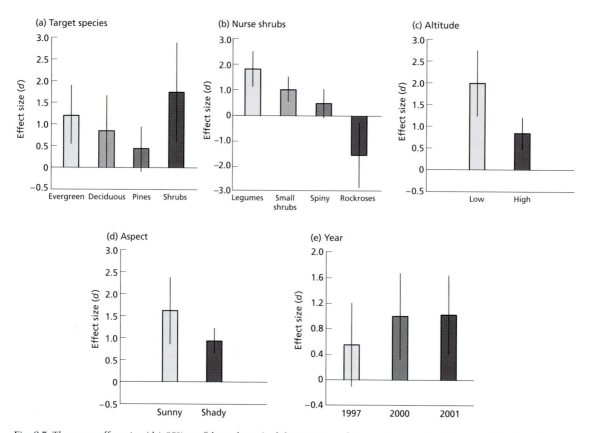

Fig. 8.7 The mean effect size ($d \pm$ 95% confidence limits) of the presence of pioneer 'nurse' shrubs for survival of seedlings of target mid- and late-successional Mediterranean forest species. Effect size is the difference in survival probability after 1 year in the presence and absence of nurse shrubs divided by the mean survival probability in their absence. Thus positive values indicate *facilitation* and larger positive values indicate a stronger facilitative effect. A negative value indicates the opposite of facilitation – survival is reduced in the presence of the (in this case inappropriately named) nurse plant. (After Gómez-Aparicio et al., 2004.)

most other shrubs, no doubt provide another benefit by reducing the effects on the target seedlings of the scorching Mediterranean summer sun. Rockroses (*Cistus* spp.) were unique in providing no 'nursing' benefit, and this reflects their production and release of 'allelopathic' chemicals into the soil, substances that have evolved to provide a competitive benefit by reducing success of neighboring plants.

The size of benefit provided by nurse shrubs also depended on local abiotic conditions. The advantage was highest at low altitudes and on sunny slopes, where lower rainfall and higher temperatures result in intense summer droughts that, in the absence of nurse plants, strongly limit seedling survival and growth (Figure 8.7c, d). Finally, it is clear that the facilitation effect of nurse shrubs was least noticeable in 1997 – a year that was uncharacteristically wet and thus provided conditions amenable to seedling survival even in the absence of nurse plants.

When pioneer species are facilitators of successional change the management prescription is to leave them in place, the converse of the case described in Section 8.2.3 where early species slowed succession to climax vegetation.

In the case of nurse plants that are legumes, the associated bacteria serve to increase nitrogen availability to target plants. Fungi may also increase nutrient availability in situations where plant nutrients are a limiting factor. Most plants do not have roots as such, they have mycorrhizas – intimate mutualisms between fungi and root tissue. The fungal networks in mycorrhizas capture nutrients from the soil, which are transported into the roots to the benefit of the plants and, in return, the fungi gain access to the photosynthetic products of the plant.

Tropical dry forests in places like the Yucatan Peninsula of Mexico have been converted to agriculture and pasture, with less than 10% of mature forest remaining. And as a result of a rapidly increasing human population, accidental fires have increased in frequency, tending to convert much of the remaining primary forest to secondary forest with less biodiversity and lower conservation value. It would be impracticable to undertake large-scale restoration to mature forest, but small patches can be improved to augment and link the remaining forest fragments. Allen et al. (2005) planted seeds of six important climax trees in a site that had been recently subject to a severe fire. They wanted to determine whether seedling success would be facilitated by mycorrhizal fungi. Thus, seeds were planted in three experimental treatments: in steam-sterilized soil (initially lacking mycorrhizal fungi), in soil inoculated with mycorrhizal fungi that are characteristic of early forest succession (principally *Glomus* spp.) or in soil inoculated with mycorrhizal fungi from late succession (including species of *Scutellospora*, *Gigaspora* and *Acaulospora*). In each case, seedling growth was enhanced by fungal inoculation of the soil, and in most cases mycorrhizal species associated with late-successional soils facilitated growth most (Figure 8.8). It is clear that the common plant nursery practice of sterilizing soil is not always appropriate when community restoration is the objective.

Pollination is another kind of mutualism with implications for restoration time-tables and reinstatement of proper community functioning. Hay meadows represent some of the most species-rich plant communities in the UK, but well over 90% have disappeared as a result of agricultural intensification. You saw earlier that meadows are being restored by returning to much less intensive patterns of grazing and mowing, as well as by planting desirable species (Section 2.4.2). Forup and Memmott (2005) compared plant–pollinator interactions in two ancient British hay meadows and in two restored hay meadows to discover whether patterns of insect visitation and pollen transport had been regained in the restoration process. In each meadow they identified and counted all insect-pollinated flowers, noted which insects visited which flowers and identified to plant species the pollen grains attached to the insects' bodies. Their analysis included 42 plant species and 85 flower-visiting insect species, dominated by two-winged flies (Diptera) and bees and wasps (Hymenoptera) with somewhat fewer beetles (Coleoptera), bugs (Hemiptera) and butterflies and moths (Lepidoptera).

The results are shown as flower-visitation webs in Figure 8.9. All meadows were similar, although a marginally higher proportion of all conceivable flower–insect links is realized in ancient meadow. There was no difference in the proportion of flower species visited, and visited plants all had more than a single species of insect visitor. Furthermore, there were no differences between old and restored meadows in the total amount of pollen transported or the average number of grains per insect. Despite differences in plant community composition among the meadows, the

Fig. 8.8 Mean seedling heights achieved during the first year by six species of forest tree when planted in steam-sterilized soil (no mycorrhiza – dotted lines), or in soil inoculated with early-successional (dashed lines) or late-successional (solid lines) mycorrhizae. The presence of mycor-rhizae significantly increased seedling growth after 1 year and in four of the species it was late-successional mycorrhizae that had the strongest facilitation effect. (After Allen et al., 2005.)

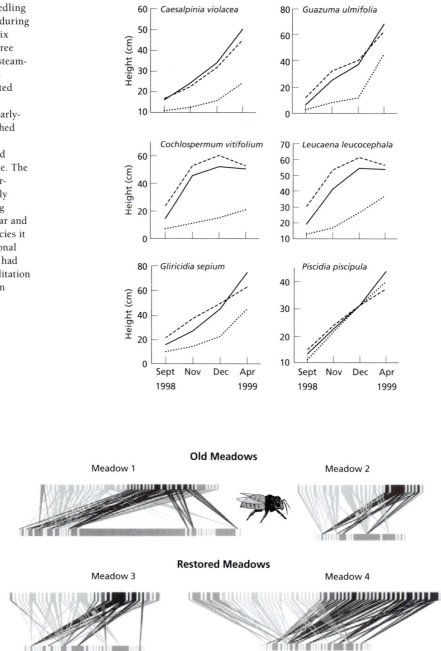

Fig. 8.9 Flower visitation webs for four neighboring hay meadows in southwest England, two of which retain their ancient plant community while the others have been actively restored. Plant species are shown as rectangles at the base of each web, insects are shown above, and interactions between them are shown as connecting lines. The relative abundance of each species is indicated by the width of its rectangle, and the frequency of each interaction is indicated by the width of the line. Hymenoptera (bees and wasps) are shown in black. (After Forup & Memmott, 2005.)

pollination web has successfully reassembled during the restoration process. In other words, restitution of the facilitation role played by pollinators posed no problem for the meadow restoration timetable.

8.2.6 Invoking enemy-interaction theory

Herbivores can be a potent influence on succession if they preferentially consume plants from a particular successional stage. They can be expected to speed up succession if they find pioneer species most tasty unless, of course, these are plants that normally facilitate later species. On the other hand, succession may be slowed if herbivores have their greatest negative influence on later species. Thus, removal of insect predators of seeds during an old-field succession led to dominance of the meadow goldenrod (*Solidago altissima*) after only 3 years, when this species normally appears about 5 years into the timetable (Carson & Root, 1999). This happened because release from seed predation allowed the goldenrods to outcompete earlier colonists more quickly.

Vertebrate herbivores can be just as influential as insects. Thus, Lai and Wong (2005) improved the growth and survival of seedlings of thick-leaved oak (*Cyclobalanopsis edithiae*), a late-successional species of the original primary forest of Hong Kong, by preventing access to browsing mammals (Figure 8.10). In their study the total cost per surviving seedling was reduced from US$6.76 in the control (access to browsers) to US$4.05, despite the cost of providing the tree guard (and weed mat).

Within the patchy mosaic of forest in the Missouri region of the USA, 'Ozark glades' are rocky outcrops where pH, nutrient concentrations and soil moisture are low, limiting plant productivity and producing characteristic prairie-like patches in the forest. Historically, these glades were maintained by fire cycles that prevented tree encroachment. But decades of fire suppression have resulted in the intrusion of fire-intolerant species such as Eastern red cedar (*Juniperus virginiana*). Current restoration practices, which include cutting down the cedars and burning the waste in 'brush' piles, seem sensible enough, but the brush piles represent habitat for

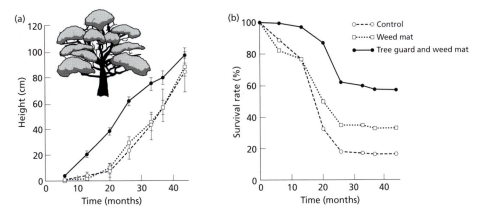

Fig. 8.10 Protection from browsing mammals by a plastic sleeve ('tree guard') increased both (a) seedling growth and (b) survivorship of thick-leaved oak in restored Hong Kong forest. The provision of a hessian 'weed mat', to reduce competition with pioneer plants, improved survivorship but to a lesser extent than when combined with a tree guard. (After Lai & Wong, 2005.)

Northern fence lizards (*Sceloporus undulatus*) that may themselves be beneficial in glade restoration. Because *Sceloporus* consume herbivores such as grasshoppers, Van Zandt et al. (2005) hypothesized that the presence of the lizards in and around brush piles might result in a trophic cascade, reducing herbivore damage to native plants. Surveys of six Missouri glades showed that lizard activity occurred mainly close to habitat structures, where they reduced grasshopper abundance by 75% and plant damage to several glade plant species by over 66% (Figure 8.11). Future glade restoration can benefit by taking into account the top-down effects of predators on grazers in this glade food web.

8.3 Managing succession for harvesting

In contrast to restoration managers who look for ways to help successions run their course, farmers and gardeners work hard to fight succession – by planting desired species and weeding out unwanted competitors. In their attempt to maintain the characteristics of an early-successional stage – growing a productive annual grass – arable farmers are forced to resist the natural succession to herbaceous perennials, and beyond, to shrubs and trees. This is a never-ending battle. In other cultures, though, agricultural 'gardening' poses fewer problems in the way that succession is confronted (Section 8.3.1). Others use controlled burning to create a successional mosaic of benefit to hunters (Section 8.3.2).

8.3.1 *Benzoin 'gardening' in Sumatra*

Benzoin is an aromatic resin, used for flavoring, incense and medicinal products, which is tapped from the bark of tropical trees in the genus *Styrax*. This has been going on for centuries and benzoin still provides a significant income to many

Fig. 8.11 (a) Lizard activity in Ozark glades declines with distance from structures such as brush piles, leading to (b) reductions in their prey, grazing grasshoppers, close to the structures. (c) Percentage herbivore damage to the two forbs *Echinacea paradoxa* and *Rudbeckia missouriensis* (but not to the sedge *Carex* sp.) was lower close to the structures, a result mirrored (d) in transplant experiments involving the herb *Aster oblongifolius* and the grass *Schizachyrium scoparium*. Standard error bars are shown. (After Van Zandt et al., 2005.)

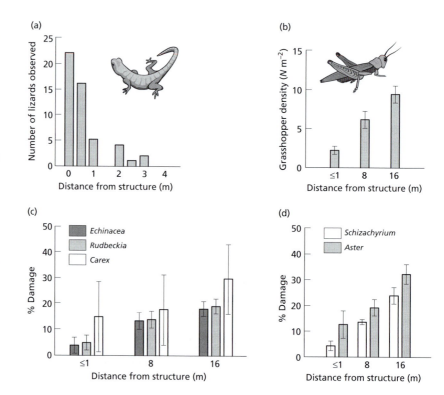

villagers in Sumatra. They plant benzoin gardens (*S. paralleloneurum*) after clearing small patches (0.5–3.0 ha) of the understory of montane broadleaf forest. Two years later farmers remove branches from the larger trees to allow light to reach the saplings, leaving the wood to decay on the forest floor. Annual tapping of benzoin begins after 8 years. Yields usually decline after 30 years of tapping but resin may be harvested for as long as 60 years before the garden is left to return to the forest.

Garcia-Fernandez et al. (2003) identified three categories of garden: G1 was the most plantation-like, with intensive thinning and high densities of *S. paralleloneurum* trees, and G3 was the most forest like. Total tree species richness was high in plots of 'primary' (pristine) and 'secondary' forest (30–40 years after gardening had ceased) but also in most of the gardens themselves, except for the most intensely managed cases where richness was lower (but still with a remarkably high average of 26 tree species) (Figure 8.12a). As predicted by succession theory (Box 8.1), climax species typical of mature forest were most common in primary forest and there was a more even mix of pioneer and mid-successional tree species in secondary forest and in the least intensively managed gardens (G3) (Figure 8.12b). Gardens with an intermediate or high intensity of management were dominated by mid-successional trees (mainly because benzoin trees are in this category).

Indigenous people are usually aware of a wide range of uses for forest plants. Figure 8.12c shows the representation in the garden and forest plots of trees in each of four classes: no known use (12%), subsistence use (food, fiber or medicine) (42%), local market use (23%) and international market use (23%). The international category dominates in intensively managed gardens (i.e. benzoin and its products) while trees in the subsistence and local market categories were well represented in less intensively managed gardens and in primary and secondary forest.

Although benzoin gardening requires competing vegetation to be trimmed back, tree species richness remains high. This traditional form of forest gardening maintains a diverse community that recovers rapidly when tapping ceases. It represents a sustainable balance between development and conservation.

8.3.2 *Aboriginal burning enhances harvests*

Fire is used as a management tool in a number of situations, including the maintenance by gamekeepers of prime game bird habitat in Scottish moorland. Fire is an important management tool also for the Australian aborigine clan who own the Dukaladjarranj area of Arnhem Land (Figure 8.13a). Burning, to provide green forage for game animals, is planned by custodians (aboriginal people with special responsibilities for the land) and focuses on dry grasses on higher ground, before moving progressively to moister sites as the dry season runs its course. Each fire is small in extent and of low intensity, producing a patchy mosaic of burned and unburned areas and thus a variety of habitats at different successional stages. Toward the end of the dry season, when the weather is very hot and dry, burning ceases except where it can be carefully controlled (such as the reburning of previously burnt areas).

In a study carried out by indigenous people and professional ecologists, Yibarbuk et al. (2001) lit experimental fires to assess their impact on the vegetation and animals. They discovered that burned sites attracted large kangaroos (Figure 8.13b) and other game animals, and yams and other important plant foods remained abundant. These results would have hardly been a surprise to the indigenous

Fig. 8.12 (a) Tree species richness in different tree size classes (Dbh is diameter at breast height) in three categories of benzoin garden (G1, most intensely managed; G2, intermediate; G3, least intensively managed) and in secondary forest (SF; 30–40 years after abandonment of benzoin gardens) and in primary (undisturbed) forest (PF). (b) Percentage of individual trees in three successional categories. (c) Percentage of individual trees in various utility categories. Each data point is based on three replicate 1-ha plots (standard errors shown). (After Garcia-Fernandez, 2003.)

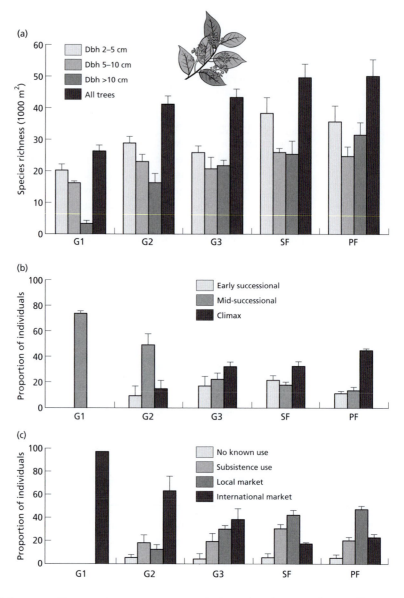

collaborators. Fire-sensitive vegetation in decline elsewhere, such as *Callitris intratropica* woodlands and shrub-dominated heathlands, remained well represented in the study area. The ecologists were surprised to note that the Dukaladjarranj area compares favorably with the Kakadu National Park, a conservation area with high vertebrate and plant diversity. Dukaladjarranj contains several rare species, a number of others that have declined in unmanaged areas and, furthermore, exotic plant and animal invaders are rare.

The aboriginal regime, with its many small, low intensity fires, contrasts dramatically with the typical modern pattern of intensive, uncontrolled fires near the end of the dry season. These blaze across vast areas, sometimes more than 1 million hectares, of western and central Arnhem Land that are unoccupied and unmanaged,

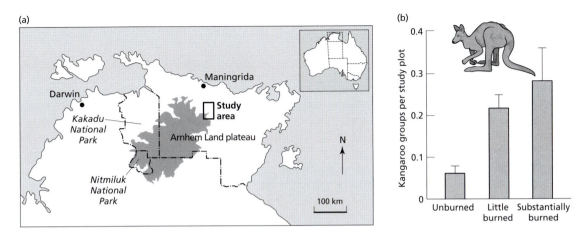

Fig. 8.13 (a) Location of the aboriginal fire management study area near the northeastern end of the Arnhem Plateau in the Northern Territory of Australia; the position of two National Parks is also shown. (b) Mean number (and standard error) of kangaroo groups sighted during a helicopter survey of 0.25 km² plots with different recent burning histories. (After Yibarbuk et al., 2001.)

and regularly make their way into Kakadu and Nitmiluk National Parks (Figure 8.13a). It seems that continued aboriginal occupation of the study area and traditional fire management practices limit the accumulation of fire-promoting grass species and plant litter. In the absence of this excellent fuel, the probability is much lower of the massive fires that can eliminate the most fire-sensitive vegetation. A return to indigenous-style burning seems to hold promise for restoration and conservation of threatened species and habitats in these Australian landscapes and provides clues for the management of fire-prone areas elsewhere.

8.4 Using succession to control invasions

Given the contrasting features of different successional stages it would not be surprising to find that invaders are hindered (or helped) according to the stage that a succession has reached. I address this topic for both grassland (Section 8.4.1) and forest settings (Section 8.4.2)

8.4.1 Grassland

Early-successional agricultural weeds can be characterized as r-selected, with short lives, rapid growth and abundant seed production (Box 3.1). This pioneer life-history stategy depends on high resource availability, and such weeds are unlikely to do well in the resource-poor, competitive environments that characterize late-successional grassland. Restoration of prairie grasslands may therefore be expected to reduce the problem of weed invasion and provide a cost-effective alternative to traditional weed control techniques (Chapter 6).

Blumenthal et al. (2003) examined the cumulative effects of restoration on weed populations after 7 years of restoration of tallgrass prairie in Minnesota. Density and biomass of weeds were compared among plots with: (i) no restoration; (ii) prairie seed addition; and (iii) 'full restoration' involving tilling and raking together with prairie seed addition. In comparison to unrestored sites, full restoration reduced weed biomass by 94%, total weed density by 76%, and the densities of four important weed species (*Berteroa incana*, *Elytrigia repens*, *Euphorbia maculata* and *Setaria*

Fig. 8.14 (a) Mean weed biomass, (b) mean weed density and (c) light penetration to ground level 7 years after the commencement of seed-only and full restoration of tallgrass prairie, in comparison to unrestored grassland. (After Blumenthal et al. 2003.)

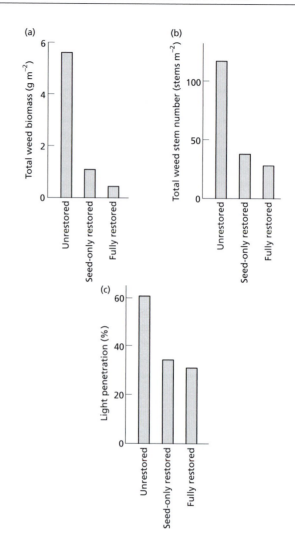

glauca) (Figure 8.14). Full restoration reduced available light and it seems likely that weed invasion was limited by decreasing availability of the crucial radiant energy resource. Prairie seed addition alone had no significant effect on weed biomass but it did reduce weed density by 45%.

In a related study, Blumenthal et al. (2005) augmented soil concentrations of the limiting plant nutrient nitrogen (by adding urea). This reduced the negative effect of restoration on weed invasion, indicating that a lack of mineral resources during late-successional stages also helps to account for poor weed performance in restored grasslands.

8.4.2 *Forest*

Invaders seem to do particularly well in disturbed situations, and one-to-one competition studies often indicate that exotic species are more highly competitive than their native counterparts. Given that early-successional communities have, by definition, been recently disturbed (Box 8.1), you might suppose that invaders will be more prominent in early- than late-successional forests. If this is so, then restoration

Fig. 8.15 Changes in (a) percentage exotic plant cover and (b) percentage of species that are exotic in each of 10 old-field successions over a 40-year period. The patterns for native species are, of course, the inverse of these but note that because cover is summed across overlapping species canopies, total cover can exceed 100%. The patterns are similar whether the starting point of the succession was a hay field or a row crop. (After Meiners et al., 2002.)

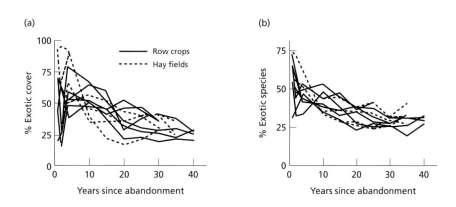

of late-successional communities may be an effective invader-control strategy (as for the grassland example in Section 8.4.1). To address this question, Meiners et al. (2002) analyzed a remarkable database of 40 years of accumulated results in ten separate old-field successions in New Jersey, USA.

Exotic species initially comprised more than 50% of plant cover, but during succession there were significant declines in abundance and richness of exotic species and increases in native species (Figure 8.15). And this occurred in the absence of any management intervention. The loss of exotics seems to be related to forest canopy closure. I have already noted how annual and biennial herbaceous species usually decline as woody cover increases during old-field succession. Many of the pioneer species in the present study are in fact invaders, so it seems that in more resource-limited, later-successional environments the native shrubs and trees outcompete the exotic pioneers. But all in the forest garden is not rosy. Some shade-tolerant invaders are showing signs of increasing after 40 years of *succession*. The biennial *Alliaria petiolata*, for example, is becoming more obvious in situations where understory diversity is low. And exotic shrub honeysuckles (*Lonicera maackii* and *L. tartarica*) are increasing in abundance and may move further into gaps left by the death of early-successional native trees. Simple successional changes may be enough to see off exotic pioneers, but shade-tolerant invaders seem to require more active management.

8.5 Managing succession for species conservation

Successional theory is relevant to the fate of endangered species too. Some require an early-successional stage, some a mosaic of patches at different stages, and others rely on a late stage (Sections 8.5.1–8.5.3). Sometimes the success of a conservation strategy depends on knowledge of the role of herbivores (enemy interaction theory – Section 8.5.4) or of species that smooth the path of succession (facilitation theory – Section 8.5.5; see Box 8.1).

8.5.1 *When early succession matters most – a hare-restoring formula for lynx*

For more than a century the common perception has been that Canada lynx (*Lynx canadensis*) inhabit remote North American primary forest unoccupied by people. Indeed, a federal judge ruling in 1997 concerning the status of lynx inferred that such forest is a prerequisite for them (Hoving et al., 2004). However, lynx are known to be specialist predators of snowshoe hares (*Lepus americanus*) whose habitat is in early-successional dense shrubland or immature forest.

To sort out this paradox, successional attributes were compared in landscapes in northern Maine, USA, where lynx were present or absent according to surveys of tracks in the snow (Hoving et al., 2004). High hare densities were found to be associated with densely regenerating forest following complete removal (clear cutting) of trees. Lynx are most likely to occur in landscape units (each 100 km^2) that contain a large proportion of regenerating forest, least likely to occur where there has been recent clear-cut or partial forest felling, and are neither positively nor negatively associated with mature forest. And snowshoe hares showed precisely the same pattern.

Clear cutting is beneficial to lynx in the long term (but not for the first several years) because it produces dense forest regeneration with abundant snowshoe hares. This pattern of forest harvest may well mimic large-scale natural disturbances such as fire or occasional insect outbreaks that kill trees over large areas. On the other hand, a recent trend towards partial harvesting, of the type that maintains forest composition at a relatively stable state (Section 7.3.2), seems to favor neither hares nor lynx and may actually jeopardize the threatened lynx population. Forest harvesting will need to be re-planned so that suitably sized patches of regenerating forest (from clear-cut areas) are available in the landscape.

8.5.2 Enforcing a successional mosaic – first aid for butterflies

Just as lynx need snowshoe hares, butterflies need suitable host plants for their caterpillars to feed on. Things are not as they once were in piñon-juniper woodland (*Pinus edulis, Juniperus monosperma*), the most common vegetation in New Mexico and Arizona, USA. During the last century or so livestock grazing and periodic drought have reduced herbaceous ground cover, interrupting the fire regime and leading to dense forest that lacks the original more open mosaic patches with grassy understories and rich soils. Kleintjes et al. (2004) assessed a management regime called overstory reduction and slash mulching (ORSM) to restore the former habitat mosaic, with expected benefits for host plants and their butterflies.

ORSM involves removing small trees (<20 cm diameter at their base) and main branches of larger trees, applying the resulting slash as a fine surface layer of organic matter (mulch), particularly to areas of eroded soil. Four years after this treatment both abundance and species richness of butterflies were significantly higher than in untreated woodland. The open patches in the restored habitat mosaic had greater ground cover of grasses and forbs, including five of the ten most common caterpillar

Fig. 8.16 (a) Abundance (±95% confidence limits) and (b) species richness of butterflies are increased 4 years after overstory reduction and slash mulching (ORSM = 'Treated', in comparison to 'Control') to create open habitat patches in piñon-juniper woodland in New Mexico. The surveys were carried out on two occasions in mid-summer. (After Kleintjes et al., 2004.)

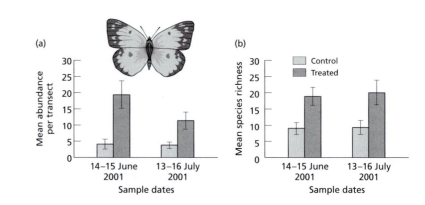

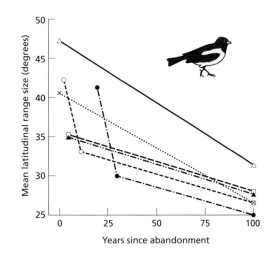

host plants. For example, the success in open patches of the legume *Lotus wrightii* benefited several butterfly species such as the orange sulfur (*Colias eurytheme*) and western green hairstreak (*Callophrys affinis*).

8.5.3 *When late succession matters most – range finding for tropical birds*

Species can differ dramatically in the size of their geographical ranges. Dunn and Romdal (2005) compared the range sizes of bird species characteristic of different successional stages of tropical forests between northern Mexico and central Peru. They found on average that species with the largest average latitudinal ranges occurred in early-successional (cleared) areas, intermediate ranges were associated with young secondary forests, smaller ranges were linked to old secondary forests and the smallest ranges of all were associated with mature forests (Figure 8.17). At one end of the spectrum, bird species from early-successional stages more often extend outside the tropics and are thus more likely to be conserved by conservation programs directed at other biomes. On the other hand, mature forest specialists may be especially at risk of extinction because they are both habitat-restricted and have small range sizes. They deserve special consideration.

8.5.4 *Controlling succession in an invader-dominated community*

When endangered animal species are associated with a particular successional stage, their conservation depends on intervention to maintain habitat at an appropriate point in the successional timetable. An intriguing example is provided by a giant New Zealand insect, the weta *Deinacrida mahoenuiensis* (Orthoptera; Anostostoma-tidae), a species believed extinct after being formerly widespread in forest habitat. The weta was rediscovered in the 1970s in an isolated patch of gorse (*Ulex euro-paeus*). Ironically, in New Zealand gorse is an introduced weed that farmers spend much time and effort attempting to control. Its dense, prickly sward provides a refuge for the giant weta against other introduced pests, particularly rats (*Rattus* spp.), but also hedgehogs (*Erinaceus europaeus*), stoats (*Mustela erminea*) and possums (*Trichosurus vulpecula*), which readily captured wetas in their original forest home.

The Department of Conservation purchased this important patch of gorse from the landowner who insisted his cattle should still be allowed to overwinter in the reserve. Conservationists were unhappy about this but the cattle proved to be part of the weta's salvation. By opening up paths through the gorse, cattle provide entry for feral goats (*Capra hircus*) that browse the gorse, producing a dense hedge-like sward and preventing the gorse habitat from succeeding to a stage inappropriate to the wetas. This story involves a single endangered native insect together with a whole suite of invaders (gorse, rats, goats, etc.) and introduced domestic animals (cattle). Before the arrival of people in New Zealand, the island's only land mammals were bats, and New Zealand's endemic fauna has proved to be extraordinarily vulnerable to the mammals that arrived with people. However, by maintaining gorse succession at an early stage, the grazing goats provide a habitat in which the weta can escape the attentions of mammalian predators.

8.5.5 *Nursing a valued plant back to cultural health*

The very identity of cultures often depends to some extent on native flora and fauna. Take the Mohawk Nation, for example, where great value is placed on the sweetgrass *Anthoxanthum nitens* for its use in basketry, a cultural practice that evolved from early utilitarian forms to intricate and ornate basket work. Sweetgrass is a mid-successional species, typically occurring with other grasses, herbs and shrubs. The species once grew along the Mohawk River valley in eastern New York State, near to the farming community known as Kanatsiohareke, but the nearest gathering site now is 325 km away. The local vegetation is now a late-successional grassland community dominated by exotic species.

Shebitz and Kimmerer (2005) evaluated the restoration potential of sweetgrass using four experimental treatments: (i) sweetgrass alone (weeded to remove competition from other grassland plants); (ii) together with existing old-field vegetation; (iii) weeded and planted with hairy vetch (*Vicia villosa*) as a nurse plant (Section 8.3.5); and (iv) weeded and planted with ryegrass (*Lolium multiflorum*) as a nurse plant. Hairy vetch is an annual legume with the desirable properties that it does not persist through succession and has a nitrogen-fixing ability that may enhance grass growth. Ryegrass, another annual, also readily establishes, has weed-suppressing properties and might therefore serve as a nurse crop, but its tendency to persist may threaten local biodiversity.

Sweetgrass biomass, height, reproduction rate and survivorship were greatest in plots that were weeded to eliminate competition and where hairy vetch was present. Planting sweetgrass with hairy vetch also generated characteristics desired by basketmakers, such as high abundance and tall blades. The annual ryegrass, on the other hand, reduced sweetgrass growth and reproduction – the term 'nurse crop' clearly does not apply in this case. Hairy vetch can thus be used to restore the native sweetgrass into its place in old-field succession at Kanatsiohareke and to bring back a significant cultural tradition.

Summary

Allogenic and autogenic succession

Most natural communities are in a constant state of flux. In the case of allogenic successions, the shifting pattern of species composition depends on a slowly changing physical environment, while interactions among the species themselves are of secondary importance. Autogenic successions, on the other hand, begin as a result of disturbance (an external physical factor) but the sequence in community composition results primarily from interactions among the species.

Primary and secondary successions

Autogenic successions can be either primary or secondary. A primary succession occurs where a newly exposed landform is uninfluenced by any previous community – examples occur on volcanic lava flows and on freshly formed sand dunes. Secondary successions, on the other hand, occur where the vegetation has been partially or completely removed, but where well-developed soil and seeds remain. Examples include the loss of trees through disease, high winds, fire or felling. The disturbances responsible for starting a succession vary in both spatial and temporal scale, often producing a mosaic of patches at different successional stages.

Underlying mechanisms

Early-successional plants have traits such as high fecundity, small seeds, effective dispersal and rapid growth when resources are abundant. Late-successional species have the opposite traits, including an ability to grow, survive and compete when resources are scarce. Early species persist either because their dispersal ability and high fecundity permits them to establish in recently disturbed sites before late-successional species arrive (competition–colonization trade off) or because rapid growth in unshaded conditions allows them to temporarily outcompete late-successional species even if these arrive at the same time (successional niche theory). Some early species change the abiotic environment in ways that make it easier for later species to establish and thrive (facilitation theory). On the other hand, herbivory can slow or stop a successional sequence in its tracks (enemy-interaction theory).

Managing succession for restoration

When planning the restoration of natural communities, managers need to be aware of the spatial and temporal scale of disturbance and the timetables of natural recovery of plant and animal communities. They also need to understand which underlying mechanism applies in their particular case.

In many parts of the world fires are now less frequent but more intense, affecting larger areas than was the case before human influence. By taking active steps to reduce the likelihood of forest or grassland fires, highly flammable 'fuels' have been allowed to build up so that when a fire is started it burns long and hard. Effective management requires a return to more natural fire patterns. In other cases, the aim is to restore land previously under agriculture and managers can wait for natural succession to run its course. But recovery can often be speeded up by sowing a species-rich mixture of seeds of desirable plants adapted to the prevailing conditions. The same is true for restoration of tidal salt marshes or of communities on old mine sites.

Managing succession for harvesting

To maintain the characteristics of an early-successional stage – growing an annual crop – farmers resist the natural succession to herbaceous perennials, shrubs and trees. In other cultures, though, agricultural 'gardening' is more in tune with successional processes. The aboriginal managers of remote landscapes in Australia and managers of grouse moors in Scotland both use controlled burning to create a successional mosaic of benefit to the game that hunters prize.

Using succession to control invasions

Early-successional agricultural weeds have a pioneer life-history stategy and are unlikely to do well in the resource-poor, competitive environments that characterize late-successional grassland. Thus, restoration of prairie grasslands can reduce the problem of weed invasion. In forests, too, invaders generally do well in disturbed situations (in other words, in early-successional settings). The restoration of late-successional forest communities can thus be an effective control strategy against some invasive species.

Managing succession for conservation

Some endangered species require an early-successional stage, some a mosaic of patches at different stages, and others rely on a late stage. Sometimes the success of a conservation strategy depends on knowledge of the role of herbivores or of species that facilitate succession. The plans of conservation managers who ignore successional patterns will often be doomed to failure.

The final word

Jack is no less upset about losing his house to fire (Section 8.1) but perhaps his perspective has broadened a little. 'OK, so nature is in a constant state of flux, and fires and other disturbances happen. The lesson seems to be that the regular setting of small, controlled fires would prevent the build up of fuel that can cause the occasional raging inferno. But that seems risky in itself – miscalculate and more homes will be lost. I'm beginning to think that the appropriate motto for us forest dwellers is "if you can't stand the heat, stay out of the kitchen".' Thinking of forest management in a different context, Jack has spotted a paradox. 'When the aim is sustainable harvest of trees, it seems obvious that selective removal of trees of particular ages is the way to go. But for the sake of the Canada lynx, clear cutting of large areas seems vital to open up early-successional areas rich in the lynx's prey.'

Consider the principles of sustainable harvesting (Section 7.3.2) and the needs of the lynx (Section 8.5.1). Outline the elements of a forest management plan that responds to both.

References

Allen, M.F., Allen, E.B. & Gomez-Pompa, A. (2005) Effects of mycorrhizae and non-target organisms on restoration of a seasonal tropical forest in Quintana Roo, Mexico: factors limiting tree establishment. *Restoration Ecology* 13, 325–333.

Blumenthal, D.M., Jordan, N.R. and Svenson, E.L. (2003) Weed control as a rationale for restoration: the example of tallgrass prairie. *Conservation Ecology* 7(1): 6. Available online: http://www.consecol.org/vol7/iss1/art6

Blumenthal, D.M., Jordan, N.R. & Svenson, E.L. (2005) Effects of prairie restoration on weed invasions. *Agriculture, Ecosystems and Environment* 107, 221–230.

Carson, W.P. & Root, R.B. (1999) Top-down effects of insect herbivores during early succession: influence on biomass and plant dominance. *Oecologia* 121, 260–272.

Connell, J.H. (1978) Diversity in tropical rainforests and coral reefs. *Science* 199, 1302–1310.

Craft, C., Megonigal, P., Broome, S. et al. (2003) The pace of ecosystem development in constructed *Spartina alterniflora* marshes. *Ecological Applications* 13, 1417–1432.

Cummings, J., Reid, N., Davies, I. & Grant, C. (2005) Adaptive restoration of sand-mined areas for biological conservation. *Journal of Applied Ecology* 42, 160–170.

Davis, A.L., van Aarde, R.J., Scholtz, C.H. & Delport, J.H. (2002) Increasing representation of localized dung beetles across a chronosequence of regenerating vegetation and natural dune forest in South Africa. *Global Ecology and Biogeography* 11, 191–209.

Davis, A.L.V, van Aarde, R.J., Scholtz, C.H. & Delport, J.H. (2003) Convergence between dung beetle assemblages of a post-mining vegetational chronosequence and unmined dune forest. *Restoration Ecology* 11, 29–42.

Dunn, R.R. & Romdal, T.S. (2005) Mean latitudinal range sizes of bird assemblages in six Neotropical forest chronosequences. *Global Ecology and Biogeography* 14, 359–366.

Forup, M.L. & Memmott, J. (2005) The restoration of plant–pollinator interactions in hay meadows. *Restoration Ecology* 13, 265–274.

Fuhlendorf, S.D. & Engle, D.M. (2004) Application of the fire-grazing interactions to restore a shifting mosaic on tallgrass prairie. *Journal of Applied Ecology* 41, 604–614.

Garcia-Fernandez, C., Casado, M.A. & Perez, M.R. (2003) Benzoin gardens in North Sumatra, Indonesia: effects of management on tree diversity. *Conservation Biology* 17, 829–836.

Gómez-Aparicio, L., Zamora, R., Gómez, J.M., Hódar, J.A., Castro, J. & Baraza, E. (2004) Applying plant facilitation to forest restoration: a meta-analysis of the use of shrubs as nurse plants. *Ecological Applications* 14, 1128–1138.

Hoving, C.L., Harrison, D.J., Krohn, W.B., Jakubas, W.J. & McCollough, M.A. (2004) Canada lynx *Lynx canadensis* habitat and forest succession in northern Maine, USA. *Wildlife Biology* 10, 285–294.

Kamijo, T., Kitayama, K., Sugawara, A., Urushimichi, S. & Sasai, K. (2002) Primary succession of the warm-temperate broad-leaved forest on a volcanic island, Miyake-jima, Japan. *Folia Geobotanica* 37, 71–91.

Kleintjes, P.K., Jacobs, B.F. & Fettig, S.M. (2004) Initial response of butterflies to an overstory reduction and slash mulching treatment of a degraded pinon-juniper woodland. *Restoration Ecology* 12, 231–238.

Lai, P.C.C. & Wong, B.S.F. (2005) Effects of tree guards and weed mats on the establishment of native tree seedlings: implications for forest restoration in Hong Kong, China. *Restoration Ecology* 13, 138–143.

Lee, C.-S., You, Y.-H. & Robinson, G.R. (2002) Secondary succession and natural habitat restoration in abandoned rice fields of central Korea. *Restoration Ecology* 10, 306–314.

Martinez-Garza, C. & Howe, H.F. (2003) Restoring tropical diversity: beating the time tax on species loss. *Journal of Applied Ecology* 40, 423–429.

Meiners, S.J., Pickett, S.T.A. & Cadenasso, M.L. (2002) Exotic plant invasions over 40 years of old field successions: community patterns and associations. *Ecography* 25, 215–223.

Pywell, R.F., Bullock, J.M., Hopkins, A. et al. (2002) Restoration of species-rich grassland on arable land: assessing the limiting processes using a multi-site experiment. *Journal of Applied Ecology* 39, 294–309.

Redi, B.H., van Aarde, R.J. & Wassenaar, T.D. (2005) Coastal dune forest development and the regeneration of millipede communities. *Restoration Ecology* 13, 284–291.

Rees, M., Condit, R., Crawley, M., Pacala, S. & Tilman, D. (2001) Long-term studies of vegetation dynamics. *Science* 293, 650–655.

Shebitz, D.J. & Kimmerer, R.W. (2005) Reestablishing roots of a Mohawk community and a culturally significant plant: sweetgrass. *Restoration Ecology* 13, 257–264.

Van Zandt, P.A., Collins, E., Losos, J.B. & Chase, J.M. (2005) Implications of food web interactions for restoration of Missouri Ozark glade habitats. *Restoration Ecology* 13, 312–317.

Wang, G.-H. (2002) Plant traits and soil chemical variables during a secondary vegetation succession in abandoned fields on the Loess Plateau. *Acta Botanica Sinica* 44, 990–998.

Warren, R.S., Fell, P.E., Rozsa, R. et al. (2002) Salt marsh restoration in Connecticut: 20 years of science and management. *Restoration Ecology* 10, 497–513.

Yibarbuk, D., Whitehead, P.J., Russell-Smith, J. et al. (2001) Fire ecology and aboriginal land management in central Arnhem Land, northern Australia: a tradition of ecosystem management. *Journal of Biogeography* 28, 325–343.

9 Applications from food web and ecosystem theory

Food webs link the community of species together in predator–prey, parasite–host and competitive interactions. They also provide the ecosystem pathways along which flow energy and nutrients. Managers need to adopt a broad ecosystem perspective to react to problems as diverse as human disease transmission, excessive nutrient enrichment of water bodies, and even the loss of protection provided by intact ecosystems against floods and storms.

Chapter contents

9.1 Introduction   230
9.2 Food web theory and human disease risk   234
9.3 Food webs and harvest management   236
    9.3.1 Who gets top spot in the abalone food web – otters or humans?   236
    9.3.2 Food web consequences of harvesting fish – from tuna to tiddlers   238
9.4 Food webs and conservation management   239
9.5 Ecosystem consequences of invasions   240
    9.5.1 Ecosystem consequences of freshwater invaders   240
    9.5.2 Ecosystem effects of invasive plants – fixing the problem   241
9.6 Ecosystem approaches to restoration – first aid by parasites and sawdust   242
9.7 Sustainable agroecosystems   245
    9.7.1 Stopping caterpillars eating the broccoli – so that people can   245
    9.7.2 Managing agriculture to minimize fertilizer input and nutrient loss   245
    9.7.3 Constructing wetlands to manage water quality   247
    9.7.4 Managing lake eutrophication   248
9.8 Ecosystem services and ecosystem health   249
    9.8.1 The value of ecosystem services   249
    9.8.2 Ecosystem health of forests – with all their mites   252
    9.8.3 Ecosystem health in an agricultural landscape – bats have a ball   253
    9.8.4 Ecosystem health of rivers – it's what we make it   254
    9.8.5 Ecosystem health of a marine environment   255

Key concepts

In this chapter you will

see that some species (strong interactors and keystone species) are more influential in food webs than others

recognize that managers need to understand food web relations to combat disease risk, manage harvests, restore ecosystems and design reserves

---

understand that human pressures alter the way energy and nutrients flow through ecosystems – restoration should take this into account

realize that the most damaging invaders are often those that upset energy and nutrient dynamics

recognize that when agricultural fertilizers are wastefully applied, they pollute the waterways that drain the land

appreciate that ecosystems provide people with many valuable services (such as forest products, fish, crop pollination and climate control)

understand that managers need indicators of ecosystem health to help them identify ecosystems in trouble

9.1 Introduction

Aretha had been spending as much time as possible in the outdoors and was enjoying her retirement after a lifetime in banking. She considered herself a lucky individual until she was involved in two events in quick succession, one personal and the other witnessed by millions around the world. *'A few weeks after hiking along a woodland trail, one side of my face became partially paralyzed. Then a little later I developed painful bouts of arthritis in my knees.'* Aretha's physician ran some tests and diagnosed Lyme disease, an infection spread by ticks and picked up as she walked through the forest. Antibiotics killed the bacteria responsible but a feeling of lethargy persisted, so Aretha decided to take a holiday in Thailand. The sun and sea were doing the trick until the coast was struck by a dreadful tsunami (tidal wave). *'I still have nightmares, plagued by images of the people who were crushed and drowned while I watched helplessly from my hotel on the hill.'* It hardly bears thinking about.

This chapter is concerned, like the last, with the community level of organization, focusing on the food webs that link species together. Studies that unravel the complex interactions in food webs can provide key information to managers on issues as diverse as minimizing human disease risk (such as Aretha's Lyme disease – Section 9.2), managing harvests (Section 9.3) and planning conservation measures (Section 9.4). When communities are considered in their physicochemical setting, we move to the ecosystem level of organization – and the focus then is on the way energy and matter pass through the food web. Coupled with an understanding of food web theory, this ecosystem approach has much to offer managers when they confront invasions (Section 9.5), plan restoration strategies (Section 9.6), design sustainable agroecosystems and remedy the effects of excess nutrients leaking into waterways (Section 9.7). The theory relating to food webs, and the way that energy and nutrients move through them, is summarized in Box 9.1. For more detailed treatments check out Begon et al. (2006).

Ecosystems provide people with many valuable services (Section 9.8.1). You have already come across some relatively straightforward ecosystem services, including crop pollination by wild bees that inhabit forest fragments (Section 4.5.4) and the disposal of dead bodies by vultures in India (Section 5.1). More surprising are services such as the protection against tsunamis provided by intact offshore coral reefs – these absorb some of the wave's power. According to the American Geophysical Union, illegal coral mining off the southwest coast of Sri Lanka allowed far more onshore destruction from the recent Pacific-wide tsunami than occurred in nearby areas whose coral reefs were intact. You saw in Section 1.2.2 that ecosystem services can be divided into several categories. Here it is evident that exploitation of a *pro-*

visioning service (coral crushed to create road surface) can result in the loss of a *regulating service*.

Each ecosystem service has a value that can be set against the economic returns of activities that put them in jeopardy. Thus, in purely economic terms, there are credible grounds for protecting nature. However, it is not straightforward to recognize ecosystem decline or to know that, by some management action, ecosystem health has been regained. For this reason, managers need indicators of ecosystem health that provide a short cut to identifying ecosystems in trouble (Sections 9.8.2–9.8.5).

Box 9.1 Food webs and pathways of energy and nutrients

Direct and indirect effects in food webs

A predator feeding on a prey species has its most obvious (and direct) impact on the prey population. A similarly straightforward effect in the case of interspecific competition is for one competitor species to negatively impact another that feeds on the same limited resource. However, the influence of a species often spreads much further than this, producing a series of indirect effects throughout the community food web. Thus, for example, the effects of a carnivore on its herbivore prey may also be felt by any plant population fed upon by the herbivore, by other predators and parasites of the herbivore, and by other herbivores with which our species competes. The description of food webs is a painstaking affair, involving a detailed taxonomic inventory and analysis of who feeds on whom. But a huge amount of information is then boiled down into a simple flow diagram such as the one in Figure 9.1. The food web is often a starting point for understanding the consequences of species loss (e.g. removing predators to assist an endangered species) or species gain (e.g. reintroducing a desirable species or assessing the risk posed by the invasion of an undesirable one).

Strong interactors and keystone species

Certain species are more tightly woven into the fabric of the food web than others. A species whose removal would produce a large change in density (or extinction) of at least one other species may be described as a *strong interactor*. Some strong interactors, if they were to be removed, would lead to marked changes spreading throughout the food web – these *keystone species* have an impact that is disproportionately large relative to their abundance (Power et al., 1996). The first use of the term keystone species was for the starfish *Pisaster ochraceus* (Paine, 1966, and see

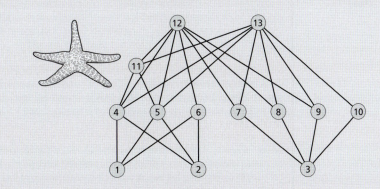

Fig. 9.1 Food web for a rocky shore in Washington State, USA. Species higher in the web feed on species lower down, as indicated by the lines. 1, fine particulate organic detritus in the water column; 2, free-living microscopic planktonic plants; 3, algae growing on the rocks; 4, acorn barnacles; 5, the mussel *Mytilus edulis*; 6, the goose barnacle *Pollicipes* sp.; 7, chitons; 8, limpets; 9, the topshell *Tegula* sp.; 10, the periwinkle *Littorina* sp.; 11, the snail *Thais* sp.; 12, the starfish *Pisaster ochraceus*; 13, the starfish *Leptasterias* sp. (After Briand, 1983.)

Figure 9.1). The starfish benefits a suite of inferior competitors (algae growing on the rocks together with a range of invertebrates) by preferentially consuming the dominant competitors (barnacles and, in particular, mussels). When the starfish are removed, the inferior competitors disappear from the scene. Managers need to keep in mind not only that a species of interest will have both direct and indirect interactions in the food web, but also that some are far more influential than others.

Trophic cascades and top-down and bottom-up control of food webs

Many examples have come to light where a carnivore (or parasite), through its influence on a herbivore, is responsible for relaxing the impact of the herbivore on its plant food, whose biomass increases as a result. The idea can be extended to a food chain with four links – in this case the top predator may reduce an intermediate predator, relaxing predation on a herbivore, leading to enhanced grazing and a reduction in plant biomass. These patterns are known as trophic cascades and the species at the top of the food chains can reasonably be called keystone species. One way to reduce nuisance algal blooms in lakes is to manipulate a cascading food chain to benefit the herbivores.

In trophic cascades, the dominant control on biomass in the food web is exerted from the *top down*. In particular, whether plant biomass is high or low is a direct consequence of the impact of a top predator. But this is not always the case. It is by no means unusual for plant biomass to be controlled *bottom up* by the supply of the resources needed for photosynthesis – light, carbon dioxide, water and nutrients such as phosphorus and nitrogen. Everyone will be aware that the application of fertilizer usually has the effect of increasing plant productivity because previously limiting nutrients become more readily available. This is bottom-up control of plant biomass. An alternative way to reduce nuisance algal blooms in lakes is thus to identify and eliminate the delivery of excess nutrients by improved sewage treatment or reduction of agricultural runoff.

Ecological energetics

Net primary productivity (NPP) is the rate of production of new biomass by plants. Plant communities can be expected to be most productive where sunlight is in good supply, and where sufficient water and plant nutrients are available to fuel plant growth. Some of the energy locked in plant biomass passes along food chains involving herbivores and their consumers – collectively known as the *grazer system*. But because herbivores are never 100% efficient at consuming NPP, some plant biomass dies and finds its way as *dead organic matter* (DOM) into the base of another component of the food web – the *decomposer system*, with its bacteria and fungi, detritivorous animals and their consumers. Note that each trophic grouping (herbivores, carnivores, detritivores, etc.) is less than 100% efficient at assimilating what is consumed (some is lost as feces, which also enter the DOM compartment) and less than 100% efficient at turning what is assimilated into new biomass (some energy is lost as respiratory heat). The rate of production of new biomass by organisms other than plants is known as *secondary productivity*.

Taking all these points into account, Figure 9.2 illustrates some gross differences in the flux of energy through contrasting ecosystems. First note that the decomposer system is responsible for the majority of secondary productivity, and of respiratory heat loss, in almost all the world's ecosystems. The grazer system holds little sway in terrestrial communities (Figure 9.2a,b) because of low herbivore consumption and assimilation efficiencies, and it hardly exists at all in many small streams and ponds because NPP is so low (Figure 9.2d). These small water bodies depend for their energy on dead organic matter that enters from the surrounding terrestrial environment. The grazer system plays its strongest role in plankton communities because a large proportion of NPP is consumed alive and it is assimilated at quite high efficiency (Figure 9.2c). Certain species may play key roles in determining the overall productivity and pattern of energy flux in an ecosystem. Thus, management sometimes relies on understanding the ecosystem role of endangered keystone native species, or of potential invaders that could fundamentally change the flux of energy and nutrients (below).

Nutrient dynamics

Living organisms extract chemicals from their environment, hold on to them for a while, then lose them again. Ecosystem ecologists are interested in how the biota accumulates, transforms and

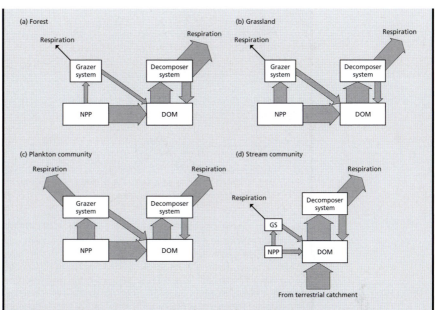

Fig. 9.2 General patterns of energy flow for: (a) a forest; (b) a grassland; (c) a marine plankton community; (d) the community of a stream or a small pond. The relative sizes of the boxes and arrows are proportional to the relative magnitudes of compartments and flows. DOM, dead organic matter; NPP, net primary production. (After Begon et al., 2006.)

moves chemical elements between the various living and nonliving (abiotic) compartments of the ecosystem. Some abiotic compartments are in the *atmosphere* (carbon in carbon dioxide, nitrogen as gaseous nitrogen), some in the rocks of the *lithosphere* (calcium, potassium) and some dissolved in the *hydrosphere* – the water of soil, stream, lake or ocean (nitrogen in dissolved nitrate, phosphorus in phosphate).

Nutrient elements, of which carbon, phosphorus and nitrogen are of critical importance, are available to plants as simple inorganic molecules or ions and are incorporated during photosynthesis into complex organic carbon compounds in biomass. Ultimately, however, when the carbon compounds are metabolized to CO_2 (via the plant's own respiratory activity or after use by herbivores, bacteria, fungi or detritivores) energy is lost as heat and the mineral nutrients are released again in simple inorganic form. Another plant may then absorb them, and so an individual atom of a nutrient element may pass repeatedly through the ecosystem. By its very nature, each unit of energy in a high-energy compound can be used only once before being lost as heat, whereas chemical nutrients can be used again and again. However, nutrient cycling is never perfect and a terrestrial ecosystem may lose nutrients via drainage to an aquatic ecosystem. Natural ecosystems tend to be more nutrient-tight than plantation forestry with its clear-felling approach (Figure 9.3) or agroecosystems. Managers of waterways need to be wary of the input of excess nutrients because this can fuel undesirably high primary productivity by phytoplankton, causing the water to become turbid and eliminating large rooted plants, and eventually, via decomposition of the excessive primary production, reducing oxygen levels and killing invertebrates and fish. This process is known as *eutrophication*.

Ecological stoichiometry – a complex tale of nutrient dynamics
Ecological stoichiometry is concerned with the consequences for ecological interactions of the relative availability of different elements – and particularly the ratio of carbon to nitrogen (Elser & Urabe, 1999). The rate at which dead organic matter decomposes depends strongly on its biochemical composition. This is because bacterial and fungal tissues have very high nitrogen

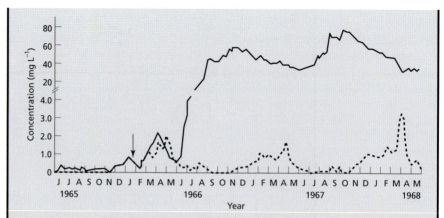

Fig. 9.3 Results of a classic large-scale experiment at Hubbard Brook, New Hampshire, USA, where all the trees in a forested catchment area were cut down (mimicking clear felling by foresters). The arrow shows when deforestation occurred. The concentrations of nitrate ions in stream water draining from the deforested catchment (solid line) were considerably greater than in a neighboring control catchment that remained forested (dashed line). This happened because deforestation fractured the within-forest nutrient cycling by uncoupling decomposition losses of nitrogen from its re-uptake by vegetation. (After Likens & Borman, 1975.)

contents, indicative of high requirements for this nutrient. The stochiometric ratio of carbon:nitrogen (C:N) in bacteria and fungi is about 10:1. Terrestrial plant material, on the other hand, has considerably higher ratios, ranging from 19:1 to 315:1 (Enriquez et al., 1993). Put another way, dead plant material has relatively much lower amounts of nitrogen than the microbes require. This means that the plant material, when it dies, can support only a limited biomass of decomposer organisms and the pace of decomposition is slowed. It turns out that there is a critical C:N ratio in dead plant material of 30:1 above which bacteria and fungi are N-limited – under these circumstances they withdraw ammonium and nitrate ions from the soil, competing with plants for these resources (Daufresne & Loreau, 2001). But when the C:N ratio is below 30:1, the microbes are C-limited and the process of decomposition leads to an increase of inorganic N in the soil, which may in turn increase N-uptake by plants. Stoichiometric knowledge can be used to manipulate the carbon content of the soil in a way that reduces nitrogen and facilitates the restoration of a native plant community.

9.2 Food web theory and human disease risk

If left untreated Lyme disease can damage the heart and nervous system and lead to a type of arthritis – tens of thousands of people are affected around the world each year. The disease is caused by the bacterium *Borrelia burgdorferi*, which is carried by ticks in the genus *Ixodes*. In their 2-year life cycle the ticks pass from egg to larva to nymph to adult, taking blood meals from a succession of vertebrate hosts.

Eggs, which generally do not carry the Lyme bacterium, are laid in spring and after hatching the larva takes a single blood meal from a vertebrate host. If the host (a small mammal, bird or reptile) carries the bacterium, this may be transmitted to the larval tick, which then remains infective for the rest of its life (as nymph and adult). The larval tick drops from its host and molts into an overwintering nymphal stage. In its second year the nymph seeks another host (small mammal, bird or human) in spring/early summer for a further single blood meal. This is the really risky stage for humans for three reasons – large numbers of people (including Aretha) are hiking in forests and parks, the nymphs are small and difficult to detect, and up to 40% of them carry the bacterium. The nymph drops to the ground where it molts into an adult before taking a final blood meal from a third host (often a larger mammal such as a deer) and laying its eggs. This food web thus involves two parasites (bacterium and tick) and a wide variety of host species. But wait, there's more! The parasite–host web is itself embedded in a larger forest food web where oak trees can play a key role.

The most abundant host of ticks in the eastern USA is the white-footed mouse (*Peromyscus leucopus*). Jones et al. (1998) recognized that acorns (oak seeds) are a preferred food of the mice, and that acorns are produced in very large numbers from time to time (known as *mast years*). To simulate a mast year they added acorns to the forest floor. Mice numbers increased the following year and, of more significance, the percentage of infected nymphal black-legged ticks (*Ixodes scapularis*) increased 2 years after acorn addition. Despite the complexity of this food web, it may be possible to predict high-risk years (in relation to acorn production) and take steps to educate and alert hikers to the danger.

It is important to realize that the potential mammal, bird and reptile hosts of ticks show great variation in the efficiency with which they transmit the Lyme bacterium to the tick. The white-footed mouse is by far the most efficient, but a plethora of other host species harbor the bacterium but rarely transmit it. Squirrels, for example, are abundant and provide blood meals to ticks but only a small percentage of these acquire the bacterium. This means that squirrels, and other species that do not transmit the bacterium efficiently, act to 'dilute' the risk of disease to humans. With this in mind, you might expect that high species richness of potential hosts would result in less human disease because the high transmission efficiency of white-footed mice is diluted by the presence of a variety of less 'efficient' species (LoGuidice et al., 2003). A related prediction is that small forest fragments, which contain far fewer potential host species but where white-footed mice remain abundant, will provide a greater risk of human infection than large tracts of undisturbed forest. If this is the case, forest fragmentation, whose impact on biodiversity I have already noted (Sections 3.4 and 4.5.1), might also be responsible for increasing the risk of disease transmission. The prediction is supported by results from New York State (Figure 9.4), which show an exponential decline in the percentage of nymphs infected by the Lyme bacterium with increasing size of forest patches.

The importance of understanding human disease transmission from wild animal populations, particularly rodents, cannot be overestimated: we can add to Lyme disease a long list of nasties including Lassa fever, bubonic plague and various hemorrhagic fevers. Mosquitos also carry many pathogens that are transmitted when the female mosquito takes a human blood meal. Invasions by foreign mosquitos are not unusual and some bring a pathogen with them, such as yellow fever transmitted by *Aedes aegypti* – both arrived in the Americas on slave-ships in the sixteenth and seventeenth centuries. Others transmit a native pathogen that was previously transmitted by native mosquitos (e.g. malaria in Brazil became transmitted by *Anopheles gambiae* after this arrived from Africa in 1930–40), and yet others arrive first but transmit a pathogen that turns up later (e.g. dengue fever, which arrived in infected humans in Hawaii in 2001, but is now transmitted by the earlier arriving *Aedes albopictus*). On a somewhat brighter note, Juliano and Lounibos (2005) note that an

Fig. 9.4 The Lyme bacterium *Borrelia burgdorferi*, transmitted by ticks and carried by hosts such as the white-footed mouse, causes disease in humans. Density of nymphs of the tick *Ixodes scapularis* that are infected with the bacterium responsible for human Lyme disease increases dramatically in small forest fragments where white-footed mice are common, but other less efficient disease transmitters are absent. (After Allan et al., 2003.)

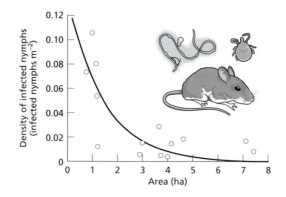

invading mosquito might conceivably reduce human disease risk if it is a less efficient disease vector than native species.

Whether or not a new mosquito becomes invasive depends on where it fits into the food web of the small water bodies that mosquitos inhabit – and whether it is more or less competitive for food, or vulnerable to predation, than the native species (Juliano & Lounibos, 2005). Thus, although the invading *Aedes albopictus* can outcompete the native North American mosquito *Ochlerotata triseriatus* in the absence of predators, the invader is more vulnerable than the native to insects that feed on mosquito larvae in their tree hole habitat. In fact, because the native mosquito hatches earlier, it achieves a size at which it can itself prey upon the smaller larvae of the invader. For these reasons *A. albopictus* fails to displace *O. triseriatus*. Biosecurity managers can glean clues about the likely harm (or good) of a potential mosquito invader through knowledge of its disease vector efficiency and food web relations.

9.3 Food webs and harvest management

9.3.1 *Who gets top spot in the abalone food web – otters or humans?*

Harvests of abalone (gastropods in the family Haliotidae) are prone to collapse through overfishing. Because adult abalones are relatively sedentary, protection of broodstock in no-take marine areas may promote export of planktonic larvae and enhance the harvested populations outside the reserves (see Section 7.2.4). But note that the function of marine protected areas is most often to conserve biodiversity. The question arises whether protected areas can serve both harvest management and biodiversity objectives. The answer is yes in some cases, but not apparently where abalone and sea otters are concerned.

A keystone species in coastal habitats along the Pacific coast of North America is the sea otter (*Enhydra lutris*), hunted almost to extinction in the eighteenth and nineteenth centuries but increasingly widespread as a result of protected status. Sea otters eat abalones and while otters were rare, valuable fisheries for red abalone (*Haliotus rufescens*) developed. Now there is concern that the fishery will be unsustainable in the presence of the resurgent sea otter.

Fanshawe et al. (2003) compared the population characteristics of abalone in sites that varied in harvest intensity and sea otter presence: two sites had no sea otters and had been 'no-take' abalone zones for 20 years or more; two sites lacked sea otters but allowed recreational fishing; and four sites were 'no-take' zones that contained sea otters. The aim was to determine whether marine protected areas can help make the abalone fishery sustainable even when all links in the food web (including the otters) are fully restored.

Sea otters and recreational harvest influence red abalone populations in similar ways but the effects of sea otters are much more pronounced. Abalone populations in protected areas have much higher densities than areas with otters, while harvested areas generally have intermediate densities (Figure 9.5a). In addition, there are differences in the size of abalone (Figure 9.5b) – 63–83% of individual abalones in protected areas are larger than the legal harvesting limit of 178 mm, compared with 18–26% in harvested areas and less than 1% in sea otter areas. Clearly the effects of otters on the size class structure of their abalone prey mean that almost none are left for human harvest. Finally, in the presence of sea otters the abalones are mainly restricted to crevices where they are least vulnerable to predation (Figure 9.5c), and most difficult for fishers to extract. Multiple-use protected areas are never likely to be feasible where a desirable top predator feeds intensively on prey targeted by a fishery. Fanshawe's team recommends separate single-purpose categories of protected area. But this may not work in the long term either. Maintenance of abalone no-take areas when sea otters are expanding their range will eventually require culling of otters, something that may not be politically acceptable.

Fig. 9.5 The influence of human harvest and otter predation history on key features of the abalone population at sites on the Californian coast: (a) mean abalone density; (b) shell length; (c) percentage in crevices, where they are least vulnerable to predation. (Redrawn from Fanshawe et al., 2003.)

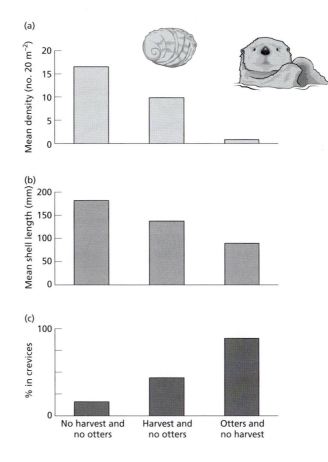

9.3.2 *Food web*
consequences of
harvesting fish –
from tuna to tiddlers

When dealing with harvest management (Chapter 7) I treated fin fisheries as if they existed in splendid isolation as single-species fisheries. In reality, of course, any fish population exists in a network of competitive and predator–prey interactions so that a fishery-induced reduction in its density must have direct and indirect consequences that spread through the food web. In this sense the fishing industry can be viewed as a keystone predator that affects the ocean food web from the top down, reducing the abundance of favored large fish, and through a reduction in top predators allowing a proliferation of smaller species further down the food chain.

Ecosystem-based fisheries models that take the reality of food webs into account, while not replacing single-species approaches, are now very much part of the fisheries manager's tool box. Figure 9.6 shows just how intricate an ocean food web can be. And another reality complicates the picture still further – different fishing methods, by preferentially catching particular species, can have different effects in

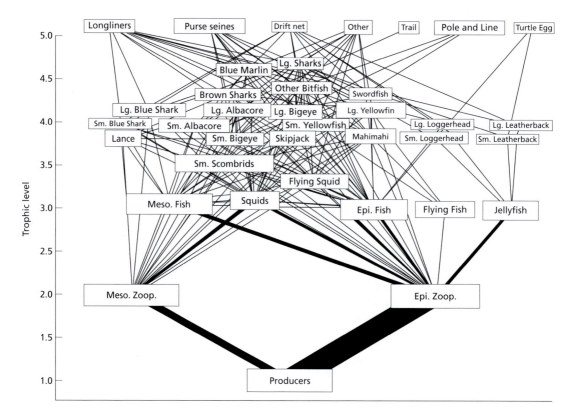

Fig. 9.6 Ocean food web for the central north Pacific, where line thickness indicates energy flow. Phytoplankton, the 'producers', are eaten by zooplankton (Zoop) both in the epipelagic zone (Epi – near surface open water) and mesopelagic zone (Meso – deeper water with little light penetration). Large and small (Lg, Sm) classes are distinguished for some big taxa. Each taxon has a vertical position in the diagram that shows its position in the food chain. Producers have a trophic level of 1, herbivores 2 and predators that eat only herbivores 3. But note that mixed diets lead to trophic level values that are not whole numbers and fish can have trophic levels anywhere from 2 to a maximum of about 5. At the top of this food web are shown six fishing methods currently or historically deployed. The purse seine fleet targets skipjack and yellowfin tuna while longliners harvest bigeye and yellowfin tuna, but both also catch many other species. Pole-and-line fishers and troll fishers go after skipjack and yellowfin but take less by-catch. A relatively short lived driftnet fishery targeted squid and albacore until 1993. 'Other' gears include hand nets and small-meshed gill nets. (After Hinke et al., 2004.)

Fig. 9.7 Analysis of trophic level averaged for all fish species in the fisheries catch statistics for the western central Atlantic. Open circles are for the USA, and closed circles are for all other areas. Note in both cases how mean trophic level has been declining since 1950 as a result of overfishing of the larger, most preferred species. These patterns are examples of a worldwide trend of 'fishing down'. (From Pauly & Palomares, 2005.)

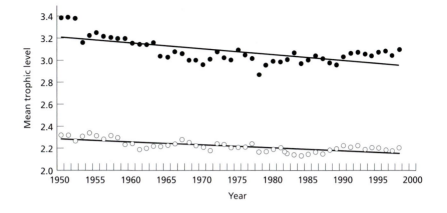

the same food web. Simulation models, involving reductions in one fishing method at a time, indicate that restrictions on shallow-set longline gear would be the step most likely to allow recovery of top predators in this food web.

A worldwide pattern, mirroring what has happened in the north Pacific, is a decline in the mean trophic level of fish in fisheries catches (Figure 9.7). In other words, by overfishing the largest and most desirable species, which in itself allows an increase in biomass of smaller species, fisheries are increasingly coming to rely on species from much lower down the food chain (i.e. the prey of the previously targeted fish). This general pattern is known as 'fishing down marine food webs' (Pauly & Palomares, 2005). Sustainability in fisheries can be expected to require management of the use of gears to allow recovery of the larger species and stability in the composition of the catch.

9.4 Food webs and conservation management

Knowledge of the intricacies of food webs can also guide managers who seek to conserve species at risk. The kokako (*Callaeas cinerea*), an endangered bird, is now restricted to about 15 populations in forest fragments in the North Island of New Zealand. Nesting success is generally poor because of predation of eggs and chicks by introduced mammals – ship rats (*Rattus rattus*), which arrived on the first sailing ships, brushtail possums (*Trichosurus vulpecula*) brought from Australia to start a fur trade, and stoats (*Mustela erminea*) introduced in a vain attempt to control rabbits. The diet of kokako includes fruits and foliage, items also consumed by possums and rats. The causal links in this food web are illustrated in Figure 9.8. Population modeling has shown, first, that reduction of any one mammalian predator is not sufficient to improve the fledging success of kokako. At the other extreme, reducing all three predators provides the strongest benefit for kokako fledging success – something that is hardly surprising. But of most note is the finding that reduction of just ship rats and possums produces very nearly as positive an outcome as reducing all three predators. This is because stoats prey on ship rats so that reduction of rats reduced stoat success, and the reduction in both predators was beneficial to the birds.

The advice to managers, then, is to reduce all three predators or, if funds are limited (which they invariably are), target the rats and possums. But don't waste money trying to reduce just one of the predators.

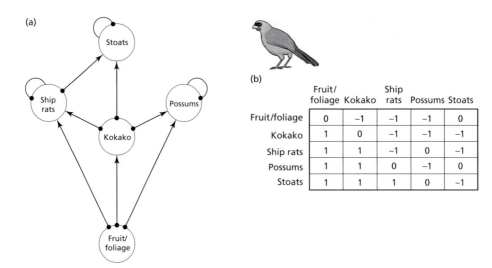

	Fruit/foliage	Kokako	Ship rats	Possums	Stoats
Fruit/foliage	0	−1	−1	−1	0
Kokako	1	0	−1	−1	−1
Ship rats	1	1	−1	0	−1
Possums	1	1	0	−1	0
Stoats	1	1	1	0	−1

Fig. 9.8 (a) The kokako food web showing positive (in direction of arrows) and negative (in direction of black dots) links among the food web components. Thus, for example, an increase in kokako has a positive effect on number of ship rats (which prey on kokako), while an increase in ship rats has a negative impact on kokako numbers. Closed loops starting and ending at the species represent density dependence, where change in population size is negatively related to current population size (refer to Box 5.1). (b) The matrix provides another way of describing the food web in a qualitative way. Note that an increase in kokako, ship rats or possums (refer to columns beneath these species) has a negative effect on fruit and foliage; this is because all three feed, at least partially, on fruit and foliage. Stoats, on the other hand, do not. All three predators, however, feed on kokako, and stoats also feed on ship rats. (After Ramsey & Veltman, 2005.)

9.5 Ecosystem consequences of invasions

You have seen in previous chapters how invaders can affect populations of native species with which they interact. On a larger ecological scale, it turns out that invaders sometimes modify the way energy and nutrients move through the native ecosystems.

9.5.1 *Ecosystem consequences of freshwater invaders*

Just as sea otters alter the behavior of their abalone prey (Section 9.3.1), invading brown trout (*Salmo trutta*) in New Zealand change the behavior of herbivorous invertebrates that graze algae on the streambed; daytime activity is significantly reduced in the presence of trout (Townsend, 2003). Brown trout use vision to capture prey, while the native fish they have replaced (*Galaxias* spp.) rely on mechanical stimuli. This means that darkness provides a refuge against trout predation (analogous to the crevices that abalone use when under threat). It is not surprising that the introduced fish should have direct effects on *Galaxias* distribution or invertebrate behavior, but its influence also cascades to the algae at the base of the food web. Three treatments were established in experimental stream channels – no fish, *Galaxias* present, or trout present, at naturally occurring densities. After 12 days, algal biomass was highest where trout were present, partly because of a reduction in grazer biomass (Figure 9.9) but also because of a reduction of grazing (reduced feeding during the day) by the remaining grazers. This trophic cascade had further profound effects. The enhanced algal biomass led to a higher rate of capture of radiant energy in photosynthesis (annual net primary production was six

Fig. 9.9 Total invertebrate biomass and algal biomass (chlorophyll *a*) (±SE) in an experiment performed in summer in replicated artificial channels placed in a small New Zealand stream. N, no fish; G, native *Galaxias* sp. present; T, brown trout present. Trout cause a trophic cascade by reducing grazing by invertebrates, thus allowing algal biomass to accumulate. (After Townsend, 2003.)

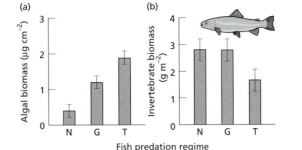

times greater in a trout stream than in a neighboring *Galaxias* stream; Huryn, 1998). This, in turn, resulted in more efficient uptake of nitrogen from the overlying water and, thus, lower nitrogen losses downstream (Simon et al., 2004). Thus, the two most important elements of ecosystem functioning – energy and nutrient flux – were altered by the invading brown trout.

The related rainbow trout (*Oncorhyncus mykiss*) has invaded fishless lakes in North America and a similar increase has been recorded in plant biomass, this time the phytoplankton that occur in open water. To some extent the cause is similar, with a fish-induced reduction in grazing (by zooplankton). But the main reason for increased primary production is that the fish feed on invertebrates living on the lake bed and then, via their excretion, transfer phosphorus (the nutrient that limits plant production in the lake) into the open water where the phytoplankton live (Schindler et al., 2001).

Invading zebra mussels (*Dreissena polymorpha*) in North American lakes (Section 4.4.1) also exert a profound ecosystem effect. As super-efficient filter-feeders that live on the lake bed, these can remove up to 25% of phytoplankton cells from the open water each day, slowing energy flow from lower to higher trophic levels. They have a further stoichiometric effect (Box 9.1). Zebra mussels excrete soluble waste products at low nitrogen to phosphorus ratios that favor the growth of cyanobacterial blooms – these are less preferred food of zooplankton, further reducing the transfer of energy between trophic levels (Conroy & Culver, 2004).

Biosecurity managers need to pay particular attention to animal invaders that have a novel method of resource acquisition (such as brown trout in New Zealand) or that link previously unlinked ecosystem compartments (such as rainbow trout and zebra mussels in North America) (Simon & Townsend, 2003).

9.5.2 Ecosystem effects of invasive plants – fixing the problem

One class of plant invaders with particularly pervasive ecosystem effects are nitrogen-fixing woody plants, including *Myrica faya* in Hawaiian forest, *Acacia* spp. in South African shrubby fynbos communities and *Lupinus arboreus* in Californian coastal prairies (Corbin & D'Antonio, 2004). These plants, with their symbiotic bacteria that are capable of fixing atmospheric nitrogen into dissolved inorganic form, alter patterns of nutrient flux and dramatically increase soil nitrogen content. Not only are they highly successful as invaders, but even after removal they can leave a long-term legacy of high soil and litter nitrogen – features that may favor other invaders at the expense of native species. Managers need to take such legacies into account when planning restoration. Thus, for example, burning may reduce soil nitrogen stocks to a greater extent than mechanical removal of the invader plants.

In other cases it is soil water balance that is affected by an invader. Californian grasslands were once dominated by perennial grasses such as wheatgrass (*Pseudo-roegnaria* sp.). Then, after European settlement, the area became dominated by exotic annual grasses such as hare barley (*Hordeum murinum*) and exotic annual forbs (nongrasses) such as tumble mustard (*Sisymbrium altissimum*). Most recent of all has been invasion by the deeply rooted and late-maturing forb, the yellow star thistle (*Centaurea solstitialis*), an invader whose notoriety has already been noted (Section 1.2.5). In an attempt to understand the star thistle's success, Enloe et al. (2004) created experimental plots with three communities that were dominated, respectively, by perennial native grass, annual exotic grasses and star thistle.

Soil moisture measurements were made regularly for 4 years at depths down to 1.5 m in the three community types. The exotic grassland maintained the highest water contents whereas star thistle dramatically reduced soil moisture (Figure 9.10a). The perennial grassland showed an intermediate pattern. It seems that the competitive status of star thistle is due, at least in part, to its ability to reduce soil water to levels at which other plants fail. The star thistle has an exceptionally long life cycle for an annual plant. The seeds germinate after the first autumn rain, then during winter the seedling forms a rosette that develops a deep taproot by spring. This enables star thistles to exploit soil water at greater depths than their competitors (Figure 9.10b). The plants bolt to a height of 1 m or more in late spring and flower and seed during summer and early autumn, well after annual grasses have died back. Attempts to restore perennial grassland could include the introduction of deeply rooted summer forbs or shrubs to help suppress star thistle by extracting water from further down the soil profile. On the other hand, if star thistle is targeted for removal, and annual grassland is regained, you might expect increases in soil water content and, consequently, of water runoff to streams. Intermittent streams might even be turned into permanent ones. The ecosystem effects of some invasive plants can extend very far indeed.

9.6 Ecosystem approaches to restoration – first aid by parasites and sawdust

Ecological restoration can also benefit from knowledge of food web interactions or of ecosystem processes such as nutrient dynamics.

Species-rich meadows are now uncommon in agricultural landscapes in Europe because decades of fertilizer application and intensive mowing or grazing have allowed a few species to competitively exclude others. You have already seen that one limitation to meadow recovery is slow colonization of restored sites by seeds of many of the species, something that can be helped by sowing appropriate mixtures (Section 3.2.1). A more insidious problem, especially where high soil nutrient concentrations persist, is for competitively superior species to exclude less competitive species that would be represented if only nutrient concentrations were lower. A possible solution is to manipulate the food web so that the competitive status of the species is evened out.

One way to enhance coexistence of more species is to cut and/or graze at key moments to open space and allow seeds of less competitive species to find a place in the sun. It is well established that grazers can reduce biomass of plant species to such an extent that the most competitive do not become abundant enough to exclude their weaker brethren – a process known as *exploiter mediated coexistence*. A less

Fig. 9.10 (a) Soil water content, averaged across soil depths, during the 4 years of a study involving plots of Californian native perennial grassland (open squares, ±SE), exotic annual grassland (open circles) and star thistle-dominated communities (closed circles). The star thistles exploit soil water and reduce soil moisture to a greater extent than their grassland counterparts. (b) Soil water content, averaged across the 4 years, in relation to depth in the soil profile. Because of their deep taproots, star thistles can reduce water content to a greater depth than the grasses. Symbols as in (a). (After Enloe et al., 2004.)

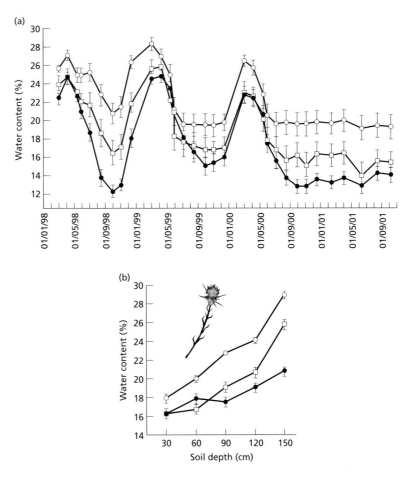

well-known form of exploiter mediated coexistence occurs when parasites exert the leveling effect. *Rhinanthus minor*, an annual plant, is capable of its own limited photosynthesis but is known as a *hemiparasite* because it typically taps into the photosynthetic products of other plants by building connections with their roots. Pywell et al. (2004) reasoned that the presence of the hemiparasite might facilitate recovery to species-rich grassland via exploiter-mediated coexistence. To test this hypothesis in a species-poor grassland, they established experimental plots in early autumn with various densities of *Rhinanthus minor*. Two years later, after the hemiparasite populations had become established, they sowed a seed mixture of ten native wildflower species that had been lost from the grassland as a result of intensive agriculture. After two further years, plant plots where the hemiparasite was more abundant achieved lower height, because of suppression of growth of the parasitized plants (Figure 9.11a). This led, by year 3, to the desired increase in grassland species richness because competitive exclusion had been circumvented (Figure 9.11b). An understanding of exploiter-mediated coexistence holds promise for future meadow restoration efforts.

Fig. 9.11 Relationships between frequency of occurrence of the hemiparasite *Rhinanthus minor* and (a) mean sward height (height achieved by the grassland community) and (b) species richness of plants per experimental plot. The presence of the hemiparasite leads to lower sward height, because of reduced success of the parasitized plants, and the following year to increased species richness because of suppression of competitive exclusion by the dominant species. (After Pywell et al., 2004.)

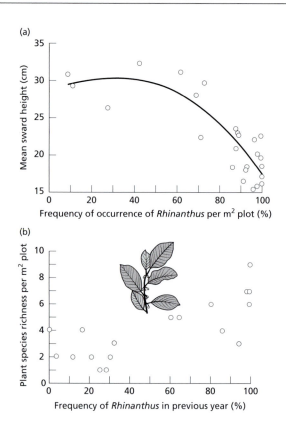

As explained in Box 9.1, there is a critical stoichiometric ratio of carbon : nitrogen in the soil of 30 : 1 above which bacteria and fungi are nitrogen-limited, and under these circumstances they withdraw nitrogen from the soil, reducing the concentration of this critical plant nutrient. Since invasive plants frequently do best when nitrogen availability is high, you might expect that native species would be favored if the C : N ratio could be increased from its natural level of about 10 : 1 by the addition of carbon to the soil. This is precisely what Blumenthal et al. (2003) did in replicate plots of tallgrass prairie established with 10 weeds and 11 tallgrass prairie species in Minnesota, USA. Carbon was added to the soil partly as sucrose (to supply readily available carbon – 5.9% of the mix) and the rest as sawdust (slower for microorganisms to metabolize). Fourteen levels of carbon addition were performed in late summer (four replicates of each) ranging from zero to more than 3000 g of carbon per square meter. As predicted, soil nitrate levels the following year were negatively related to carbon addition (Figure 9.12a). Weed biomass, in total and for a number of individual weeds (including *Sinapis arvensis* and *Solanum nigrum*), declined with carbon addition, no doubt because of reduced nitrate availability (Figure 9.12b). The opposite pattern was found for total biomass of native prairie plants (Figure 9.12c) and for a number of individual species (including *Sorghastrum nutans* and *Monarda fistulosa*). Carbon addition may be a useful tool for restoring grassland communities as long as undesirable weeds have high nitrogen demands and are suppressing less nitrogen-demanding native species.

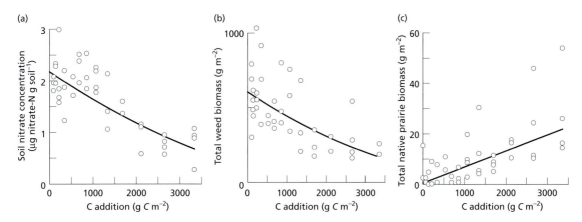

Fig. 9.12 Relationships between amount of carbon added to the soil the previous year and (a) soil nitrate concentration, (b) biomass of all weed species combined and (c) biomass of all native prairie plants combined. (After Blumenthal et al., 2003.)

9.7 Sustainable agroecosystems

9.7.1 Stopping caterpillars eating the broccoli – so that people can

A trophic cascade, in which top predators reduce herbivore populations and thus benefit the plant trophic level, can provide a useful and natural pest control mechanism for certain crops. The tangle web spider, *Nesticodes rufipes*, is the most abundant of several predatory species found on crops of broccoli (*Brassica oleracea*) in Hawaii. In addition, two species of bird, the red-crested and northern cardinals (*Paroaria coronata* and *Cardinalis cardinalis*), have been observed to feed on caterpillars of various butterflies and moths, including the cabbage butterfly (*Artogeia rapae*), which is the most abundant pest of broccoli crops. Four treatments were established in a series of experimental broccoli plots: (i) spiders were excluded by daily hand removal (birds only); (ii) both birds and spiders were present; (iii) birds were excluded by cages that permitted access to spiders (spiders only); and (iv) birds were excluded and spiders removed daily.

Access to birds (with or without spiders) led to the biggest reduction in caterpillar density, minimal defoliation damage (because of reduced herbivore activity) and large crop weights (Table 9.1). This evidence of a strong trophic cascade provides good reason for protecting or enhancing bird populations around the crop. On their own, spiders had a smaller effect, but they did reduce caterpillar density to a degree, with some benefit in reducing defoliation and increasing broccoli biomass. Perhaps their effect should be called a trophic trickle.

9.7.2 Managing agriculture to minimize fertilizer input and nutrient loss

Intensive agriculture is frequently associated with pollution, by phosphorus or nitrogen, of the waterways that drain the land. This happens when there is a mismatch between the application of fertilizer (rich in P or N or both) and its use by the intended recipients – crop plants, forest trees or pasture (and, subsequently, grazing stock animals). Excess nitrate that finds its way into drinking water may pose a health hazard, potentially contributing to the formation of carcinogenic nitrosamines. The US Environmental Protection Agency recommends a maximum concentration of 10 mg/liter. Moreover, nitrogen and other plant nutrients that leach into the groundwater, and thence into rivers, lakes, estuaries and oceans, can profoundly affect aquatic food webs and ecosystem functioning. For now, let's consider

Table 9.1 Consequences of manipulation of predation by birds and spiders for cabbage butterflies (*Artogeia rapae*) and broccoli plants. A trophic cascade is evident. The herbivores are significantly reduced in density and crop damage lessened when birds are allowed access (with or without spiders). On their own, spiders produce a smaller effect. Plant dry weight is for above-ground biomass and does not include the flower head. (Data from Hooks et al., 2003.)

Treatment	Cabbage butterflies		Broccoli plants	
	Large larvae per plant $\pm$ SE	Pupae per plant $\pm$ SE	% of plants with >50% defoliation at 25 days	Plant dry weight (g) at 54 days $\pm$ SE
Birds only	0.18 $\pm$ 0.18	0	0	153.5 $\pm$ 6.7
Birds plus spiders	0.44 $\pm$ 0.24	0	0	166.2 $\pm$ 9.7
Spiders only	3.62 $\pm$ 1.15	4.2 $\pm$ 0.85	9.1	139.2 $\pm$ 9.5
No birds or spiders	4.70 $\pm$ 0.46	5.0 $\pm$ 0.22	72.7	91.2 $\pm$ 7.7

ways to minimize the unintended loss of nutrients from the land. I will focus on nitrogen, but the problems are much the same for phosphorus and other limiting mineral nutrients.

Most of the fixed nitrogen in natural communities is present in vegetation and in the organic fraction of the soil. As organisms die they contribute to nitrogen in soil organic matter, eventually transformed into nitrate ions that are leached by rainfall down through the soil profile. Both decomposition and nitrate production are generally fastest in summer, when natural vegetation is growing most quickly. Nitrate ions may then be re-absorbed by the growing plants as fast as they are produced, so that few ions are leached out of the plants' rooting zone and lost from the terrestrial ecosystem. As noted in Box 9.1, natural vegetation is most often a 'nutrient-tight' ecosystem.

By contrast, there are several reasons why nitrates leach more readily from agricultural land and plantation forests. For one thing, agricultural land for part of the year carries little or no living vegetation to absorb nitrates – and for many years in the forest cycle, biomass is below its maximum. Then again, crops and managed forests are usually monocultures that may capture nitrates only from specific rooting depths, whereas natural vegetation often has a diversity of rooting systems and depths. Third, the practice of burning straw, crop stubble or forestry waste returns the organic nitrogen in a pulse back to the soil as nitrate, at a time when plants are unavailable to reabsorb it. Fourth, fertilizer is usually applied just once or twice a year rather than in the continuous stream that occurs in natural vegetation. Finally, there are the subtle consequences of stoichiometry (Box 9.1): when agricultural land is used for grazing animals, their metabolism speeds up the rate at which carbon is respired, reducing the C:N ratio, and thus increasing nitrate formation and leaching.

There are a number of tools to minimize fertilizer loss from the land (thus saving money) to the water (where a useful resource becomes an irritating pollutant). Farmers might aim to maintain ground cover of vegetation year-round, practice mixed cropping rather than monoculture and take care to return organic matter to the soil. The overriding objective should be to match nutrient supply to crop demand. Modern 'controlled-release' fertilizers hold much promise in this regard (Mosier et al., 2002).

Farm animals are responsible for further problems of nutrient loss from the land. Cattle, pigs and poultry are the three major nitrogen contributors in agriculture feedlots. The nitrogen-rich waste from poultry can be dried to provide a readily transportable, inoffensive and valuable fertilizer for crops and gardens. By contrast, cattle and pig excreta is 90% water and smells bad. A unit of 10,000 pigs produces as much pollution as a town of 18,000 people. In many parts of the world, the law increasingly restricts the discharge of animal waste into waterways. The simplest practice converts pollutant into fertilizer and returns the material to the land as semisolid manure or sprayed on as slurry, diluting its concentration to something similar to a more 'primitive' and sustainable type of agriculture. When it comes to nitrate pollution of waterways, one of the biggest culprits is farm specialization where forage crops are grown in one area, but stock is fattened on the other side of the country. This means that fertilizer must be used to make up the shortfall when plants are reaped and transported to the stock, whose excreta can hardly be shipped all the way back to the farm of origin. In the USA, for example, only 34% of the nitrogen excreted in animal waste is returned to cropped fields (Mosier et al., 2002) – much of the rest eventually finds its way into streams and rivers. A change in practice to one where animal feed crops and stock fattening occur in the same area would certainly reduce nutrient loss to waterways. Of course, there is an even more pervasive problem that intimately involves us all – the nutrients released as human excreta are rarely taken back to be placed on the land where they originated.

9.7.3 Constructing wetlands to manage water quality

The excess input of nutrients from sources such as agricultural runoff and sewage has caused many 'healthy' *oligotrophic* lakes (low nutrients, low plant productivity with abundant water weeds, and clear water) to switch to a *eutrophic* condition where high nutrient inputs lead to high phytoplankton productivity (sometimes dominated by bloom-forming toxic species), making the water turbid, eliminating large plants and, in the worst situations, leading to anoxia and fish kills. This process of *cultural eutrophication* of lakes has been understood for some time. But it was only recently that people noticed huge 'dead zones' in the oceans near river outlets, particularly those draining large catchment areas such as the Mississippi in North America and the Yangtze in China. The nutrient-enriched water flows through streams, rivers and lakes, and eventually to estuary and ocean where the ecological impact may be huge, killing virtually all invertebrates and fish in areas up to 70,000 km^2 in extent. The United Nations Environment Program (UNEP) has reported that 150 sea areas worldwide are now regularly starved of oxygen as a result of decomposition of algal blooms fueled particularly by nitrogen from agricultural runoff of fertilizers and sewage from large cities (UNEP, 2003). Ocean dead zones are typically associated with industrialized nations and usually lie off countries that subsidize their agriculture, encouraging farmers to increase productivity and use more fertilizer.

The only way to alleviate problems in the world's oceans is by careful management of terrestrial catchment areas to reduce agricultural runoff of nutrients and to treat sewage to remove nutrients before discharge (known as tertiary treatment). The vegetation zones between land and water, such as wetland areas (consisting of swamps, ditches and ponds) and riparian forest along the banks of streams, can also be beneficial because the plants and microorganisms remove some of the dissolved nutrient as it filters through the soil.

Unfortunately, riparian and wetland communities have usually been cut down or drained to provide a greater area for agricultural production. Riparian vegetation and wetlands can sometimes be restored to a seminatural state. An alternative is 'treatment wetlands', which are constructed, planted and flow-controlled to maximize removal of pollutants from wastewater. How much of a difference might wetland restoration/construction be expected to make? It seems that at a maximum, wetlands in temperate areas might remove 1000–3000 kg ha^{-1}yr^{-1} of nitrogen and 60–100 kg ha^{-1}yr^{-1} of phosphorus from water flowing through them. Estimates for stream catchment areas in southern Sweden, which are a major source of nitrate enrichment of the Baltic Sea, indicate that to remove 40% of the nitrogen currently finding its way into the sea, a system of wetlands covering about 5% of the total land area would need to be recreated (Figure 9.13).

9.7.4 *Managing lake eutrophication*

Lake eutrophication, where phosphorus is often the principal culprit, can be reversed by either chemical or biological means. In the first case, reduction of P inputs may be combined with an intervention such as chemical treatment to immobilize P in the sediment – in many cases recovery to a more oligotrophic state occurs within 10–15 years (Jeppesen et al., 2005). In essence, this is *bottom-up* control (Box 9.1) of nutrient availability, which reduces phytoplankton productivity (increasing water clarity) and successively reduces biomass in the higher trophic levels of zooplankton and fish.

The aim of biological control – known as *biomanipulation* – is also to reduce phytoplankton density and increase water clarity, but it does this via an increase in grazing by zooplankton that comes about after the biomass of zooplanktivorous fish has been actively reduced (by fishing them out or by increasing piscivorous fish biomass). The outcome is the same, but the process by which it is achieved is *top-down* control of a cascade in the food web.

Lathrop et al. (2002) attempted to biomanipulate the relatively large and deep eutrophic Lake Mendota in Wisconsin, USA. Their approach was to increase the

Fig. 9.13 The locations of 148 wetlands under construction along tributaries of the Rönneå River in southern Sweden. If these are built to occupy 5% of the total land area, a 40% reduction can be expected in agricultural nitrogen input to the Baltic Sea. (From Verhoeven et al., 2006, based on Arheimer & Wittgren, 2002.)

density of two species of piscivorous fish – walleye (*Stizostedion vitreum*) and northern pike (*Esox lucius*). In total, more than 2 million fingerlings of the two species were stocked beginning in 1987 (Figure 9.14a) and total piscivore biomass responded rapidly, stabilizing at 4–6 kg ha^{-1}. The biomass of zooplanktivorous fish declined, as a result of increased predation by the piscivores, from 300–600 kg ha^{-1} prior to biomanipulation to 20–40 kg ha^{-1} in subsequent years. The consequent reduction in predation pressure on zooplankton (Figure 9.14b) led, in turn, to a switch from small zooplankton grazers (*Daphnia galeata mendotae*) to the larger and more efficient *Daphnia pulicaria*. The increased grazing pressure had the desired effect of reducing phytoplankton density and increasing water clarity (Figure 9.14c).

9.8 Ecosystem services and ecosystem health

Having considered a range of impacts of human activities on ecosystem functioning, attention can now be turned to the question of how to 'cost' these impacts or, more particularly, how to put a value on the ecosystem service(s) that have been lost (Section 9.8.1). Beyond this, managers have another requirement to help them identify and prioritize ecosystems in trouble – easily measured indicators of the 'ecosystem health' of terrestrial, freshwater and marine ecosystems (Sections 9.8.2 – 9.8.5).

9.8.1 *The value of ecosystem services*

Some human effects are more significant than others for the welfare of nature and of humans, but until recently it was not possible to set benefits and costs against each other so that managers can identify priorities for action. The private economic benefits of exploiting a forest for timber or a coral reef for fish (or for coral) are usually clear. But the public benefits associated with the pristine ecosystem (or of less intensive exploitation) are harder to quantify. Considerable ingenuity is required to assign value to benefits that accrue to people from natural ecosystems – *ecosystem provisioning services* such as wild food and forest products and *ecosystem regulating services* such as maintenance of the chemical quality of natural waters, buffering of human communities against floods and droughts, protection and maintenance of soils, regulation of climate, and breakdown of organic and inorganic wastes.

The most straightforward approach is to compare the value of retaining a particular habitat in a relatively undisturbed condition as opposed to intensively exploiting it. Going beyond the mere calculation of private benefit to incorporate the dollar values of diverse public benefits of ecosystem services, Balmford et al. (2002) analyzed five case studies. Two of these concern tropical forest (Figure 9.15a,b). High-impact (and unsustainable) logging in Malaysia produces the highest private benefits (to the loggers), but after taking ecosystem services into account (nontimber forest products, flood protection, carbon accumulation in vegetation reducing atmospheric carbon dioxide and counteracting global warming – Chapter 11), the economic value of high-impact forestry is actually 14% lower than for sustainable, low-impact logging. In the case of Cameroon, conversion to plantation (oil palm and rubber) was compared with conversion to small-scale agriculture or retention of low-impact forestry. The value of all ecosystem services combined is highest under sustainable forestry while, at the opposite extreme, plantation conversion actually makes a net loss when both private benefit and ecosystem services were included in the analysis.

Analysis of a mangrove ecosystem in Thailand showed that the private benefit from shrimp farming shrinks almost to nothing when the economics take into

Fig. 9.14 (a) Fingerlings of two piscivorous fish stocked in Lake Mendota; the major biomanipulation effort started in 1987 (vertical dotted line). (b) Estimates of zooplankton biomass consumed by zooplanktivorous fish per unit area per day. The principal zooplanktivore fish were *Coregonus artedi*, *Perca flavescens* and *Morone chrysops*. Note that consumption of zooplankton is reduced because the piscivorous fish reduce densities of the zooplanktivorous fish. (c) Mean and range during summer of the maximum depth at which a Secchi disk is visible (a measure of water clarity); dotted vertical lines are for periods when the large and efficient grazer *Daphnia pulicaria* was dominant. This grazing zooplankton species was much more prominent after biomanipulation had allowed zooplankton to increase in density, and *D. pulicaria* plays a large role in reducing the density of phyto-plankton so that water clarity increases (Secchi disk visible at greater depth). (From Begon et al., 2006, after Lathrop et al., 2002.)

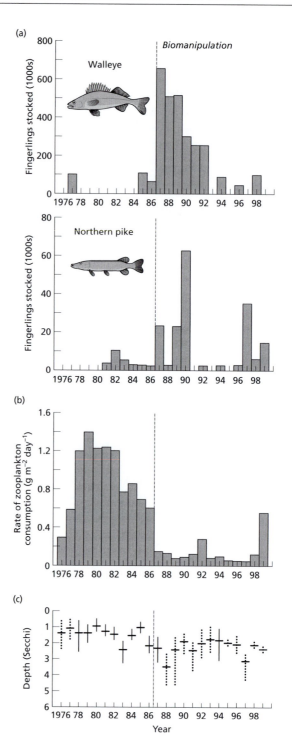

Fig. 9.15 The economic value of retaining or converting natural habitats expressed in US$ per hectare, each estimated over a 30- to 50-year period. In each case, the less intensive exploitation regimes are shown as light histograms and the heaviest exploitation as dark histograms. (a) Tropical forest in Malaysia; (b) tropical forest in Cameroon; (c) mangrove in Thailand; (d) coral reef in the Philippines; (e) wetland in Canada. (After Balmford et al., 2002.)

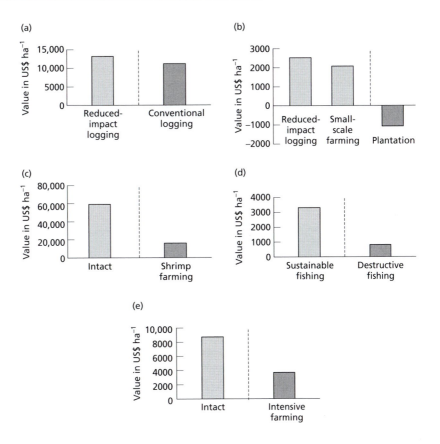

account the loss of ecosystem services from timber and nontimber products, charcoal, offshore fisheries and the storm protection associated with the intact ecosystem (Figure 9.15c). And the picture is similar for coral reefs in the Philippines. Despite high initial benefits from destructive fishing (involving dynamite), the combined benefits from sustainable fishing and tourism, when added to the reef's ecosystem service of coastal protection from storms, far outweigh any short-term gain (Figure 9.15d). Had there been more intact reefs around coasts hit by the 2004 Pacific-wide tsunami, who can say how many lives and livelihoods would have been saved (Section 9.1).

Finally, the draining of freshwater marshes for agriculture often produces private benefit (sometimes, as in this case, because of drainage subsidies provided by the government). However, ecosystem services from intact wetland include recreational opportunities (hunting, trapping and angling) and the uptake of nutrients as these move from land to rivers and beyond; when the dollar values of these are taken into account the overall economic value of intact wetland exceeds converted land by about 60% (Figure 9.15e).

Most often a number of species contribute to each ecosystem service and it is important to remember that some may make a more important contribution than others, something that managers need to recognize when assessing threats. I described earlier the contribution made to watermelon pollination by wild bees that

Fig. 9.16 Relative contributions of species to ecosystem services (species shown in order of the size of their contribution). (a) Cumulative watermelon pollination of 11 native bee species in California. (b) Cumulative contribution to carbon storage of more than 40 tree species in tropical rainforest at Chiapas in Mexico. (After Balvanera et al., 2005.)

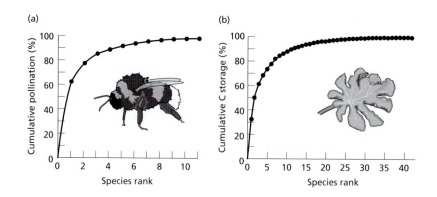

live in neighboring patches of intact forest (Section 4.5.4). In the case of organic watermelon farms close to intact forest, various species of native bee make a contribution to pollination – calculated as the product of bee abundance and pollen grains transported per individual. When ranked in order of contribution it becomes clear that just one species is responsible for more than 60% of the pollination ecosystem service (Figure 9.16a). Knowledge of its niche and habitat requirements would be paramount when countering any threats that may become apparent. Similarly, between them the forest trees of a protected Mexican rainforest store about 94 tonnes of carbon per hectare. But just 13% of the tree species contribute 90% of carbon storage (being more abundant and/or bigger and/or with denser wood) (Figure 9.16b). An emerging threat to one or more of these key species (perhaps a disease) could have a profound effect on the C-storage ecosystem service.

9.8.2 Ecosystem health of forests – with all their mites

Many ecosystems around the world have been degraded by human activities. Using an analogy with the human condition, managers frequently describe ecosystems as 'unhealthy' if their community structure (species richness, species composition, food web architecture), or their ecosystem functioning (productivity, nutrient dynamics, decomposition rate), has been fundamentally upset by human pressures. Aspects of ecosystem health are sometimes reflected directly in human health. These include nitrogen content in groundwater and thus drinking water, toxic algal outbreaks in lake and ocean, and transmission of Lyme disease in oak forest fragments. But any measure of ecosystem health must also reflect whether valued ecosystem services remain intact.

Management strategies can be usefully framed in terms of *pressure* (human action), *state* (resulting community structure and ecosystem functioning) and management *response* (Figure 9.17). Just as physicians use indicators in their assessment of human health (body temperature, blood pressure, etc.), ecosystem managers need ecosystem health indicators when prioritizing ecosystems for action and, just as important, to enable them to determine whether their interventions have succeeded.

The Ponderosa pine forest (*Pinus ponderosa*) of the western USA illustrates the relationship between pressure, state and response (Yazvenko & Rapport, 1997). A variety of human influences are at play but the most important *pressure* has been fire suppression. Ponderosa pine forest evolved in a situation where periodic fires occurred naturally but, with human occupation, attempts to suppress fire have caused the *state* of the forest to shift towards lower productivity and higher tree

Fig. 9.17 Links between *pressure* caused by human activities, *state* in terms of community composition and ecosystem processes, and *management* response to reduce the pressure. Adverse effects on ecosystems sometimes involve processes with clear value in human terms – ecosystem services such as recreational opportunities, water quality, natural flood control, harvestable wildlife and general biodiversity. (From Begon et al., 2006.)

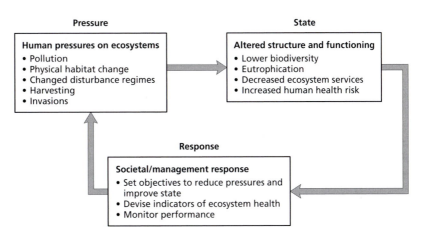

mortality, changed patterns of nutrient cycling, and more frequent outbreaks of tree pests and diseases. These properties may serve as indicators of ecosystem health – successful restoration (*response*) will be evident when the indicator trends reverse.

Ruf and Beck (2005) note the potential value of predatory soil mites as indicators of forest health. This is a very diverse group of organisms with, for example, 800 species in Europe in total, and up to 60 to be found at a given site. The predatory mites occupy a unique position in the soil food web, straddling the three pathways of energy and nutrients – from primary production (of plant roots), as well as fungal-based and bacteria-based decomposer chains (Figure 9.18). As such, they are well placed as integrators of the degrading effects of human pressures to serve as ecosystem health indicators. For example, characteristic sets of predatory mite species dominate the community at different levels of soil acidity.

9.8.3 *Ecosystem health in an agricultural landscape – bats have a ball*

Predatory mites could equally well serve as ecosystem health indicators in agricultural soils. But at a rather different scale, bats might be used to the same end (Wickramasinghe et al., 2003). Intensification of agriculture has placed increasing pressures on the agroecosystem, including unintended effects of fertilizers, herbicides and pesticides on nontarget species, as well as the loss of traditional farmland habitats such as hedges. Recognizing that organic farming prohibits the use of agrochemicals, and that organic farmers are more likely to leave hedges and riparian vegetation intact, bat activity and bat species richness were assessed in southern Britain in 24 carefully matched pairs of farms, one organic and the other conventional in each pair. There was no statistical difference in bat species richness, with a total of 14 of the 16 British bat species observed on organic farms compared to 11 on conventional farms. In most cases, however, the organic farm received more visits from bats and recorded more feeding activity (measured by acoustic surveys of feeding 'buzzes') than its conventional counterpart. Overall, bat numbers were 61% higher and feeding activity 84% higher on organic farms, probably reflecting the greater preponderance of hedges and other linear habitats along which bats fly and where their insect prey tend to be most common. In addition, agrochemical use on conventional farms may reduce abundance of both terrestrial insects and, through deterioration in water quality, of the adults of aquatic insects upon which bats prey.

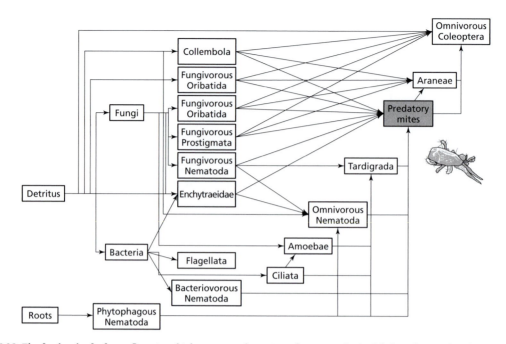

Fig. 9.18 The food web of a forest floor, in which energy and nutrients flow towards the left from living plant biomass (roots) and dead organic matter (detritus). Mites feature as fungivores/detritivores in the fungus-based chain (Oribatida and Prostigmata) as well as predatory mites. Other important soil organisms include nematodes, springtails (Collembola), pot worms (Enchytraeidae), flagellates, ciliates, amoebae, tardigrades, and the predatory spiders (Araneae) and omnivorous beetles (Coleoptera). Note how predatory mites are high on all three food chains – whether based on living plants (grazer system) or on detritus/fungus or detritus/bacteria (decomposer systems) (Box 9.1). (After Ruf & Beck, 2005, and Berg et al., 2001.)

9.8.4 *Ecosystem health of rivers – it's what we make it*

River health has been measured in a number of ways, from assessment of abiotic evidence of changed *state* (e.g. nutrient concentrations and sediment loads), through community composition (of fish, invertebrates or algae) to ecosystem functioning (such as rate of decomposition in rivers of leaves falling from overhanging vegetation). Some health indexes include more than one indicator, while in other cases managers rely on a single measure. In New Zealand, for example, river managers use the macroinvertebrate community index (MCI, Stark, 1993). Based on the presence or absence of certain types of river invertebrates that differ in their ability to tolerate pollution, healthy streams with abundant species that are intolerant of pollution attain high scores of 120 or more, whereas unhealthy streams have values as low as 80 or less. Land development in the river's catchment area is often the major *pressure*. Figure 9.19a shows the relationship for tributaries between MCI and the percentages of their catchment areas that have been developed for pasture or urban growth. The positive relationship between *pressure* and the *health index* gives confidence that the measure has value when identifying unhealthy rivers for remedial action (*response*) and for monitoring recovery.

It is worth remembering that the concept of ecosystem health is often, or partly, a social construct. A healthy ecosystem is one that humans believe to be healthy, and different groups may legitimately hold different ideas. Anglers may consider a

Fig. 9.19 Relationships between percentage development (for pasture and urban use) of the catchment area of tributaries of the Kakaunui River and (a) the Macroinvertebrate Community Index (MCI), commonly used by river managers in New Zealand, and (b) a Maori Cultural Stream Health Measure (CSHM). (After Townsend et al., 2004.)

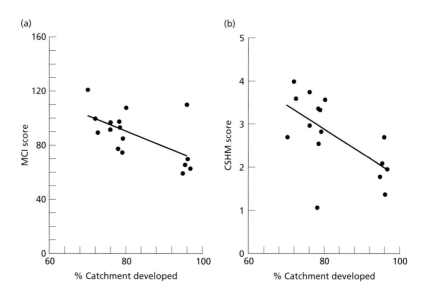

river healthy if it contains big fish of desired species; parents if their children do not get sick when swimming; conservationists if native species are well represented. The Kakaunui River in New Zealand is within the territory of a Maori group who developed an index that represented their perceptions of river health. Their Cultural Stream Health Measure (CSHM) includes components that reflect the perceived extent to which the surrounding catchment area, the riparian zone, the banks and the streambed have been impacted by humans. The CSHM (Figure 9.19b) is strongly correlated with the MCI despite the fact that it included no invertebrate component. Both indexes have merit in reflecting the effect of human *pressures* on the *state* of the ecosystem.

You have seen how particular taxonomic groups may lend themselves to use as ecosystem health indicators, including predatory soil mites, bats and stream invertebrates. A somewhat surprising alternative might be to use parasites as indicators. Marcogliese (2005) points out, for example, that the species richness of metazoan parasites of American eels (*Anguilla rostrate*) declines with level of acidification in North American streams (associated with acid rain caused by power generation). Moreover, the percentage infestation of another fish (spottail shiner – *Notropis hudsonius*) by myxozoan parasites increases below an urban effluent outflow. Parasites might prove good ecosystem health indicators, either because they are especially sensitive themselves to heavy metals or exotic chemicals, or because they have intermediate hosts that decline (or increase) as a result of the pressure concerned.

9.8.5 *Ecosystem health of a marine environment*

My final example comes from Tolo Harbor in northeast Hong Kong, an almost landlocked marine water body with an area of 52 km^2. The most significant physical and chemical *pressures* since the 1970s have been caused by increased urbanization (urban runoff and sewage), reservoir development (affecting patterns of freshwater input), industrialization (pollutants) and livestock rearing. Industrial pollutants and nutrient concentrations rose steeply from the early 1970s, more or less eradicating ecosystem services such as the provision of seafood and of clean water for swimming. Xu et al. (2004) have proposed a suite of ecosystem health indicators that encompass physical *state* (e.g. turbidity), chemical *state* (e.g. dissolved oxygen near the seabed), biological *state* (e.g. mangrove biomass, phytoplankton biomass, fecal coliform numbers) and the *state* of ecosystem services (e.g. seafood productivity and quality, feasibility of marine farming, safety of contact recreation). The management *response* has been the implementation of the Tolo Harbor Action Plan, which aimed to significantly reduce pollutants and excess nutrients entering the rivers that flow into Tolo Harbor. It is gratifying to note that the majority of state indicators have improved a little during the 1990s but there is still a way to go before regaining the full suite of ecosystem services that existed in 1970.

Summary

Food webs

The influence of a predator population on its prey, or of one competitor on another, is easy to envisage. But the influence of a species often spreads via a series of indirect effects throughout the food web. The food web is often the starting point for predicting the consequences of the arrival of an invader, for planning the removal of predators of an endangered species or for devising a pest control strategy. Managers need to be aware that some species (keystone) are much more influential in the food web than others.

Top-down and bottom-up control of food webs

A carnivore or parasite, through its influence on a herbivore, may reduce consumption of plants whose biomass increases as a result. Thus, plant biomass may sometimes be controlled *top down*. In other cases plant biomass is controlled *bottom up* by the supply of resources needed for photosynthesis – light, carbon dioxide, water and nutrients, such as phosphorus and nitrogen. Nuisance blooms of algae in lakes may be controlled by increasing predation on herbivores (top down) or by reducing the delivery of excess nutrients (bottom up) via improved sewage treatment or reduction of agricultural runoff.

Energy and nutrient dynamics in ecosystems

Plant communities are most productive where sunlight, water and nutrients are in good supply. Some of the energy locked in plant biomass passes along food chains to herbivores and their consumers – the grazer system. A proportion of plant biomass dies and moves as dead organic matter into the decomposer system, with its bacteria and fungi, detritivorous animals and their consumers. Living organisms extract chemicals from their environment, hold on to them for a while, then lose them again. Ecosystem ecologists are interested in how the biota moves chemical elements between the various living and nonliving compartments of the ecosystem. Certain species play key roles in determining patterns of energy and nutrient flux, and biosecurity managers need to pay particular attention to invaders that have a novel method of resource acquisition or that change the links between ecosystem compartments. Ecological stoichiometry is concerned with the relative availability of different elements – particularly the ratio of carbon to nitrogen. Stoichiometric knowledge can sometimes be used to manipulate soil carbon content to facilitate the restoration of a native plant community.

Human disease risk

Lyme disease (and other tick-transmitted diseases) is part of a complex food web involving ticks, small mammals, birds, reptiles, large mammals and oak trees. The transmission of malaria (and other diseases carried by mosquitoes) is similarly affected by a variety of species with which the mosquitoes interact. Based on an understanding of these interactions, managers can sometimes reduce disease risk.

Harvest management

Harvesting one species invariably affects others in the food web and it can be risky to treat a marine fishery as though the target species exists in isolation. In reality, any fishery has direct and indirect consequences, generally reducing the abundance of favored large fish and, through a reduction in top predators, allowing a proliferation of smaller species further down the food chain. By overfishing the largest and most desirable species, fishers increasingly rely on species from much lower down the food chain – known as 'fishing down marine food webs'.

Managing agriculture to minimize fertilizer input and nutrient loss

Intensive agriculture leads to pollution, by phosphorus or nitrogen, of the waterways that drain the land when there is a mismatch between the application of fertilizer and its use by the intended recipients (crop plants, plantation trees). Approaches to minimize fertilizer loss from the land include maintenance of ground cover of vegetation year-round, mixed cropping and careful return of organic matter to the soil. In addition, restoration of the vegetation zones between land and water (wetlands and riparian vegetation) can reduce inputs to the aquatic ecosystem because the plants and microorganisms remove some of the dissolved nutrient as it filters through the soil.

Ecosystem services and ecosystem health

Benefits accrue to people from natural ecosystems – *ecosystem provisioning services* such as wild food and forest products and *regulating services* such as high water quality and maintenance of soils. It is important for managers to be able to put a value on ecosystem services that are put at risk by human economic activities. Activities that are economic from the point of view of private benefit often turn out to be uneconomic when public benefits are also taken into account.

Many ecosystems have been degraded by human activities. Managers describe ecosystems as 'unhealthy' if their community structure or ecosystem functioning has been fundamentally upset. To help managers identify and prioritize ecosystems in trouble, easily measured indicators of 'ecosystem health' have been developed.

The final word

As a retired banker Aretha understands classical economics pretty well, but she was surprised to find that such a wide range of ecosystem services are now having dollar values placed on them. She was struck by the thought that often one ecosystem service has to be traded off against another. *'Harvesting abalone (a provisioning service) and conserving the sea otter (arguably a cultural service) are both great objectives, but they cannot be achieved in the same place at the same time. A balancing act is required. In the same vein, the negative effects on ecosystem functioning of exotic trout need to be balanced against the improved recreational opportunities they provide.'* Then again, she noted at least one 'win–win situation' – where two or more ecosystem services benefit from the same action. *'On a personal note, maybe I wouldn't have caught Lyme disease if larger tracts of forest and greater mammal diversity had been preserved – and that would have provided a conservation benefit too.'*

How many similar 'win–win situations' can you think of?

References

Allan, B.F., Keesing, F. & Ostfeld, R.S. (2003) Effect of forest fragmentation on Lyme disease risk. *Conservation Biology* 17, 267–272.

Arheimer, B. & Wittgren, H.B. (2002) Modelling nitrogen retention in potential wetlands at the catchment scale. *Ecological Engineering* 19, 63–80.

Balmford, A., Bruner, A., Cooper, P. & 16 others (2002) Economic reasons for conserving wild nature. *Science* 297, 950–953.

Balvanera, P., Kremen, C. & Martinez-Ramos, M. (2005) Applying community structure analysis to ecosystem function: examples from pollination and carbon storage. *Ecological Applications* 15, 360–375.

Begon, M., Townsend, C.R. & Harper, J.L. (2006) *Ecology: from individuals to ecosystems*, 4th edn. Blackwell Publishing, Oxford.

Berg, M., de Ruiter, P., Didden, W., Janssen, M., Schouten, T. & Verhoef, H. (2001) Community food web, decomposition and nitrogen mineralization in a stratified Scots pine forest soil. *Oikos* 94, 130–142.

Blumenthal, D.M., Jordan, N.R. & Russelle, M.P. (2003) Soil carbon addition controls weeds and facilitates prairie restoration. *Ecological Applications* 13, 605–615.

Briand, F. (1983) Environmental control of food web structure. *Ecology* 64, 253–263.

Conroy, J.D. & Culver, D.A. (2004) Do dreissenid mussels affect Lake Erie ecosystem stability processes? *American Midland Naturalist* 153, 20–32.

Corbin, J.D. & D'Antonio, C.M. (2004) Effects of exotic species on soil nitrogen cycling: implications for restoration. *Weed Technology* 18, 1464–1467.

Daufresne, T. & Loreau, M. (2001) Ecological stoichiometry, primary producer–decomposer interactions, and ecosystem persistence. *Ecology* 82, 3069–3082.

Elser, J.J. & Urabe, J. (1999) The stoichiometry of consumer-driven nutrient recycling: theory, observations, and consequences. *Ecology*, 80, 735–751.

Enloe, S.F., DiTomaso, J.M., Orloff, S.B. & Drake, D.J. (2004) Soil water dynamics differ among rangeland plant communities dominated by yellow star thistle (*Centaurea solstitialis*), annual grasses, or perennial grasses. *Weed Science* 52, 929–935.

Enriquez, S., Duarte, C.M. & Sand-Jensen, K. (1993) Patterns in decomposition rates among photosynthetic organisms: the importance of detritus C:N:P content. *Oecologia* 94, 457–471.

Fanshawe, S., VanBlaricom, G.R. & Shelly, A.A. (2003) Restored top carnivores as detriments to the performance of marine protected areas intended for fishery sustainability: a case study with red abalones and sea otters. *Conservation Biology* 17, 273–283.

Hinke, T., Kaplan, I.C., Aydin, K. et al. (2004) Visualizing the food-web effects of fishing for tunas in the Pacific Ocean. *Ecology and Society* 9(1), 10. Available at: http://www.ecologyandsociety.org/vol9/iss1/art10.

Hooks, C.R.R., Pandey, R.R. & Marshall, W.J. (2003) Impact of avian and arthropod predation on lepidopteran caterpillar densities and plant productivity in an ephemeral agroecosystem. *Ecological Entomology* 28, 522–532.

Huryn, A.D. (1998) Ecosystem level evidence for top-down and bottom-up control of production in a grassland stream system. *Oecologia*, 115, 173–183.

Jeppesen, E., Sondergaard, M., Jensen, J.P. & 27 others (2005) Lake responses to reduced nutrient loading – an analysis of contemporary long-term data from 35 case studies. *Freshwater Biology* 50, 1747–1771.

Jones, C.G., Ostfeld, R.S., Richard, M.P., Schauber, E.M. & Wolff, J.O. (1998) Chain reactions linking acorns to gypsy moth outbreaks and Lyme disease risk. *Science* 279, 1023–1026.

Juliano, S.A. & Lounibos, L.P. (2005) Ecology of invasive mosquitoes: effects on resident species and on human health. *Ecology Letters* 8, 558–574.

Lathrop, R.C., Johnson, B.M., Johnson, T.B. et al. (2002) Stocking piscivores to improve fishing and water clarity: a synthesis of the Lake Mendota biomanipulation project. *Freshwater Biology* 47, 2410–2424.

Likens, G.E. & Bormann, F.G. (1975) An experimental approach to New England landscapes. In: *Coupling of Land and Water Systems* (A.D. Hasler, ed.), pp. 7–30. Springer-Verlag, New York.

LoGiudice, K., Ostfeld, R.S., Schmidt, K.A. & Keesing, F. (2003) The ecology of infectious disease: effects of host diversity and community composition on Lyme disease risk. *Proceedings of the National Academy of Science, USA* 100, 567–571.

Marcogliese, D.J. (2005) Parasites of the superorganism: are they indicators of ecosystem health? *International Journal for Parasitology* 35, 705–716.

Mosier, A.R., Bleken, M.A., Chaiwanakupt, P. et al. (2002) Policy implications of human-accelerated nitrogen cycling. *Biogeochemistry* 57/58, 477–516.

Paine, R.T. (1966) Food web complexity and species diversity. *American Naturalist* 100, 65–75.

Pauly, D. & Palomares, M-L. (2005) Fishing down marine food web: it is far more pervasive than we thought. *Bulletin of Marine Science* 76, 197–211.

Power, M.E., Tilman, D., Estes, J.A. et al. (1996) Challenges in the quest for keystones. *Bioscience* 46, 609–620.

Pywell, R.F., Bullock, J.M., Walker, K.J., Coulson, S.J., Gregory, S.J. & Stevenson, M.J. (2004) Facilitating grassland diversification using the hemiparasitic plant *Rhinanthus minor*. *Journal of Applied Ecology* 41, 880–887.

Ramsey, D. & Veltman, C. (2005) Predicting the effects of perturbations on ecological communities: what can qualitative models offer? *Journal of Animal Ecology* 74, 905–916.

Ruf, A. & Beck, L. (2005) The use of predatory soil mites in ecological soil classification and assessment concepts, with perspectives for oribatid mites. *Ecotoxicology and Environmental Safety* 62, 290–299.

Schindler, D.E., Knapp, K.A. & Leavitt, P.R. (2001) Alteration of nutrient cycles and algal production resulting from fish introductions into mountain lakes. *Ecosystems* 4, 308–321.

Simon, K.S. & Townsend, C.R. (2003) The impacts of freshwater invaders at different levels of ecological organization, with emphasis on ecosystem consequences. *Freshwater Biology* 48, 982–994.

Simon, K.S, Townsend, C.R., Biggs, B.J.F., Bowden, W.B., & Frew, R.D. (2004) Habitat-specific nitrogen dynamics in New Zealand streams containing native or invasive fish. *Ecosystems* 8, 777–792.

Stark, J.D. (1993) Performance of the Macroinvertebrate Community Index: effects of sampling method, sample replication, water depth, current velocity, and substratum on index values. *New Zealand Journal of Marine and Freshwater Research* 27, 463–478.

Townsend, C.R. (2003) Individual, population, community and ecosystem consequences of a fish invader in New Zealand streams. *Conservation Biology* 17, 38–47.

Townsend, C.R., Tipa, G., Teirney, L.D. & Niyogi, D.K. (2004) Development of a tool to facilitate participation of Maori in the management of stream and river health. *Ecology and Health* 1, 184–195.

UNEP (2003) *Global Environmental Outlook Year Book 2003*. United Nations Environmental Program, GEO Section, PO Box 30552, Nairobi, Kenya.

Verhoeven, J.T.A., Arheimer, B., Yin, C. & Hefting, M.M. (2006) Regional and global concerns over wetlands and water quality. *Trends in Ecology and Evolution* 21, 96–103.

Wickramasinghe, L.P., Harris, S., Jones, G. & Vaughan, N. (2003) Bat activity and species richness on organic and conventional farms: impact of agricultural intensification. *Journal of Applied Ecology* 40, 984–993.

Xu, F.L., Lam, K.C., Zhao, Z.Y., Zhan, W., Chen, Y.D. & Tao, S. (2004) Marine coastal ecosystem health assessment: a case study of the Tolo Harbor, Hong Kong, China. *Ecological Modelling* 173, 355–370.

Yazvenko, S.B. & Rapport, D.J. (1997) The history of Ponderosa pine pathology: implications for management. *Journal of Forestry* 95, 16–20.

10 Landscape management

Ecological landscapes are patchy and contain a variety of habitats. Some ecological problems can be addressed at the scale of a single patch – a solitary agricultural field or marine reserve. But others require managers to focus at the much larger scale of multiple patches and habitats in regional landscapes and waterscapes.

Chapter contents

10.1	Introduction	262
10.2	Conservation of metapopulations	267
	10.2.1 The emu-wren – making the most of the conservation dollar	267
	10.2.2 The wood thrush – going down the sink	268
	10.2.3 The problem with large carnivores – connecting with grizzly bears	269
10.3	Landscape harvest management	270
	10.3.1 Marine protected areas	270
	10.3.2 A Peruvian forest successional mosaic – patching a living together	271
10.4	A landscape perspective on pest control	272
	10.4.1 Plantation forestry in the landscape	272
	10.4.2 Horticulture in the landscape	273
	10.4.3 Arable farming in the landscape	274
10.5	Restoration landscapes	274
	10.5.1 Reintroduction of vultures – what a carrion	275
	10.5.2 Restoring farmed habitat – styled for hares	276
	10.5.3 Old is good – willingness to pay for forest improvement	276
	10.5.4 Cityscape ecology – biodiversity in Berlin	277
10.6	Designing reserve networks for biodiversity conservation	277
	10.6.1 Complementarity – selecting reserves for fish biodiversity	279
	10.6.2 Irreplaceability – selecting reserves in the Cape Floristic Region	279
10.7	Multipurpose reserve design	280
	10.7.1 Marine zoning – an Italian job	280
	10.7.2 A marine zoning plan for New Zealand – gifts, gains and china shops	283
	10.7.3 Managing an agricultural landscape – a multidisciplinary endeavor	283

Key concepts

In this chapter you will

see that populations often exist as linked subpopulations in a metapopulation – conservation then requires attention to the size and number of subpopulations and interlinking corridors

realize that the effectiveness of a pest control program may be compromised by the make-up of the surrounding landscape matrix

---

appreciate that because species often use a variety of habitats, a landscape perspective may be
   needed for reintroduction programs
identify biodiversity hotspots with exceptional concentrations of endemic species undergoing
   exceptional loss of habitat
understand that regional species richness (gamma richness) consists of within-habitat richness
   (alpha) plus additional species added as further habitats are included (beta richness)
recognize the importance of social and economic dimensions of landscape management, particularly where multipurpose reserves are concerned

10.1 Introduction

Omar is usually objective about the documentary films he has to review as part of his course, but he became unusually absorbed in this one – 'The Great Dance: a hunter's story' (directed by Craig and Damon Foster). In the words of a Kalahari bushman 'tracking is like dancing because your body is happy'. The film takes us inside the hunters' minds, and we discover how they themselves 'put on' the mind of the animal they are tracking, how much they understand and respect their quarry, how they must help it die as quickly and painlessly as possible. The film recounts their four hunting techniques – the use of arrows tipped with poison, the tracking of cheetah who can be chased from freshly killed prey, the scaring of vultures in desperate times to feed from rotting flesh, and, most extraordinary of all – 'the hunt by running'. *'I could hardly believe that when the weather is really hot a small group of bushmen will run for hours after an antelope, never letting it settle. Remarkably, the animal becomes exhausted before the bushmen, who spear it and carry the prime meat back to the tribe.'* The film records one chase through desert scrub that lasts for 6 hours before the bushman's voice tells us '. . . but now they have run into the wildlife reserve. We must leave them, or else we'll be put in gaol.' Omar feels really angry that conservation managers have not found a way to accommodate a lifestyle that has lasted 30,000 years.

In previous chapters I have treated most environmental issues as if I was peering at them through a microscope, asking questions about the match between environment and organisms in relatively homogeneous and small-scale settings. I say 'relatively' because no environment, even when viewed with a 'microscope', is entirely homogeneous. The approach I have used is generally adequate when planning pest control in a farmer's field, designing a fishing regime in a particular marine area, conserving koalas in a particular forest, or setting aside a single nature reserve. But ecological problems are not always confined to a single habitat patch. The population of an endangered species may exist as a metapopulation of subpopulations in physically separate patches, but with some interchange of individuals among them (Section 10.2). The fish in a marine protected area are likely to contribute to the population dynamics of the fishery in a different way than fish at large in the sea (as already touched on in Chapter 7) (Section 10.3). The effectiveness of a pest control plan may be compromised by the nature of the surrounding countryside (Section 10.4). And the restoration or protection of maximum biodiversity demands a network of reserves that, between them, incorporate the most species (Sections 10.5, 10.6). Sometimes the aim is even more ambitious and complicated – to design a multipurpose reserve that simultaneously incorporates different objectives (Section 10.7), such as hunting and conservation in the Kalahari Desert. In all these cases, managers must put the microscope back in its box and take up a telescope – but let's call it, more aptly, a 'macroscope' (Brown, 1995). In other words, the scale of consideration needs to be increased to include a much larger landscape. Some important general principles of landscape ecology are described in Box 10.1 before I proceed to a systematic exploration of ecological applications at the landscape level.

Box 10.1 Landscape theory

Patches, patch dynamics and landscapes

Habitats are invariably patchy in the distribution of abiotic conditions, resources and enemies. This patchiness affects the performance of individual organisms, the distribution of populations and the composition of communities. Box 8.1 noted how communities, when viewed at a large enough scale, often consist of a mosaic of patches at different successional stages. The patch dynamics concept of communities views the habitat as a set of patches, with patches being disturbed and recolonized by individuals of various species. Landscape ecology takes an even broader view, being concerned with the way individuals, populations and communities behave in the patchwork of habitats that occur in a region.

Metapopulations

A population that exists as a discrete entity is made up of individuals distributed over an area of suitable habitat, where rates of birth and immigration add individuals to the population and rates of death and emigration subtract them. This is the simple view of population dynamics. Where a species exists as a metapopulation (a collection of subpopulations linked by dispersal), each subpopulation might also be considered, in isolation, to behave according to the simple view of dynamics. But, by contrast, the dynamics of the metapopulation as a whole are mainly determined by the rate of extinction of individual subpopulations, and the rate of colonization – by dispersal from existing subpopulations – of habitable but uninhabited patches, creating or restoring a subpopulation in that part of the landscape (Hanski, 1999). Note the intrinsic importance in metapopulation dynamics of dispersal and migration (dealt with in detail in Chapter 4). The metapopulation concept has particular relevance to metapopulation conservation and to harvest management.

Metacommunities and Island Biogeography Theory

So far, most of my discussion of applications from community ecology (Chapters 8 and 9) has treated the community as a more or less discrete entity. By analogy with metapopulations, however, if communities are viewed through the 'macroscope' a metacommunity may be discerned – a set of local communities that are linked by dispersal of multiple, potentially interacting species (Leibold et al., 2004). When looked at like this, you can see that the particular community that assembles in a patch of habitat is influenced by other habitat patches and other habitats in the vicinity. The metacommunity concept is relevant to harvest management, pest control and restoration and conservation of biodiversity.

The number of species in a community is a balance between the principal process that adds species (colonization from other communities in the region) and the principal process that subtracts species (extinction). Macarthur and Wilson's (1967) 'Island Biogeography Theory' has been very influential in ecology, and is equally applicable to oceanic islands and to 'habitat islands' such as lakes, mountain tops and forest patches. In essence, the theory holds that there is a continual turnover of species – because of species extinction and colonization events – but that a balance is achieved that mainly reflects two things (Figure 10.1). The first is *island size* – larger islands have a higher equilibrium number of species because of higher colonization rates (large islands are bigger 'targets') and lower extinction rates (large islands have larger average population sizes). The second is *isolation* from colonist sources – isolated islands have lower colonization rates and therefore lower equilibrium numbers of species. See Begon et al. (2006) for a more detailed treatment of Island Biogeography Theory.

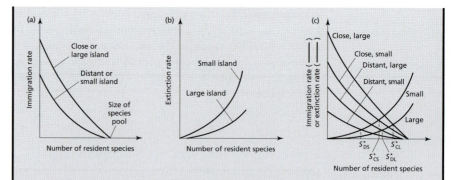

Fig. 10.1 Macarthur and Wilson's (1967) equilibrium theory of island biogeography. (a) The rate of species immigration on to an island, plotted against the number of resident species already there, for large and small islands and for close and distant islands. All immigration curves reach zero at the same point (when all members of the region's species pool have arrived), but the rate of arrival will be higher on islands close to the source of immigrants than on more remote islands, because colonizers have a greater chance of reaching an island the closer it is to the source. Immigration rates are also expected to be higher on larger islands because these provide a larger target for colonizers. (b) The rate of species extinction on an island, plotted against the number of resident species. The rate of species extinction will obviously be zero when there are no species there, and will generally be low when there are few species. But, as the number of resident species rises, the extinction rate increases at a more than proportionate rate. This is expected because with more species, competitive exclusion becomes more likely, and the population size of each species is on average smaller, making it more vulnerable to chance extinction. Extinction rates should be higher on small than on large islands because population sizes will typically be smaller on small islands. (c) The net effect of immigration and extinction can be seen when their two curves are superimposed. The equilibrium number of species where the curves cross (S^*) is the characteristic species richness for the island in question. Below S^*, richness increases (immigration rate exceeds extinction rate); above S^*, richness decreases (extinction exceeds immigration). C, close; D, distant; S, small; L, large. (From Begon et al., 2006.)

Island Biogeography Theory is relevant to biodiversity conservation because many conserved areas are surrounded by an 'ocean' of habitat made unsuitable by human activities. Three general points can be made:

1 Conservation managers often face the problem of whether to construct one large reserve or several small ones that add up to the same total area (sometimes referred to as the SLOSS (single large or several small) debate). If each of the small reserves supported the same species, then the theoretical relationship between island size and species richness suggests it would be preferable to construct a single large reserve in the expectation of conserving more species.

2 But if managers shift attention to the landscape perspective and find that the region as a whole is heterogeneous, each of the small reserves may support a different group of species and the total conserved in the small reserves is then likely to exceed that in a large reserve.

3 A key tenet of Island Biogeography Theory is that local extinctions are common events so that recolonization of habitat fragments is critical for the survival of metapopulations, and for species richness in the metacommunity as a whole. This means that particular attention must be paid to connectivity amongst fragments, and particularly the provision of dispersal corridors that increase the probability of recolonization after any local extinction.

What determines species richness?

In very general terms, the species that assemble to make up a community are determined by: (i) dispersal constraints; (ii) environmental constraints; and (iii) internal dynamics (Figure 10.2).

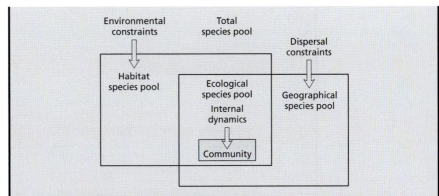

Fig. 10.2 Relationships among five types of species pools – the total pool of species in a region, the geographical pool (species able to arrive at a site), the habitat pool (species able to persist under the abiotic conditions of the site), the ecological pool (the overlapping set of species that can both arrive and persist) and the community (the pool that remains in the face of biotic interactions). (Adapted from Belyea & Lancaster, 1999, and Booth & Swanton, 2002.)

To be part of the community, in other words, a species has to be able to get there, must be able to persist in the face of abiotic conditions and find appropriate food resources (i.e. circumstances match its niche requirements – Chapter 2), and must win out in any biotic interactions with competitors, predators and parasites.

Let's explore what determines species richness by means of a conceptual model. For simplicity, assume that the resources available can be depicted as a one-dimensional continuum, R units long (Figure 10.3). The various species use the portion of this resource continuum corresponding to their *niche breadths* (n), and the average niche breadth is $\bar{n}$. Some of these niches will overlap, with an overlap o between adjacent species and an average niche overlap of $\bar{o}$

Now you can see why some communities might be richer in species than others. First, for any given values for niche breadth and overlap, a community will contain more species the greater the range of resources present (i.e. the larger the value of R – Figure 10.3a). Second, for a given range of resources R, more species can be fitted in to the community if $\bar{n}$ is smaller – in other words, if the species are more specialized in their use of resources (Figure 10.3b). Alternatively, more species can coexist if their niches overlap to a greater extent (greater $\bar{o}$ – Figure 10.3c). Finally, a community will contain more species the more fully saturated it is; conversely, it will contain fewer species when more of the resource continuum is unexploited (Figure 10.3d).

Some communities, most notably tropical forests, contain remarkably high species richness. A combination of circumstances seems responsible – very high productivity, a considerable range of resources, and a long evolutionary history producing saturated communities with large numbers of specialists. Such *biodiversity hotspots* around the world provide a particular focus for conservation action.

The partitioning of species richness at local and landscape scales
The total species richness of a region, known as gamma richness, is made up of two components, alpha and beta, that relate respectively to our small-scale, homogeneous view of communities and the larger-scale, landscape view of all local communities combined. Alpha richness is the number of species present within a particular habitat area, whereas beta richness refers to the extra species added when other habitat areas are included – thus, beta richness is the between-habitat component of overall regional richness. If every habitat area had identical species inventories, beta richness would be zero, and regional (gamma) richness would simply equal the within-habitat alpha diversity. But whenever there is heterogeneity in the distribution of species across habitats in a region, beta richness will make an important contribution to gamma richness.

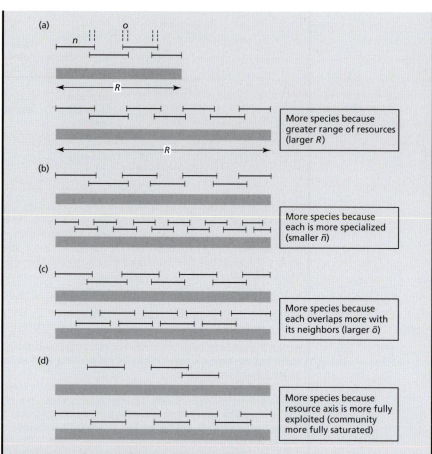

Fig. 10.3 A simple model of species richness. Each species utilizes a portion n of the available resources (R), overlapping with adjacent species by amount o. (From Begon et al., 2006, after MacArthur, 1972.)

Drawing the lines between alpha, beta and gamma richness is to some extent a matter of choice. In Gering's study of beetle species richness on trees in Ohio and Indiana, USA, alpha richness was calculated as the average number of beetles identified per site in Figure 10.4. Beta richness is then the difference between alpha richness and richness for the region as a whole (gamma). Beta richness itself had two components at increasing scales – the average beetle richness for each of three different sites within an ecoregion (ecoregions contrast in their glaciation history, topography and soil type), or the average for each of two ecoregions present in the region as a whole. Figure 10.4 shows how regional beetle species richness is partitioned between the local site scale (alpha), the between-sites scale (beta$_1$) and the between-ecoregions scale (beta$_2$). Evidently, designation of a nature reserve in a single site would safeguard fewer than 50% of the beetles. Addition of extra sites would bring the percentage up to about 70%, but addition of another ecoregion would be necessary to safeguard most species in the region.

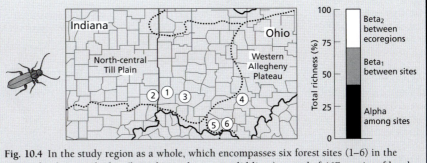

Fig. 10.4 In the study region as a whole, which encompasses six forest sites (1–6) in the states of Indiana and Ohio (boundaries shown as solid lines), a total of 467 species of beetles were identified in early summer. The histogram shows how this overall regional beetle species richness (gamma richness) is partitioned within habitats (alpha richness – average species richness per site) and between habitats in the region as a whole (beta_1 – the extra species added as extra forest sites are included, beta_2 – the further species added when two ecoregions are included (North-central Till Plain and Western Allegheny Plateau – boundaries shown as dotted lines)). (After Gering et al., 2003.)

10.2 Conservation of metapopulations

It is not unusual for endangered populations to exist as a series of subpopulations in a metapopulation. And even in cases where numbers have become so small that only a single 'subpopulation' remains, a successful first management step might well be to increase the size of the population so that it can be subdivided into a metapopulation, thus moving away from the all-eggs-in-one-basket syndrome. When it comes to predicting the persistence of metapopulations, the manager has to focus on three things – the size (and therefore vulnerability to extinction) of each subpopulation, the extent to which subpopulations are connected by emigration and immigration, and the relative favorability of the landscape within which the subpopulations are embedded. To illustrate the nature of the problem and the approaches managers may take, I present three case studies. The first concerns an endangered Australian bird and makes explicit the relationship between economics and population viability. Various management options are compared to determine not only the likely outcome for the bird but also to minimize the economic cost of achieving a viable population (Section 10.2.1). Then I turn to a North American bird in decline, and show how at a broad geographical scale managers can identify areas that deserve conservation effort and others that are lost causes. Finally, I compare the conservation gain of providing greater connectivity among existing grizzly bear reserves as opposed to simply enlarging the reserves.

10.2.1 *The emu-wren – making the most of the conservation dollar*

When a species exists as a metapopulation it is necessary to take into account the probabilities of local extinctions of subpopulations and of their recolonization by individuals from surviving subpopulations. With this in mind, Westphal et al. (2003) built a metapopulation simulation model for the critically endangered southern emu-wren (*Stipiturus malachurus intermedius*), following a procedure similar to those described in Section 5.4. The metapopulation in a South Australian upland area occurs in six remaining patches of dense swamp habitat (Figure 10.5). Emu-wrens are poor flyers and interpatch corridors of appropriate vegetation are likely

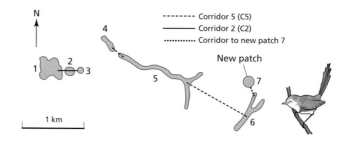

Fig. 10.5 The southern emu-wren metapopulation, showing size and location of patches (1–6 and new patch 7) and corridors (dashed, solid and dotted lines). Five different strategies were assessed in the study: Enlarge patch 2 (E2); Enlarge patch 5 (E5); Create new patch 7 plus corridor (dotted line) to patch 6 (E7); Create corridor (solid line) linking patch 2 with 1 and 3 (C2); Create corridor (dashed lines) linking patch 5 with 4 and 6 (C5). (After Westphal et al., 2003.)

to be important for metapopulation persistence. The management strategies evaluated were: (1) enlargement of existing patches; (2) linking patches via newly created corridors of suitable habitat; and (3) creating a new patch (see Figure 10.5). The economic 'cost' of the three strategies was standardized to be equivalent to 0.9 ha of newly vegetated area in each case.

The next step was to compare the effectiveness of the three different management actions, and also to compare among a variety of multi-step management scenarios – for example, first build a corridor from the largest patch to its neighbor, then, in the next time period, enlarge the largest patch, then create a new patch, and so on. (A process known as stochastic dynamic modeling was used for these analyses.) The aim was to find the strategy or scenario that reduced the 30-year extinction risk to the greatest extent.

It turns out that optimal metapopulation management decisions depend on the current state of the population. For example, if only the two smallest patches (2 and 3) are occupied, the optimal single action would be to enlarge one of them (patch 2; Strategy E2). However, when only the large patch 5 is occupied (which is more resistant to extinction because of the larger population it contains), connecting it to neighboring patches is optimal (strategy C5). The best of these fixed strategies reduced 30-year extinction probabilities by up to 30%. Even better, when chains of different actions were taken over successive time periods (e.g. strategy C5 followed by strategy E2, followed by strategy C2, and so on), the optimal scenario reduced extinction probabilities by 50–80% compared to no-management models.

These results hold a number of lessons for conservation managers. First, good decisions rely on knowledge of patch occupancy and understanding of extinction and recolonization rates. Second, the sequence of actions taken can be critical and this is where approaches such as stochastic dynamic modeling are of value (Clark & Mangel, 2000). Most important of all is the point that funds available for conservation will always be limited and tools such as these can help achieve the best returns from scarce resources.

10.2.2 *The wood thrush – going down the sink*

There is convincing evidence that certain bird species living on the edges of forest patches are more vulnerable to predation of eggs and young (by predatory mammals and birds) and nest parasitism (when birds such as the brown-headed cowbird, *Molothrus ater*, replace the host's eggs with their own). This pattern occurs because predators and brood parasites often enter the forest from surrounding agricultural habitats. Lloyd et al. (2005) estimated for wood thrushes (*Hylocichla mustelina*) how the rate of mortality (from predation and parasitism), and thus the population's fundamental rate of increase (λ; see Box 5.1), vary in forest patches of different sizes and in landscapes containing more or less agricultural land. Recall that a population will increase in size when $\lambda > 1.0$, and decrease when $\lambda < 1.0$ (Box 5.1). In this analysis of 30 landscapes across the USA, λ was often reduced below the 1.0 threshold by the combined effects of enemies, and this reduction is partly due to the degree of forest fragmentation (smaller fragments more likely to have $\lambda < 1.0$) and partly to the extent of agricultural land development in the surrounding landscape ($\lambda < 1.0$ when percentage forest cover in a 10 km radius is less than 80%). Taking all this into account, Lloyd's team produced a map of the USA showing areas where the populations are likely to decline to extinction and, more importantly, areas where wood thrush populations are currently safe (Figure 10.6). The populations in decline ($\lambda < 1.0$) can be thought of as 'sink' populations (dependent on recruits from other populations) while the safe populations ($\lambda > 1.0$) can be thought of as 'source' populations whose excess recruits help sink populations to persist. Areas where source populations predominate should be the focus of conservation efforts, because further declines here would be particularly damaging for the species as a whole. On the other hand, areas where sink populations predominate must either be targeted for remedial action or given up as lost causes.

10.2.3 *The problem with large carnivores – connecting with grizzly bears*

Wildlife reserves are islands in an ocean of more or less favorable habitat that becomes increasingly inhospitable with time because of the relentless encroachment of human-dominated landscapes. You saw in Box 10.1 how Island Biogeography Theory predicts smaller and more isolated reserves will lose species faster than larger and well-connected reserves. It is important to remember that large and

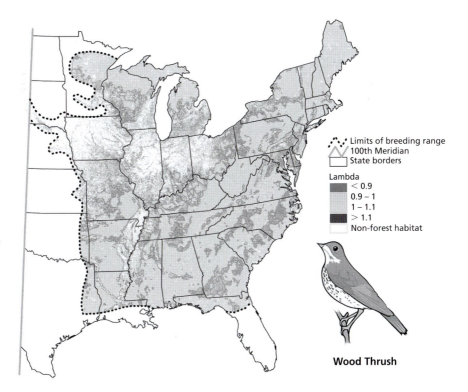

Fig. 10.6 Predicted population growth rate (λ) of wood thrush populations in relation to forest cover (colored cells) in the eastern USA. The extent of their breeding range is shown by dotted lines. Blue areas depict source populations that are currently safe ($\lambda > 1.0$), while pink areas show sink populations likely to go extinct ($\lambda < 1.0$). Safe populations are generally those in larger forest fragments and/or in landscapes containing only a small proportion of nonforest habitat. (After Lloyd et al., 2005.) (This figure also reproduced as color plate 10.6.)

Limits of breeding range
100th Meridian
State borders

Lambda
< 0.9
0.9 – 1
1 – 1.1
> 1.1
Non-forest habitat

Wood Thrush

long-lived species, such as grizzly bears (*Ursus arctos*) in the Rocky Mountains of North America, may persist for some time even after changes to the landscape around their reserves have already sealed their fate – extinction within a few generations. Thus, the particular value of studies such as that on the wood thrush (Section 10.2.2) and on grizzly bears (Carroll et al., 2005) lies in identifying probable future extinctions so that plans can be made in time. In the case of the grizzlies, Carroll's team showed that a doubling of reserve area produces a 47% increase in the probability of population persistence in highly 'developed' landscapes and a 57% increase in semideveloped landscapes. However, increasing connectivity between reserves is much more influential, increasing the probability of population persistence by 81% in undeveloped settings and by an impressive 350% in semideveloped parts of the species' range.

10.3 Landscape harvest management

Harvesting rarely takes place in a homogeneous landscape. In the ocean, some areas are heavily exploited while little or no fishing takes place in others. This is another situation where an understanding of both landscape (actually waterscape) structure and metapopulation dynamics is necessary when devising management strategies. Terrestrial landscapes can be even more patchy. Irregular disturbances produce a mixture of early successional, mid-successional and late-successional patches, each with different mixes of products that people can exploit. In this section I explore how a landscape perspective can be used to guide harvesting behavior in the ocean (Section 10.3.1) and on land (Section 10.3.2).

10.3.1 Marine protected areas

I noted in Section 7.2.4 how easy it is to imagine the potential benefits for harvest management of setting aside zero-fishing zones (marine protected areas). Thus, for species that are relatively sedentary and do not venture outside, individuals within the zero-fishing zone may be protected from fishing and achieve higher density. It is also possible that the unfished population will consist of larger and more fecund individuals, leading to a further elevation of density. If such a density increase leads to net migration from the zero-fishing zone to fished areas, then the marine protected area may contribute to sustainable exploitation of the fishery at large.

Despite the elegance of this reasoning, however, there have been very few cases where the potential benefits of marine protected areas have actually been shown to occur. And there are important gaps in our knowledge, particularly relating to patterns of dispersal of larval stages away from protected areas and towards fished zones. Robinson et al. (2005) have helped to bridge this gap by modeling the movement of 'virtual fish larvae' (i.e. theoretical 'particles' with appropriate properties) within the region as a whole. Their simulation models were run for 90 days (the average larval duration) and incorporated three-dimensional knowledge of ocean currents in late winter when high abundances of fish larvae are observed in the region. They tracked, through the ocean at large, fish larvae 'released' in each of ten coastal marine protected areas. You can see in Figure 10.7, first, that the different marine protected areas do not operate as isolated systems – each can contribute and receive fish recruits from one or more other areas. This metapopulation structure may help counter local extinctions and enhance genetic diversity (a component of biodiversity – Box 10.1). Second, there is clearly a major flow of recruits away from the protected areas to other areas of the coast and the open ocean, possibly contributing to the maintenance of exploited fish populations.

10.3.2 *A Peruvian forest successional mosaic – patching a living together*

Primary forest in tropical Peru has traditionally been cut down to make way for pasture and agriculture. After some years these patches are left fallow and begin their succession to secondary forest, but later they may be cleared again. The resulting successional mosaic (defined in Box 8.1) contains patches with different harvest values, and local people can weigh up the options of further clearing or of harvesting forest products. Gavin (2004) gathered ecological and sociological data from families with rights to a mixture of agricultural land, fallow fields, young secondary forest and old secondary forest patches in Peru's Cordillera Azul region.

Of all the successional stages, old secondary forest contained the largest number of species of useful plants and animals and also more species found at no other successional stage, including large-bodied mammals such as tapirs (*Tapirus terrestris*) and armadillos (*Priodontes maximus*). Fallow fields and young forests yielded mainly small-bodied birds and rodents, as well as fishes. Food (both animals and plants), wood and medicines were the most valuable commodities from all vegetation types, but their relative importance varied with successional stage (Table 10.1), and their availability, use and perceived value contrasted dramatically from family to family.

If farmers in this area want to maximize land value while maintaining access to all necessary forest products, Gavin suggests they should manage their forest-clearing and fallow cycles to encourage the greatest possible range of patches of different ages. The families will take into account also that the older the forest patch the more fertile it will be when cleared, but more effort has to be expended to clear larger trees. And it has to be borne in mind that median values for agricultural products on cleared land are much higher than for forest products (e.g. $167 ha^{-1} yr^{-1} for coffee). Gavin's work highlights the range of information needed for a systematic approach to landscape harvest management.

Fig. 10.7 Simulated spread on ocean currents along the north Pacific coast of Canada of particles introduced at a depth of 2 m to represent fish larvae, from each of ten marine protected areas (outlined in white) over a 90-day period. Particles from each protected area are shown in a unique color (see key). Grey areas represent Vancouver Island and the Canadian mainland. The continental shelf is shown in pink. NK, Naikoon; BISR, Banks Island sponge reef; BY, Byers; MBSR, Middle Bank sponge reef; LS, Laredo Sound; GIB, Goose Island Bank; HK, Hakai; SI, Scott Islands; QCS, Queen Charlotte Sound; BS, Blackfish Sound. (After Robinson et al., 2005.) (This figure also reproduced as color plate 10.7.)

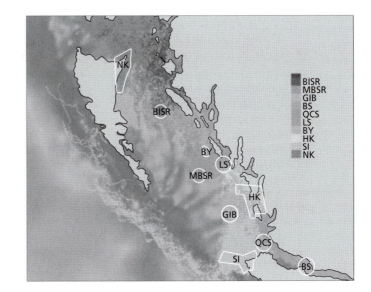

Table 10.1 Value of different harvest-use categories in three successional stages of tropical forest after clearing for agriculture. Results are for 67 households in the Cordillera Azul region of Peru. (After Gavin, 2004.)

Forest type	Use category	Number of species	Median value (US$ ha⁻¹yr⁻¹)	Maximum value (US$ ha⁻¹yr⁻¹)
Fallow fields	Total	45 animals, 138 plants	8.20	305.70
	Food	79	1.19	66.78
	Wood	28	0.00	281.63
	Medicine	67	0.10	41.90
	Weavings	10	0.00	21.50
	Adornments	23	0.00	7.66
Young forest	Total	49 animals, 109 plants	28.60	1034.80
	Food	78	2.40	121.06
	Wood	34	3.30	1000.90
	Medicine	40	0.00	88.50
	Weavings	12	0.00	110.82
	Adornments	14	0.00	5.12
Old forest	Total	81 animals, 143 plants	6.80	1183.00
	Food	115	1.02	301.84
	Wood	63	3.50	1167.20
	Medicine	48	0.10	48.20
	Weavings	21	0.20	15.50
	Adornments	47	0.00	2.79

10.4 A landscape perspective on pest control

As with other kinds of resource management, the effectiveness of pest control can be affected by the nature of the landscape in which it is carried out. In this section I discuss how the surrounding landscape matrix can augment or diminish a pest problem in forestry (Section 10.4.1), horticulture (Section 10.4.2) and arable settings (Section 10.4.3).

10.4.1 Plantation forestry in the landscape

As a result of rising demand for wood and pulp products, plantation forests have been expanding worldwide, matched by a growing concern about their vulnerability to pest damage. Being a monoculture, plantation forestry may be particularly susceptible to pest attack. You might predict that the presence of a variety of tree species in the vicinity of the plantation would foster a greater richness of natural enemies of the pests. And if this is the case, forest pest damage might be reduced. In other words, manipulation of tree species in the landscape may increase the likelihood of bringing together pests and their enemies in the metacommunity (Box 10.1).

One important pest of pine trees (*Pinus pinaster*) is the European stem borer *Dioryctria sylvestrella*, a moth caterpillar that induces trunk malformations and increases the likelihood of wind damage to the trees. The moths are attracted from a distance by a volatile chemical released from wounds caused by forestry practice (pruning). In their study of French pine plantations, Jactel et al. (2002) measured rates of borer infestation in stands of pine trees and found fewer attacks when the trees were bordered by a mixture of broad-leaved tree species (Figure 10.8a). As predicted, there was an increase in borer attack the greater the distance to the nearest broad-leaved stand, and this was matched by a decrease in parasitism of the pest by certain natural enemies (the parasitoid wasps *Macrocentrus sylvestrellae* and *Venturia robusta*) (Figure 10.8b,c). An understanding of how the configuration of

Fig. 10.8 In two French pine plantations the percentage of trees attacked by stem-boring caterpillars is lower (a) when individual pine stands are bordered by a mixture of broad-leaved trees and (b) when broad-leaved trees are closer to the pines in question. (c) Mirroring these patterns, parasitoid attacks on the stem borers are more prevalent when broad-leaved trees are nearby. Parasitoid activity was gauged by purposely causing standard damage in five pine trees in each stand to encourage stem borers to invade, and subsequently counting the number of parasitized stem borers present. (After Jactel et al., 2002.)

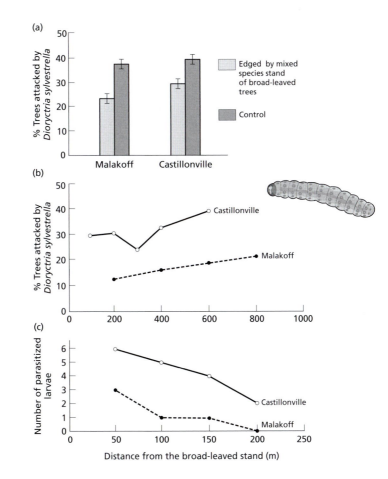

plantation and other trees in the landscape can reduce pine damage can be put to use when designing future plantings.

10.4.2 *Horticulture in the landscape*

An analogous study has involved Brussels sprouts in the Netherlands (Bianchi et al., 2005). Caterpillars of the moth *Mamestra brassicae* are significant pests of this crop but a variety of natural enemies can help depress pest populations. *M. brassicae* eggs are eaten by staphylinid and carabid beetles and parasitized by *Trichogramma* wasps, while the caterpillars are consumed by beetles and birds and parasitized by braconid wasps. To discover the relationship between landscape factors and pest damage, standard egg batches were placed in 42 fields of Brussels sprouts. Predation rates on eggs and caterpillars increased with the extent of woody habitat in the landscape as a whole, whereas parasitism rates increased with area of pasture. Once again you can see how landscape configuration may influence the probability that predators and parasitoids will co-occur with target pests in a metacommunity associated with a monoculture. Structurally complex landscapes with substantial amounts of woody habitat and pasture hold the most promise for sustainable pest control by natural enemies.

10.4.3 *Arable farming in the landscape*

While most people are unambiguous in their views about insect pests (get rid of them!) weeds are another kettle of fish. From the point of view of farm productivity the fewer weeds the better. But when it comes to biodiversity, farming practices that enhance the richness of noncrop plants (which farmers usually call weeds) have much to offer. Organically farmed wheat fields in Germany, for example, contain a greater richness of weeds per field (30–66 species) than conventionally farmed fields where herbicides are used routinely (13–40 species) (Roschewitz et al., 2005). Of more interest in the context of landscape management, however, is the finding that heterogeneity of land use around the wheat fields strongly influences the richness of weeds they contain. More complex landscapes have greater proportions of field margin, fallow land, grassland and gardens in the 1-km circle surrounding the wheat field, while simple, homogeneous landscapes have very large proportions of arable land (around 95%). It turns out that an increase in landscape complexity produces a greater gain in weed species richness in conventional than organic fields, to the extent that richness (alpha, beta and gamma – see Box 10.1) is similar in both farming systems when the surrounding landscape is complex (Figure 10.9).

10.5 Restoration landscapes

Many species use more than a single habitat. So when it comes to reintroducing species that were formerly common in an area (the vultures in Section 10.5.1), or modifying patterns of land management to benefit a species in decline (the hares in

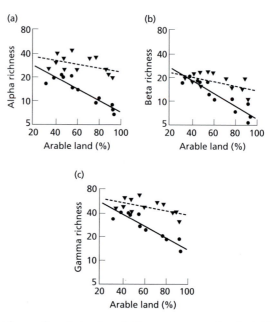

Fig. 10.9 Species richness of weeds in wheat fields, grown conventionally (with intensive use of fertilizers and pesticides – circles and solid lines) or organically (triangles and dashed lines), in relation to percentage arable land in the area surrounding the field (1 km radius). Wheat fields in the simplest landscapes are surrounded by about 95% arable land. Lower values for percentage arable land correspond to more complex landscapes with greater proportions of grassland, gardens, etc. (a) Alpha richness is the average number of weed species in four plots in each field. (b) Beta richness is the between-plot component of richness. (c) Gamma richness is the total richness of all four plots combined. (After Roschewitz et al., 2005.)

Section 10.5.2), you really need to know just how they make use of the landscape at large.

It is one thing to decide where to release vultures or how to manage farmland for the benefit of hares, but quite another to work out how to implement and finance the management recommendations. Such decisions will sometimes be made by government conservation agencies on the basis of expert advice and available funds, and without reference to public opinion or desires. In other situations, voluntary agreements with landowners may be sufficient. Where more persuasion is needed, government subsidies may be used to encourage people to act in appropriate ways. In further cases, sociological and economic instruments may be used to gauge the public's readiness to pay for conservation gains. Management of forest in Oregon, USA, provides a case in point (Section 10.5.3). Section 10.5.4 describes how even the landscapes of cities contain considerable biodiversity that is well worth protecting.

10.5.1 Reintroduction of vultures – what a carrion

Bearded vultures (*Gypaetus barbatus*), which became extinct in the European Alps a century ago, have been the focus of a reintroduction program since 1986. Captive-reared individuals were released at four widely dispersed locations in the Alps from where they spread to new areas. Hirzel et al. (2004) performed niche analyses (like those discussed in Box 2.1) based on sightings in the Valais region of Switzerland (not one of the original release sites). They contrasted the period 1987–94, when mainly immature birds were in a so-called 'prospecting' phase, with 1995–2001 when sightings were of subadults in a 'settling' phase (Figure 10.10). During the prospecting phase the most important variable explaining bearded vulture distribution was the biomass of ibex (*Capra ibex*), whose carcasses are an important resource. During the settling phase, by contrast, the presence of vultures correlated with the distribution of craggy limestone zones. These provide protected nest sites, ideal

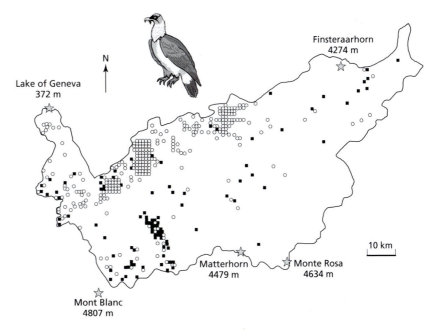

Fig. 10.10 Map of the study area in the Valais region of Switzerland, showing 1-km squares with bearded vulture sightings during the juvenile 'prospecting' phase (1987–94 – black squares) and the adult settling phase (1995–2001 – white circles). (After Hirzel et al., 2004.)

thermal conditions for soaring, and the finely structured limestone screes used for bone breaking and food storage; food abundance is now of secondary significance. The authors suggest that future reintroductions in the European Alps should recognize the diverse habitat requirements of adult vultures and be concentrated in large limestone massifs. It would be misleading to assume that a landscape that seemed ideal for prospecting juveniles would ensure success of the population as a whole.

10.5.2 *Restoring farmed habitat – styled for hares*

Sometimes the aim of land managers is to restore the landscape for the benefit of a particular species. The European hare *Lepus europaeus* provides an example. The hare's fundamental niche (Box 2.1) includes habitats created over the centuries by human activity. Hares are most common in farmed areas, but populations have declined where agriculture has become too intensive and the species is now protected in England and Wales. Vaughan et al. (2003) used a farm postal survey (1050 farmers responded) to investigate relationships between hare abundance and current land management. Their aim was to establish key features of the two most significant niche dimensions for hares, namely resource availability (crops eaten by hares) and habitat availability, and then propose management action to maintain and restore landscapes beneficial to the species.

Hares are more common on arable farms, especially on those growing wheat or beet, and where fallow land is present (areas not currently used for crops). They are less common on pasture farms, but the abundance of hares increases if 'improved' grass (ploughed, sown with a grass mixture and fertilized), some arable crops or woodland are present (Table 10.2). To increase the distribution and abundance of hares, the recommendations of Vaughan's team include the provision on all farms of year-round cover (from foxes *Vulpes vulpes*), the provision on pasture farms of woodland, improved grass and arable crops, and on arable farms of wheat, beet and fallow land.

10.5.3 *Old is good – willingness to pay for forest improvement*

A problem facing all conservation managers is the intangible nature of the value of biodiversity. Even though the populace may be broadly in favor of conservation, the absence of a publicly accepted 'value' of a management strategy makes its implementation a difficult process, especially where people are asked to vote in a State ballot. To circumvent this information gap, Garber-Yonts et al. (2004) mailed out surveys to several thousand Oregon taxpayers inviting respondents to assess a number of management options. Our focus here is on holistic forest restoration to regain a 'natural' balance of different age classes, from young stands to old growth. Those surveyed were informed about the current status of Oregon forest, and offered one of three choices, together with the estimated annual tax cost per household for each choice: status quo (only 5% of trees older than 150 years); management to achieve 33% old growth; or management to achieve 50% old growth forest (highest cost). Respondents indicated a higher willingness to pay for forest-age management than for any other option (such as conserving salmon habitat, endangered species protection or the establishment of biodiversity reserves). The analysis confirmed that Oregon people rate forest management particularly highly for wildlife management, recreation and the provision of clean water. Remarkably, households indicated a 'willingness to pay', on average, US$380 annually to increase the proportion of old trees from 5% to 35%.

Table 10.2 Habitat variables that determine the abundance of hares (estimated from the frequency of hare sightings) analyzed separately for arable and pasture farms. (After Vaughan et al., 2003.)

Variable	Variable description	Arable farms	Pasture farms
Wheat	Wheat *Triticum aestivum* (no, yes)	***	–
Barley	Barley (no, yes)	**	–
Cereal	Other cereals (no, yes)	NS	–
Spring	Any cereal grown in spring? (no, yes)	*	–
Maize	Maize *Zea mays* (no, yes)	NS	–
Rape	Oilseed rape *Brassica napus* (no, yes)	**	–
Legume	Peas/beans/clover *Trifolium* sp. (no, yes)	**	–
Linseed	Flax *Linum usitatissimum* (no, yes)	NS	–
Horticulture	Horticultural crops (no, yes)	NS	–
Beet	Beet *Beta vulgaris* (no, yes)	***	–
Arable	Arable crops present (see above; no, yes)	–	**
Grass	Grassland present (no, yes)	NS	–
Type grass	Ley (nonpermanent grassland), improved, semi-improved, unimproved grassland	NS	***
Fallow	Set aside as fallow land (no, yes)	***	–
Woods	Woodland/orchard (no, yes)	NS	*

–, Analysis was not performed for variables where fewer than 10% of farmers responded.
Variables that were significantly related to frequency of hare sightings by farmers are indicated as: $*p < 0.05$; $**p < 0.01$; $***p < 0.001$; NS, not significant.

10.5.4 *Cityscape ecology – biodiversity in Berlin*

When you think about biodiversity conservation, it is the relatively pristine settings of national parks and reserves that come readily to mind. However, agricultural and forestry landscapes can also contain a wealth of species. And even our big cities contain a biodiversity that can be fostered by retaining or restoring an appropriate mixture of habitat types.

Take Berlin as an example. The city center and the agricultural outskirts of Berlin have the smallest variety of land-use types, while the intermediate zones contain the most – including large parks, urban forests and wastelands, as well as the smaller-scale gardens, lawns, paths and street borders. Species richness of plants is roughly correlated with habitat diversity across the transect (Figure 10.11). The only exception is the city center where a large and species-rich wasteland remains after a former railway station was closed and its 'development' neglected because of the political division of the city after World War II. When attention is shifted to endangered plant species (such as *Oenothera parviflora* and *Arabis hirsute*), lawns harbor more than their fair share, with gardens not far behind. Particularly high richness of endangered and other plants is associated with areas having a long history of the same kind of management – such as residential areas built between 1920 and 1930. These deserve special protection.

10.6 Designing reserve networks for biodiversity conservation

Different areas vary dramatically in the number of species they contain, some having considerably more than others (hotspots of species richness). Areas also differ in the extent to which the biota is unique (hotspots of endemism) and the extent to which the biota is endangered (hotspots of extinction, for example because of imminent habitat destruction). Figure 10.12 shows the distribution across the face of the globe of some significant biodiversity hotspots (numbers of globally threatened species of birds plus amphibians). At this scale, the identification of global hotspots

Fig. 10.11 Map showing study areas along a transect through the city of Berlin, Germany. The total number of plant species identified in each area is generally correlated with the number of habitat types (such as wasteland, park, garden, street border) with the exception of the central city, whose high biodiversity is mainly related to a single large wasteland left undeveloped after World War II. (After Zerbe et al., 2003.)

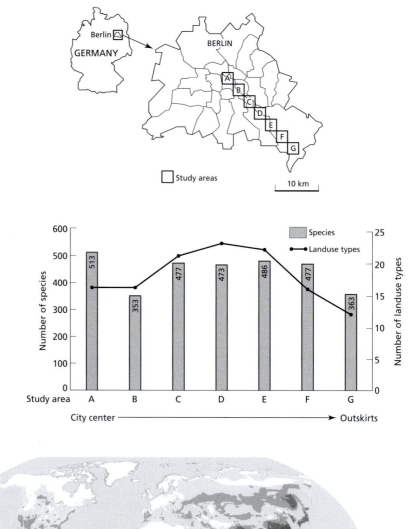

Fig. 10.12 Distribution of biodiversity hotspots, showing numbers of species of globally threatened birds plus amphibians mapped on an equal area basis (each grid cell is 3113 km^2). (After Rodrigues et al., 2006.) (This figure also reproduced as color plate 10.12.)

can guide international conservation efforts, particularly where the host country itself has few resources to devote to conservation.

At a national or regional scale, the aim of conservation is to represent the biota in a way that separates biodiversity from the processes that threaten it. Margules and Pressey (2000) recommend the following steps for systematic conservation planning:

1 *Compile data* on biodiversity and on the distribution of rare and endangered species in the planning region.

2 *Identify conservation goals* and set explicit targets for species and ecosystem types.

3 *Review existing conservation areas* to measure the extent to which quantitative goals have already been achieved.

4 *Select additional conservation areas* to augment existing reserves in a way that best achieves the conservation goals.

5 *Implement conservation actions* having decided the most appropriate form of management for each area and established an implementation timetable if resources are not available for all actions to be carried out at once.

6 *Maintain the required values of conservation areas* and monitor key indicators that will reflect management success, modifying management as required.

When selecting new biodiversity reserves (step 4), two key principles can be taken into account – *complementarity* (Section 10.6.1) and *irreplaceability* (Section 10.6.2).

10.6.1
Complementarity – selecting reserves for fish biodiversity

The basic approach in complementarity selection is to assess the biodiversity content of candidate areas and to proceed in a stepwise fashion, selecting at each step the site that is most *complementary* to those already selected in terms of the biodiversity it contains. My example concerns coastal marine fishes in Western Australia, with its extensive coastline from the tropical north to the temperate south. The results of a complementarity analysis showed that more than 95% of the total of 1855 species could be represented in just six, appropriately located, 100-km-long sections of the coast (see stars in Figure 10.13).

10.6.2
Irreplaceability – selecting reserves in the Cape Floristic Region

An approach that contrasts subtly with complementarity analysis concerns the 'irreplaceability' of each candidate area. *Irreplaceability* is defined as the likelihood of an area being required to achieve conservation targets or, conversely, the likelihood of one or more targets not being achieved if the area is not included. Cowling et al. (2003) used irreplaceability analysis as part of their conservation plan for South Africa's Cape Floristic Province. With more than 9000 species of plants this is without doubt a global hotspot.

Cowling's team followed the conservation planning steps listed in Section 10.6. Step 1, the compiling of data, is well advanced in this important region. Their goal (step 2) is that *'the natural environment and biodiversity of the region will be effectively conserved, restored wherever appropriate, and will deliver significant benefits to the people of the region in a way that is embraced by local communities, endorsed by government and recognized internationally'*. Note the emphasis placed on the acceptance of the plan by local people. The involvement of landowners and local communities was fostered by initiatives ranging from education to membership of the steering committee for the conservation plan. As part of step 2, the team identified a variety

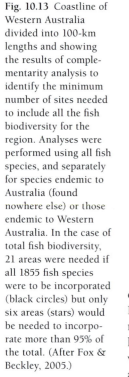

Fig. 10.13 Coastline of Western Australia divided into 100-km lengths and showing the results of complementarity analysis to identify the minimum number of sites needed to include all the fish biodiversity for the region. Analyses were performed using all fish species, and separately for species endemic to Australia (found nowhere else) or those endemic to Western Australia. In the case of total fish biodiversity, 21 areas were needed if all 1855 fish species were to be incorporated (black circles) but only six areas (stars) would be needed to incorporate more than 95% of the total. (After Fox & Beckley, 2005.)

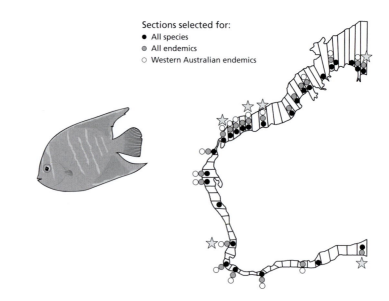

of targets including, among others, the minimum acceptable number of species of Protacea plants to be safeguarded (for which the region is famous), the minimum number of ecosystem types and even the minimum number of individuals (or populations) of large mammal species. At step 3 they assessed the extent to which targets were being met by the current set of nature reserves and then at step 4 they used an *irreplaceability* approach to guide the choice of areas to add to existing reserves that would best achieve the conservation targets (Figure 10.14). This is an excellent example of a systematic approach to conserving biodiversity in one of the world's most diverse biological regions. The ambitious aim is to achieve their goal by 2020.

10.7 Multipurpose reserve design

The design and implementation of a plan to achieve a single aim is problematic enough. Imagine how much more difficult it is to cater to the diverse goals of different groups in the community. To illustrate the problems, and the means to solve them, I discuss examples involving exploitation, recreation and conservation goals in marine settings (Section 10.7.1, 10.7.2) and agricultural production, water quality and conservation goals in a terrestrial landscape (Section 10.7.3).

10.7.1 *Marine zoning – an Italian job*

Villa et al. (2002) used a systematic approach to design one of the first marine zoning plans in Italy. They took pains to involve all the different interest groups (fishing, recreation, conservation) in defining priority areas for different uses and degrees of protection. Italian law recognizes reserves with three levels of protection: 'integral' reserves (only available for research), 'general' reserves and the less restrictive 'partial' reserves. Villa's team recognized the need to split 'integral' reserves into two categories: no-entry, no-take zones (where only nondestructive research is permitted) and public entry, no-take zones that allow visitors a full experience of the reserve, other than fishing. Permitted activities for the four categories are shown in Table 10.3.

The next step was to produce maps of 27 factors important to one or more of the interest groups. These included pollution status, fish diversity, fish nursery areas,

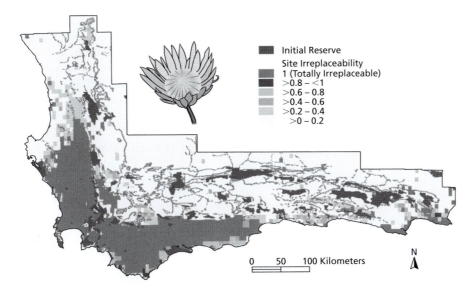

Initial Reserve
Site Irreplaceability
1 (Totally Irreplaceable)
>0.8 – <1
>0.6 – 0.8
>0.4 – 0.6
>0.2 – 0.4
>0 – 0.2

0   50   100 Kilometers

N

Fig. 10.14 Map of South Africa's Cape Floristic Region showing site *irreplaceability* values for achieving a range of conservation targets in the 20-year conservation plan for the region. Irreplaceability is a measure, varying from zero to one, which indicates the relative importance of an area for the achievement of regional conservation targets. Existing reserves are shown in blue. (After Cowling et al., 2003.) (This figure also reproduced as color plate 10.14.)

Table 10.3 Activities permitted or prohibited for each of four planned levels of protection (from left to right in order of decreasing protection) for the Asinara Island National Marine Reserve of Italy. (After Villa et al., 2002.)

Category	Activity	No-take, no-entry	Entry, no-take	General reserve	Partial reserve
Research	Nondestructive research	Aa	Aa	A	A
Sea access	Sailing	P	L	A	A
	Motor boating	P	P	L	L
	Swimming	P	P	A	A
Staying	Anchorage	P	P	L	L
	Mooring	P	L	Aa	A
Recreation	Snorkeling/scuba	P	L	Aa	A
	Guided tours	P	L	Aa	A
	Recreational fishing	P	P	L	A
Exploitation	Traditional subsistence	P	P	L	L
	Sport	P	P	P	L
	Scuba	P	P	P	P
	Commercial fishing	P	P	P	P

Key to activities: A, allowed without authorization; Aa, allowed upon authorization; L, subject to specific limitations; P, prohibited.

sites used by life-history stages of key species (e.g. limpets, sea mammals, marine birds), archaeological interest, suitability for various forms of fishing (e.g. traditional subsistence, commercial), and suitability for various recreational (e.g. snorkeling, whale watching) and tourism activities. The relative importance of each factor for each interest group was determined in planning sessions. Taking all the results into account, five higher-level maps were produced: natural value of the marine environment (NVM – aggregating all interest groups' values related to biodiversity, rarity, crucial habitats such as nursery areas; Figure 10.15a), natural value of the coastal environment (NVC – endemic coastal species including seabirds, habitat suitable

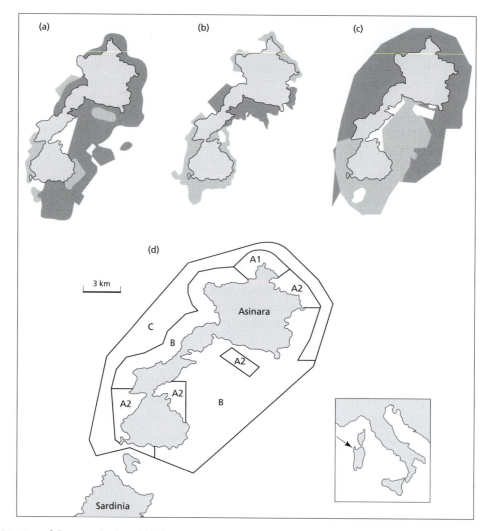

Fig. 10.15 Maps of the natural value of (a) the marine environment (NVM), (b) the coastal environment (NVC) and (c) recreational activity value (RAV) for areas around Asinara Island (island land area is shown in the center in grey). Lighter shades represent higher value. (d) Final zoning plan for the Asinara Island National Marine Reserve. A1: no-entry, no-take (the most restrictive category); A2: entry, no-take; B: general reserve; C: partial reserve (the least restrictive category). The inset map shows the location of the reserve in relation to the mainland of Italy. (From Begon et al., 2006, after Villa et al., 2002.)

for reintroduction of turtles and seals; Figure 10.15b), recreational activity value (RAV – aggregating values for all recreational activities; Figure 10.15c), as well as commercial resource value (aggregating traditional fishing sites plus other suitable areas) and ease of access value (aggregating marine access routes and harbors).

The final stage was to produce a zoning plan that provides, as far as possible, for the sustainable desires of all interest groups. Villa's team tried to avoid complex zoning that would make management and enforcement difficult. They also paid careful attention to the views of the various groups to minimize any remaining conflicts. The final plan (Figure 10.15d) has: a single no-entry, no-take zone (reflecting biological importance and relative remoteness), four entry, no-take zones to protect specific values such as endangered species (reflecting biological value but with easy access); two general reserve zones (to protect sensitive benthic assemblages, such as seagrass meadows that suffer little from permitted activities; Table 10.3); and one partial reserve zone as a buffer for adjacent reserve zones (in an area where traditional fishing practices are compatible with conservation). The plan also includes three channels that provide maximum boat access where ecological disturbance will be minimal.

10.7.2 *A marine zoning plan for New Zealand – gifts, gains and china shops*

An analogous approach has been taken to develop a management plan for the extensive Fiordland region (Section 4 in Box 1.2) in the southwest of New Zealand (Teirney, 2003). This was an entirely bottom-up effort by the local community (with no top-down direction by government agencies), which took 8 years from first meetings to the publication of a comprehensive plan. The diverse groups worked face-to-face from the beginning. The *Guardians of Fiordland's Fisheries and Marine Environment* comprise Maori, recreational and commercial fishers, tourism operators, marine scientists and environmentalists. While challenging to manage (a skilled facilitator was involved), this approach provides a model for minimizing conflict, stimulating reciprocal learning, and formulating objectives for sustainable ecosystem use that have proved difficult to achieve by top-down means.

A significant feature of the proposal was the concept of *gifts* and *gains* by the various groups. Thus the plan called for new fishing behavior: a reduction in bag limits for recreational fishers, the withdrawal of commercial fishers from the inner fiords, and a voluntary suspension of certain customary fishing rights by Maori. In addition, a number of marine reserves and protected areas were identified to protect representative ecosystems and *china shops* – areas with outstanding but vulnerable natural values. These gains in sustainability and conservation were balanced by the gift from environmentalists to refrain from pursuing their original goal of a much more extensive marine reserve program. As a result, the plan does not represent either extreme of the resource-use spectrum – preserve everything *or* exploit as a free for all. Instead a sustainable middle ground was identified, with the Maori concept of *kaitiakitanga* or guardianship at its root. The New Zealand government agreed to implement the plan in its entirety and has passed the new legislation necessary.

10.7.3 *Managing an agricultural landscape – a multidisciplinary endeavor*

Agricultural landscapes can harbor a wealth of biodiversity. But when farm production becomes too intensive and widespread, both biodiversity and ecosystem services suffer. Biodiversity is reduced because of the loss of species-rich habitat remnants and the impact of high levels of pesticides. At the same time there is an adverse effect on the provision of water of high quality for drinking and contact recreation. These ecosystem services, normally provided 'free' from a healthy landscape, can be lost because of the input of large quantities of nitrogen and phosphorus (from fertilizers applied to the land), fine sediment (from eroding land) and an increase in water-borne pathogens from farm animals that affect humans (such as the *Giardia* parasite). In addition, intensive farming may be associated with higher flooding probabilities because of the loss of vegetation that recirculates water from the soil after storm events. Flood protection is another ecosystem service associated with a healthy landscape.

The impact of agriculture depends on how much of the landscape is used for production. A single small farm – even one involving the excessive use of plough, fertilizer and pesticide – will have little effect on biodiversity and water quality in the landscape as a whole. It is the cumulative effect of larger and larger areas of intensive agriculture within a river's catchment area that depletes the region's biodiversity and reduces the quality of the water needed for other human activities. For this reason, management of agricultural landscapes needs to be carried out on a large scale. Management must also involve a wide range of disciplines, from farm production and economics to biology, chemistry and public health. This ideal approach has proved difficult to achieve.

Santelmann et al. (2004) show the way forward. They incorporated knowledge of experts in the various fields into alternative visions of a particular landscape – the catchment area of Walnut Creek in an intensively farmed part of Iowa, USA. They mapped the present pattern of land use and also created three scenarios of the way the area might look in 25 years if particular strategies are followed. They also assessed how farm income, water quality and biodiversity would be expected to change according to the three scenarios. A *production* scenario imagines what the catchment will be like if continued priority is given to corn and soybean production ('row' crops), following a policy that encourages extension of cultivation to all highly productive soils in the catchment. A *water quality* scenario envisions the catchment under a new (hypothetical) federal policy that enforces chemical standards for river and ground water, and supports agricultural practices that reduce soil erosion and flooding. Finally, a *biodiversity* scenario assumes a new (hypothetical) federal policy to increase the abundance and richness of native plants and animals. In this scenario, a network of biodiversity reserves is established with connecting habitat corridors that include the riparian (bankside) zones of rivers.

Figure 10.16 compares for the three scenarios the distribution of agricultural and 'natural' habitats in 25 years time. Not surprisingly, the Production scenario produces the most homogeneous landscape. Compared with the current situation, there is an increase in row crops (corn and soybean) at the expense of less profitable pasture and forage crops. Note that pasture and forage crops provide year-round *perennial* ground cover that is conducive to both higher water quality and biodiversity. The Water Quality scenario leads to more extensive riparian strips of natural vegetation cover and more perennial cover in total. Finally, the Biodiversity scenario has even wider riparian strips as well as prairie, forest and wetland reserves, and an increase in strip intercropping, a farming practice that is more sensitive to biodiversity because it helps increase connectivity between reserves.

The percentage change for each scenario, compared to the current situation, in economic, water quality and biodiversity terms is shown in Figure 10.17. It is hardly

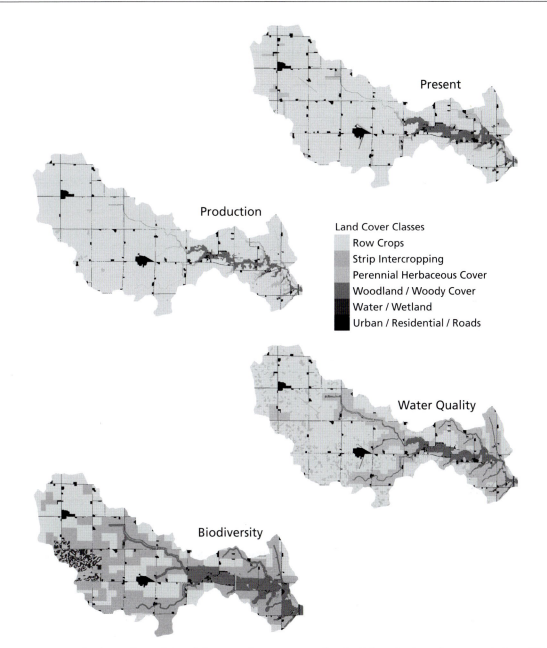

Fig. 10.16 Present landscape (top right) and alternative future scenarios for the Walnut Creek catchment area in Iowa, USA. In comparison to the present situation, note the increase in row crops at the expense of perennial cover in the Production scenario. In the Water Quality scenario, note in particular the increase in perennial cover (pasture and forage crops) and wider riparian buffers. Finally in the Biodiversity scenario, note the increase in strip intercropping, the wide riparian buffers and the extensive prairie, forest and wetland restoration reserves. (From Santelmann et al., 2004.) (This figure also reproduced as color plate 10.16.)

Fig. 10.17 Percent change in the Walnut Creek Catchment Area for each scenario (Production, Water Quality and Biodiversity, compared to the current situation) in water quality measures (sediment, nitrate concentration), an economic measure (farm income in the catchment as a whole), a measure of farmer preference for each scenario (based on farmer ratings of images of what the land cover would look like under each scenario) and two biodiversity measures (plant and vertebrate). The Biodiversity scenario ranks consistently above the Production scenario, and the Water Quality scenario ranks above the Production scenario in all but economic profitability. (After Santelmann et al., 2004.)

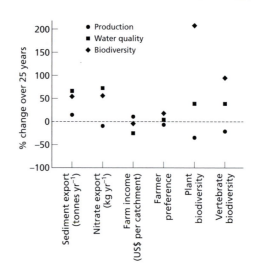

surprising that the Biodiversity scenario ranks highest for improvements in plant and animal biodiversity. Much more surprising is the finding that the land-use and management practices required by the Biodiversity scenario are nearly as profitable as current practices. The Biodiversity scenario also ranks highest in terms of acceptability to farmers (based on farmer ratings of images of what the land cover would look like under each scenario), and provides water quality improvements similar in magnitude to those in the Water Quality scenario. It seems that despite the slightly higher profitability of the Production scenario, farmers would not be unhappy with a Biodiversity strategy that provides the greatest benefits to the community at large in terms of biodiversity and ecosystem services.

Summary

Landscapes, metapopulations and metacommunities

Landscape ecology takes a broad view, dealing with the way individuals, populations and communities behave in the patchwork of habitats that occur in a region.

Where a species exists as a metapopulation (a collection of subpopulations linked by dispersal), its dynamics are strongly influenced by the rate of extinction of individual subpopulations, and the rate of colonization – by dispersal from existing subpopulations – of habitable but uninhabited patches. It is not unusual for an endangered species to exist as a series of subpopulations in a metapopulation. And even where only a single 'subpopulation' remains, the first management step may be to increase the size of the population so that it can be subdivided into a metapopulation, moving away from the all-eggs-in-one-basket syndrome.

By analogy with metapopulations, when the whole community of species is viewed at the landscape scale, a metacommunity may be discerned – a set of local communities that are linked by dispersal of multiple, potentially interacting species. 'Island Biogeography Theory', equally applicable to oceanic islands and to 'habitat islands', holds that there is a continual turnover of species – because of species extinction and colonization events – but that a balance is achieved. Larger islands tend to have a higher equilibrium number of species, while more isolated islands have a lower number. These patterns provide clues for the design of nature reserves, in terms of the importance of reserve size and connectivity among reserves in a network.

Species richness and biodiversity

Species richness is simply the total number of species present in a defined area. Species richness is a key component of *biodiversity*, but the latter can also be viewed at scales smaller (genetic diversity within species) and larger (the variety of ecosystem types present) than the species. Some communities (e.g. tropical forests) contain remarkably high species richness, reflecting high productivity, a considerable range of resources and a long evolutionary history producing saturated communities with large numbers of specialists.

Landscape harvest management

Harvesting rarely takes place in a homogeneous landscape so an understanding of landscape (or waterscape) structure and metapopulation dynamics is necessary when devising management strategies. Marine protected areas may potentially reduce the likelihood of overexploitation of a fishery. Whether they do so depends, among other things, on patterns of dispersal of larval stages away from protected areas and towards fished zones. Harvest management in forests must also take heterogeneity into account, depending on knowledge of the successional mosaic that exists, and the contrasting harvest values (wood products, bushmeat, medicinal plants) in different patches.

Pest control in the landscape

The effectiveness of pest control (whether in forestry, horticulture or pasture settings) can be affected by the nature of the landscape in which it is carried out. A greater variety of tree species or land uses in a production landscape may provide refuges for natural enemies of forest or crop pests. Thus, the details of landscape configuration can influence the probability that predators and parasitoids will co-occur with target pests in a metacommunity. From one perspective crop weeds are undesirable but, in terms of biodiversity, farming practices that enhance the richness of noncrop plants have much to offer. It turns out that heterogeneity of land use around arable fields can strongly influence the richness of weeds they contain.

Restoration landscapes

Many species use more than a single habitat, so reintroduction or protection of threatened species requires knowledge of how they make use of the landscape at large. When considering biodiversity in general, it is surprising how many species occur even in modern cityscapes. However, some aspects of landscape heterogeneity in cities are more crucial than others in determining species richness.

Designing reserve networks for biodiversity conservation or for multiple goals

Different parts of the globe vary dramatically in their species richness, and the extent to which the biota is unique or endangered. At this global scale, the identification of 'global hotspots' can guide international conservation efforts. At a national or regional scale, the steps to designing a reserve network are to compile data on biodiversity, identify conservation goals, select additional reserves to augment existing ones in a way that best achieves the conservation goals, implement conservation actions, and monitor key indicators that will reflect management success. There are two approaches when selecting new biodiversity reserves. *Complementarity* selection involves proceeding in a stepwise fashion, selecting at each step the site that is most complementary to those already selected in terms of the biodiversity it contains. *Irreplaceability* selection involves ranking potential sites in terms of the likelihood they will be required to achieve conservation targets or, conversely, the likelihood of one or more targets not being achieved if the area is not included.

It is even more difficult to design and implement plans to achieve multiple goals – sustainable exploitation, recreational use and biodiversity conservation. Arguably the most effective approach is to involve, from the outset, all the different interest groups in the planning process.

The final word

Given his choice of career as a film-maker, Omar appreciated the analogy of employing a 'macroscope' to take in the big picture when dealing with landscape issues. '*This makes very clear how mistakes might be made if managers work at too fine a scale.*' Omar also noted that the most problematic management scenarios of all are those that attempt to achieve multiple, and often conflicting, goals. '*But for me, these studies provided the most hope for the future. By involving local people and interest groups in a sympathetic way it is clear that compromises will be made (I like the idea of "gifts and gains" in Section 10.7.2). So why on earth can't something be worked out to allow the Kalahari bushmen in my documentary film to continue hunting?*'

Assemble information about the situation of the bushman tribes of southern Africa and outline a strategy that would respect their history and culture while conserving biodiversity in their landscape. Or do you think that their hunting interests are outweighed by a broader objective to maintain biodiversity?

References

Begon, M., Townsend, C.R. & Harper, J.L. (2006) *Ecology: from individuals to ecosystems*, 4th edn. Blackwell Publishing, Oxford.

Bianchi, F.J.J.A., van Wingerden, W.K.R.E., Griffioen, A.J. et al.(2005) Landscape factors affecting the control of *Mamestra brassicae* by natural enemies in Brussels sprouts. *Agriculture, Ecosystems and Environment* 107, 145–150.

Brown, J.H. (1995) *Macroecology*. University of Chicago Press, Chicago.

Carroll, C., Noss, R.F., Paquet, P.C. & Schumaker, N.H. (2005) Extinction debt of protected areas in developed landscapes. *Conservation Biology* 18, 1110–1120.

Clark, C.W. & Mangel, M. (2000) *Dynamic State Variable Models in Ecology*. Oxford University Press, New York.

Cowling, R.M., Pressey, R.L., Rouget, M. & Lombard, A.T. (2003) A conservation plan for a global biodiversity hotspot – the Cape Floristic Region, South Africa. *Biological Conservation* 112, 191–216.

Fox, N.J. & Beckley, L.E. (2005) Priority areas for conservation of Western Australian coastal fishes: a comparison of hotspot, biogeographical and complementarity approaches. *Biological Conservation* 125, 399–410.

Garber-Yonts, B., Kerkvliet, J. & Johnson, R. (2004) Public values for biodiversity conservation policies in the Oregon Coast Range. *Forest Science* 50, 589–602.

Gavin, M.C. (2004) Changes in forest use value through ecological succession and their implications for land management in the Peruvian Amazon. *Conservation Biology* 18, 1562–1570.

Gering, J.C., Crist, T.O. & Veech, J.A. (2003) Additive partitioning of species diversity across multiple spatial scales: implications for regional conservation of biodiversity. *Conservation Biology* 17, 488–499.

Hanski, I. (1999) *Metapopulation Ecology.* Oxford University Press, Oxford.

Hirzel, A.H., Posse, B., Oggier, P.-A., Crettenand, Y., Glenz, C. & Arlettaz, R. (2004) Ecological requirements of reintroduced species and the implications for release policy: the case of the bearded vulture. *Journal of Applied Ecology* 41, 1103–1116.

Jactel, H., Goulard, M., Menassieu, P. & Goujon,G. (2002) Habitat diversity in forest plantations reduces infestations of the pine stem borer *Dioryctria sylvestrella. Journal of Applied Ecology* 39, 618–628.

Leibold, M.A., Holyoak, M., Mouquet, N. et al. (2004) The metacommunity concept: a framework for multi-scale community ecology. *Ecology Letters* 7, 601–613.

Lloyd, P., Martin, T.E., Redmond, R.L., Langner, U. & Hart, M.M. (2005) Linking demographic effects of habitat fragmentation across landscapes to continental source–sink dynamics. *Ecological Applications* 15, 1504–1514.

MacArthur, R.H. (1972) *Geographical Ecology.* Harper & Row, New York.

MacArthur, R.H. & Wilson, E.O. (1967) *The Theory of Island Biogeography.* Princeton University Press, Princeton, NJ.

Margules, C.R. & Pressey, R.L. (2000) Systematic conservation planning. *Nature,* 405, 243–253.

Robinson, C.L.K., Morrison, J. & Foreman, M.G.G. (2005) Oceanographic connectivity among marine protected areas on the north coast of British Columbia, Canada. *Canadian Journal of Fisheries and Aquatic Sciences* 62, 1350–1362.

Rodrigues, A.S.L., Pilgrim, J.D., Lamoreux, J.F., Hoffman, M. & Brooks, T.M. (2006) The value of the IUCN Red List for conservation. *Trends in Ecology and Evolution* 21, 71–76.

Roschewitz, I., Gabriel, D., Tscharntke, T. & Thies, C. (2005) The effects of landscape complexity on arable weed species diversity in organic and conventional farming. *Journal of Applied Ecology* 42, 873–882.

Santelmann, M.V., White, D., Freemark, K. and 13 others (2004) Assessing alternative futures for agriculture in Iowa, USA. *Landscape Ecology* 19, 357–374.

Teirney, L.D. (2003) *Fiordland Marine Conservation Strategy: Te Kaupapa Atawhai o Te Moana o Atawhenua.* Guardians of Fiordland's Fisheries and Marine Environment Inc., Te Anau, New Zealand.

Vaughan, N., Lucas, E.-A., Harris, S. & White, P.C.L. (2003) Habitat associations of European hares *Lepus europaeus* in England and Wales: implications for farmland management. *Journal of Applied Ecology* 40, 163–175.

Villa, F., Tunesi, L. & Agardy, T. (2002) Zoning marine protected areas through spatial multiple-criteria analysis: the case of the Asinara Island National Marine Reserve of Italy. *Conservation Biology* 16, 515–526.

Westphal, M.I., Pickett, M., Getz, W.M. & Possingham, H.P. (2003) The use of stochastic dynamic programming in optimal landscape reconstruction for metapopulations. *Ecological Applications* 13, 543–555.

Zerbe, S., Maurer, U., Schmitz, S. & Sukopp, H. (2003) Biodiversity in Berlin and its potential for nature conservation. *Landscape and Urban Planning* 62, 139–148.

11 Dealing with global climate change

Human activities have been pouring greenhouse gases into the atmosphere and there is ever increasing evidence of global changes to temperature and precipitation – a huge multidisciplinary effort is now focused on what the world's climate holds in store for the future. Ecological managers need to consider the consequences of shifting habitable areas for species distributions, the location of reserves, and sustainable harvest and pest control.

Chapter contents

11.1 Introduction   291
11.2 Climate change predictions based on the ecology of individual organisms   297
    11.2.1 Niche theory and conservation – what a shame mountains are conical   297
    11.2.2 Niche theory and invasion risk – nuisance on the move   298
    11.2.3 Life-history traits and the fate of species – for better or for worse   300
11.3 Climate change predictions based on the theory of population dynamics   303
    11.3.1 Species conservation – the bear essentials   303
    11.3.2 Pest control – more or less of a problem?   303
    11.3.3 Harvesting fish in future – cod willing   304
    11.3.4 Forestry – a boost for developing countries?   305
11.4 Climate change predictions based on community and ecosystem interactions   306
    11.4.1 Succession – new trajectories and end points   306
    11.4.2 Food-web interactions – Dengue downunder   307
    11.4.3 Ecosystem services – you win some, you lose some   307
11.5 A landscape perspective – nature reserves under climate change   308
    11.5.1 Mexican cacti – reserves in the wrong place   309
    11.5.2 Fairy shrimps – a temporary setback   310

Key concepts

In this chapter you will

recognize that no model of global climate is perfect – thus the importance of checking predictions for a range of possible circumstances

understand that the ecologists' role is to take the physical predictions of climate modelers (temperature, precipitation, extreme climate events, ocean currents, etc.) and envision the consequences for species and ecosystems

grasp the importance of niche theory and life-history traits when predicting future distributions of endangered, invading, harvested and pest species

see that the consequences of changes to habitable area are all the more severe if species cannot disperse from their current range

---

recognize that the effects of rising temperature may be predicted using knowledge of consequences for population birth and death rates

understand that the ecological effects of climate change may be positive or negative from a human perspective

realize that food-web interactions can be crucial when predicting risks from global climate change – from human diseases and pests to endangered species

appreciate that nature reserves may turn out to be in the wrong places

11.1 Introduction

Legend tells of an ancient island civilization called 'Atlantis'. When the once-virtuous inhabitants became corrupted by greed and power, the gods punished them – their island was engulfed and lost forever in a single violent surge of the Atlantic Ocean.

Perhaps we should now speak of 'Pacifis', because rising sea level seems set to swallow up many low-lying islands in the Pacific Ocean. First to go will be the Carterets, six tiny horseshoe-shaped coral atolls with a population of 980. Next will be the turn of Kiribati and the Marshall Islands. Valerie is an anthropologist who has paid several visits to the Carterets. *'I imagined I would find the tropical paradise of popular imagination – a life of ease amid coconut palms. But the reality was harsh. People were often close to starvation because salt-water intrusions were killing the trees and ruining breadfruit crops. Some families left voluntarily back in the 1980s but they became caught up in political violence and returned to the islands – despite the hardship they knew they would have to endure. Now the islanders are preparing to leave for good.'* The inhabitants of the Carteret Islands may go down in history as the first to be officially evacuated as a result of global warming. The Papua New Guinea government has resolved that as soon as it can muster international financial support, 10 families at a time will be moved to Bougainville, a larger island 100 km away. By 2015 the Carterets are expected to be completely submerged.

For years, scientists and politicians have debated whether human activities, most notably the burning of fossil fuels and the consequent increase in atmospheric carbon dioxide, are responsible for observed climate change. But now even most politicians agree that we have been responsible for a global increase in temperature that has seen glaciers and icecaps melt and sea levels rise. Valerie notes the irony. *'It is the developed nations of the world who are responsible for climate change – but impoverished pacific islanders will be first to pay. In contrast to the Atlantis myth, in my view the reality is that a simple and virtuous people is being punished for the greed of an entirely different civilization on the other side of the world.'*

A huge multidisciplinary effort is now focused on what the world's climate holds in store (IPCC, 2007). Climate scientists have developed models that predict the magnitudes of future change in temperature and precipitation. Their work is then used by hydrologists to forecast patterns of flooding and runoff from land to rivers, and by ocean scientists to develop models to predict changes in sea level and ocean currents. Despite considerable uncertainty at every step in this modeling process, all results point to some dramatic changes. And changes, moreover, that would still occur through the twenty-first century even if by some magic we could immediately eliminate further human inputs to the atmosphere. This is partly because carbon dioxide in the atmosphere has a residence time of decades, but also because of thermal inertia – the idea that climate change is delayed because ocean water takes longer to heat up than air.

It is not my task to consider the political and personal changes to human activities that would eventually remedy the climatic problems we are causing. The focus, instead, will be on what ecological managers can do to mitigate the worst effects of climate change – because it is not just human populations that will be disrupted. Nature reserves already set up for key species may turn out to be in the wrong places and species currently appropriate for restoration projects may no longer succeed. Moreover, each region of the world is likely to be subject to a new set of invaders, pests and diseases. So where precisely do ecologists fit into this picture? Our role is to take the physical predictions (temperature, precipitation, extreme climate events, ocean currents, etc.) and envision the consequences for species and ecosystems. Only then can we properly work out plans to manage biodiversity.

Future species distributions can be expected to be determined in line with their niche requirements (Chapter 2), life-history features (Chapter 3) and ability to move from where they are now to where their optimal niche conditions will be in future (Chapter 4). Some of these species are ones deserving of conservation effort (Chapter 5) or merit attention for their harvest value (Chapter 6). Others are the focus of biosecurity (invaders) or pest control (Chapter 7). Sometimes our predictions about ecological change will depend on an understanding of community and ecosystem functioning, such as successional processes (Chapter 8), food-web interactions and ecosystem services (Chapter 9). And always a landscape perspective (Chapter 10) will be appropriate, because managers must envision how global climate change will reorganize regional landscapes. In other words, understanding and managing our ecological future needs to be underpinned by all the aspects of ecological theory you have met so far in this book.

Box 11.1 gives a taste of how the climate change modelers go about their business, producing predictions upon which ecological managers can base their work. This chapter then follows the logical structure of the book as a whole. First, I consider examples where managers can base their plans on our knowledge of individual organisms – their niche requirements, life-history features and dispersal powers (Section 11.2). Then I turn to the guidance provided by an understanding of population dynamics – when managers plan for the conservation of endangered species, the harvest of animals and plants, and the control of pests (Section 11.3). You will discover that an understanding of community and ecosystem processes can also guide managers (Section 11.4) and finally, at the largest scale, I will take out my macroscope again and adopt a landscape perspective of ecological management in our rapidly changing world (Section 11.5).

Box 11.1 Predicting the ecological effects of global climate change

The human activities that have changed the composition of the atmosphere, and the links between atmospheric change and climate change, were explained in Chapter 1. Here, I present some of the observational evidence of a changing climate that already exists, and then outline the modeling process used to indicate the pattern and scale of changes we can expect in future. The models of the physical scientists are simplifications of a very complex reality. It is important to consider a variety of models, each based on a different set of reasonably realistic but imperfect assumptions, to provide not a single definitive (and probably wrong) result, but rather a range of change that encompasses the probable truth of the matter. The transformation of these likely climate scenarios into ecological consequences is a further step. This is based either on what is known about current species distributions in relation to climate or, alternatively, on an understanding of the biological consequences of climate change for reproduction, growth and survival.

Observed changes to atmospheric composition and climate

We know that the atmospheric concentration of carbon dioxide increased from 280 ppm (registered for the period 1000–1750) to 368 ppm in the year 2000, representing an increase of $31 \pm 4\%$. Other greenhouse gases have also increased, most notably methane, with a $151 \pm 25\%$ increase over the same period. Associated climate changes have included an increase of $0.6 \pm 0.2°C$ in global mean surface temperature during the twentieth century, an increase in the number of hot days and a decrease in frosty days, an increase in heavy precipitation events at mid- and high northern latitudes, an increase by 5–10% in precipitation in much of the Northern Hemisphere but a decrease in other areas (e.g. north and west Africa), and an increase in summer drought in some areas (such as parts of Asia and Africa).

Observed physical environmental consequences

In response to these changes, there is good evidence in recent decades of degradation of the permafrost (permanently frozen soil in polar and mountainous areas), loss of snow cover (10% since 1960), reduction by 2 weeks in the period of ice cover of mid-high latitude lakes, glacial retreat, and thinning (by 40%) and reduction in extent (by 10–15%) of late summer/autumn Arctic sea ice. Ocean temperatures have increased and, because of thermal expansion of water together with inputs from melting ice, global mean sea level has risen at an annual rate of 1–2 mm during the twentieth century. Rather more surprising, the increased inputs of meltwater may be reducing the strength of the Gulf Stream (an ocean current that moves between Africa and the east coast of North America) (Bryden et al., 2005).

Observed ecological consequences

Alterations to mean, minimum and maximum temperature and to precipitation patterns have been responsible for a plethora of ecological changes. The growing season, the characteristic period of net primary production by vegetation, has lengthened by up to 4 days per decade in the last 40 years, especially at high latitudes in the Northern Hemisphere. In addition, the ranges of various plants, insects, birds and fish have shifted towards the poles and higher in altitude. Thus, various butterfly species have expanded northward by up to 200 km. And some alpine plants have been moving higher at a rate of 1–4 m per decade. Meanwhile plants are flowering, insects emerging, amphibians breeding and migratory birds arriving earlier, by several days per decade over the past 60 years (Walther et al., 2002).

   Profound ecological changes have also been happening in the oceans. Repeated surveys of zooplankton species (small, passively drifting animals) provide a particular insight. Note in Figure 11.1 how, in response to warmer temperatures, the distribution of species that are typical of warm temperate situations has shifted polewards in the North Atlantic, while the cold-loving subarctic species have become more confined to the far north.

Predicted changes to climate in the twenty first century

Depending on the precise assumptions that are made (about how the climate system functions, how global human population size changes, the energy policy choices that are made, and any technological advances to reduce or resorb greenhouse gases, etc.) the concentration of CO_2 is predicted to rise from 368 ppm (in 2000) to between 540 and 970 ppm by 2100, with a concomitant rise in average global surface temperature of between 1.8 and 4.0°C, but with considerable variation from place to place (Figure 11.2a). Predictions are also available for precipitation (Figure 11.2b), sea level rise (Figure 11.2c), glacial retreat and polar ice loss.

Translating future climate change into ecological consequences

Maps like those in Figure 11.2 provide ecologists with templates of temperature and precipitation (and other climatic features) onto which can be mapped future distributions of the world's biomes (tropical rain forest, savanna, arid deserts, etc; Figure 11.3) as well as agriculture and forestry. Maps of future climate are also available at regional levels. You have already seen how bioclimatic modeling of current distributions can be used to produce climate envelopes for individual species (Box 2.1). These envelopes can be superimposed onto the regional templates of predicted climate to indicate where species may occur in future (Figure 11.4).

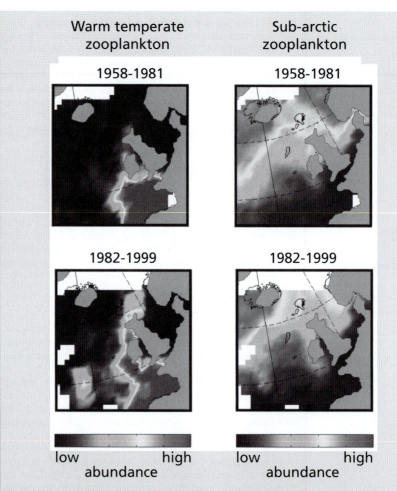

Fig. 11.1 The northerly shift by about 1000 km of zooplankton species typical of warm temperate conditions in the northeast Atlantic over the last 40 years. The scale represents the relative abundance and species richness of warm temperate species in samples taken from different geographical locations. By contrast, the colder-water species that are typical of subarctic conditions have contracted their range. (From Hays et al., 2005.) (This figure also reproduced as color plate 11.1.)

Fig. 11.2 Predicted annual changes in (a) average surface air temperature (°C), average precipitation (mm day^{-1}) and (c) average sea level rise (m) from 1960–1990 to 2070–2100, when carbon dioxide concentration in the atmosphere is expected to have doubled. These predictions are based on a coupled atmosphere–ocean general circulation model (HadCM2) of the Hadley Centre for Climate Prediction and Research, which assumes a midrange global average temperature increase of 3.2°C. (© Crown copyright 2005, Published by the Met Office Hadley Centre.) (This figure also reproduced as color plate 11.2.)

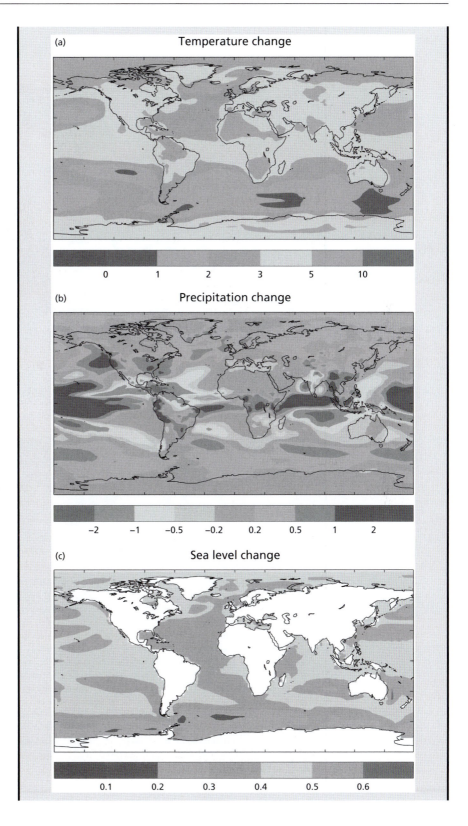

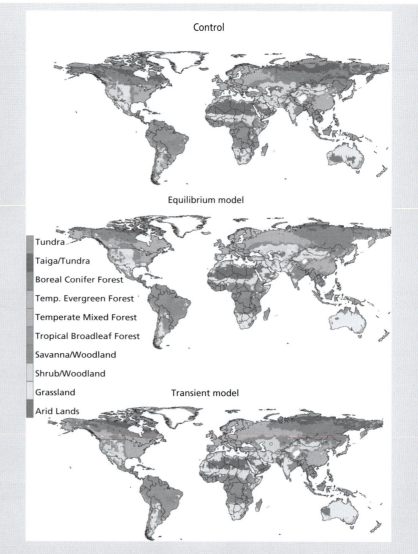

Fig. 11.3 Simulated distribution of biomes throughout the world using contrasting climate change models. Biomes are regional ecosystems characterized by distinct types of vegetation fitted to the prevailing physical conditions. (a) The 'control' output represents the current distribution of biomes. (b) Output from an 'equilibrium' model, in which the doubling of CO_2 is assumed to produce its full temperature effect instantaneously. (c) Output for a 'transient model', in which atmospheric CO_2 responds to dynamic feedback between atmosphere and oceans, achieving only about 65% of the eventual predicted temperature change at the time of CO_2 doubling. Under both scenarios, forests shift northwards to currently unforested areas. However, in temperate latitudes the transient model produces much lower increases in temperature and larger increases in precipitation than the equilibrium model, with consequent differences in the distribution of biomes in the two cases. (Tundra is a treeless plain, Boreal (= northern) Conifer Forest consists of needle-leaf trees, while Taiga/Tundra is intermediate between the two, with tundra vegetation and scattered conifer trees. Savanna is grassland with scattered trees.) (From Neilson & Drapek, 1998.) (This figure also reproduced as color plate 11.3.)

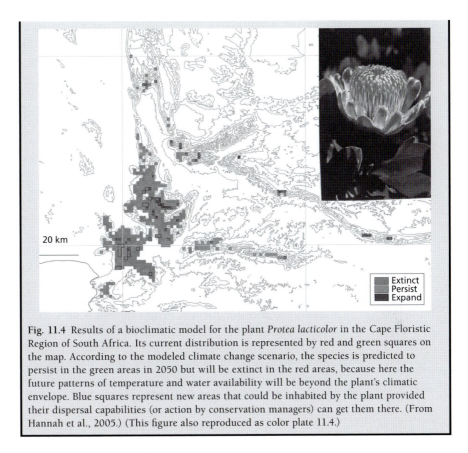

Fig. 11.4 Results of a bioclimatic model for the plant *Protea lacticolor* in the Cape Floristic Region of South Africa. Its current distribution is represented by red and green squares on the map. According to the modeled climate change scenario, the species is predicted to persist in the green areas in 2050 but will be extinct in the red areas, because here the future patterns of temperature and water availability will be beyond the plant's climatic envelope. Blue squares represent new areas that could be inhabited by the plant provided their dispersal capabilities (or action by conservation managers) can get them there. (From Hannah et al., 2005.) (This figure also reproduced as color plate 11.4.)

11.2 Climate change predictions based on the ecology of individual organisms

Mirroring the logical organization of Chapters 2–4, in this section I present examples of climate change predictions that rely on theories of the niche, life history and dispersal.

11.2.1 *Niche theory and conservation – what a shame mountains are conical*

A remarkably diverse vegetation type occurs around South Africa's Cape of Good Hope. The 'fynbos' is characterized in particular by *Protea* species such as the one shown in Figure 11.4, significant both from a biodiversity point of view and because of their economic importance in the floral trade. Extinction risk as a result of habitat loss is already high in the Cape Region, and global climate change makes the fate of many of the species even more precarious. Bioclimatic niche modeling is feasible for these species because their distributions are well studied, fine-scaled current bioclimatic data are available (together with other important environmental data such as soil type) and future climate modeling is in an advanced state.

Figure 11.4 illustrates how individual species can be expected to make both gains and losses in habitable area so that, in theory, the effect of climate change might be neutral. However, such a benign outcome is not very likely where suitable habitat has been reduced and fragmented by land-use development. A crucial question then is whether or not future habitat substantially overlaps with nature reserves. In addition, many proteas are at risk because of a strong decline in habitable area related to their expected movement to higher elevations. This happens partly because the ranges of many species are already close to absolute upper elevational limits, but also to some extent because of the 'conical' shape of mountains. As climate warms, habitable area (and thus species range) shifts into the smaller and smaller areas available at higher and higher elevations. In the case of *Protea lacticolor*, for example, there is a predicted range loss in lowland areas, species persistence in the uplands, and some new potential range at higher elevations (Figure 11.4).

Beaumont and Hughes (2002) used a similar approach to predict the effect of climate change on the distribution of 24 Australian butterfly species, which, like the proteas, are strongly influenced by temperature and moisture. Under even a moderate set of future conditions (temperature increase of 0.8–1.4°C by 2050), more than half the butterflies lose at least 20% of their ranges. One species at particular risk is *Hypochrysops halyetus* in the coastal heathland of Western Australia. It is predicted to lose 58–99% of its current range. In addition, less than 27% of its future distribution occurs in locations that are currently occupied.

These examples illustrate two important points. The first is that many species can be expected to suffer a reduction in habitable area and, because smaller areas support smaller populations, extinction risk is increased. The second point is just as significant: nature reserves may turn out to be in the wrong place in the shifting template of temperature and moisture. When selecting protected areas, managers must take account of predicted range shifts of the species to be protected. Equally, climate-related shifts in range should also be considered when selecting candidate species for habitat restoration.

11.2.2 Niche theory and invasion risk – nuisance on the move

Spiny acacia (*Acacia nilotica* subspecies *indica*) is a woody legume whose native range includes parts of Africa and extends into India. It is now on the march across Australia where it was originally introduced as an ornamental plant and to provide food and shade for domestic animals. Its spread has been dramatic and the plant is now labeled a noxious weed because it reduces pasture production, impedes access of livestock to water and makes stock mustering a difficult business. Given knowledge of conditions in its natural range, Kriticos et al. (2003) determined the species' fundamental niche (Box 2.1). This was defined in terms of optimal conditions of temperature and moisture (and lower and upper tolerance limits), as well as thresholds for cold stress, heat stress, dry stress and wet stress (water-logging). They then modeled the acacia's invasion potential under two climate change scenarios. Both assumed a middle-of-the-range 2°C temperature rise, but this was coupled with either a 10% increase or 10% decrease in rainfall, reflecting the uncertainty surrounding future precipitation in Australia.

Currently, the plant is spread widely through the range indicated by the model based on present climatic conditions (Figure 11.5a), but it is not yet in all predicted areas. When climate change is taken into account, its eventual invadable range will be much larger (Figure 11.5b,c). This is partly because of changed climatic conditions, but also because spiny acacia is expected to become more efficient in its use of water (making dryer areas inhabitable) as a result of a 'fertilizing' effect on growth caused by increased atmospheric CO_2. Thus, elevated atmospheric CO_2 concentration can have both indirect effects, via climate change, and direct effects on the performance and distribution of plants (Volk et al., 2000). Given that we now know

Fig. 11.5 Predicted distribution of the invasive spiny acacia in Australia on the basis of (a) current climate, (b) a climate change scenario with an average 2°C increase and a 10% increase in precipitation, (c) a scenario with a 2°C increase and a 10% decrease in precipitation. The predicted distributions in (b) and (c) also assume an increased efficiency of water use by spiny acacia because the increase in atmospheric CO_2 enhances its performance. (From Begon et al., 2006, after Kriticos et al., 2003.)

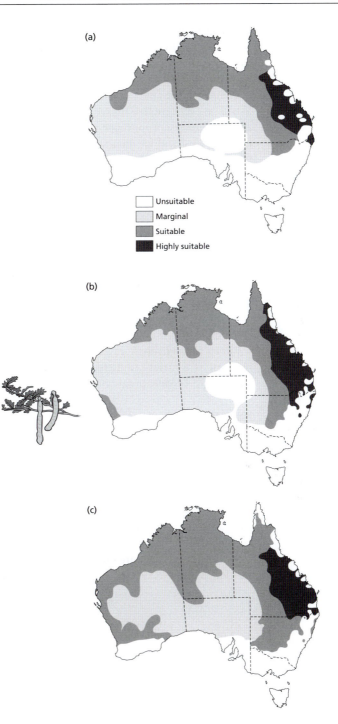

where to look, further spread of this species should be containable because trees can be physically removed, and the spread of seeds (in stock feces) can be prevented as long as animals are not moved indiscriminately. Part of the management response will be to raise public awareness of the weed and how to control it.

The Argentine ant (*Linepithema humile*), a native of South America, is now established on every continent except Antarctica. It can achieve extremely high densities and has unpleasant consequences for biodiversity (eliminating native ants and other invertebrates) and for domestic life, with its nasty tendency to swarm over human foodstuffs and even sleeping babies. A niche model was developed for the ant, based on occurrences in its native and invaded ranges and related both to climatic data (e.g. maximum, minimum and mean temperatures, precipitation, number of frost days, number of wet days) and topographic data (e.g. elevation, slope and aspect). The niche model provided a good fit with current distribution based on current climate. Next, several different climate change scenarios were modeled to produce a range of predictions of the ant's future distribution. Figure 11.6 (see also color plate 11.6) shows the average outcome of the models, indicating in red those areas predicted to improve for the ant by 2050 (increased likelihood of ant occurrence) and in blue those areas expected to worsen. The species will retract its range in tropical areas but expand into higher latitudes. Ironically, the Argentine ant looks set to do less well in its native South America than in North America and Europe.

Efforts to eradicate Argentine ants have rarely been successful. The management response is therefore to increase vigilance and biosecurity precautions in regions expected to become progressively more invadable as climate change takes hold.

11.2.3 Life-history traits and the fate of species – for better or for worse

You learnt in Chapter 3 how *r*-selected species (with a combination of traits that allow them to multiply rapidly and produce large numbers of progeny) are often good invaders, while *K*-selected species (with life histories that enable them to

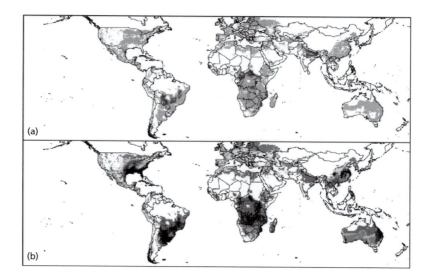

Fig. 11.6 Predicted changes to distribution of the Argentine ant between now and 2050. The map represents the average predictions of four global climate change scenarios. Red areas are those predicted to improve for Argentine ants, whereas blue areas are predicted to worsen for the species. (After Roura-Pascual et al., 2004.) (This figure also reproduced as color plate 11.6.)

survive where there is intense competition) are more likely to be candidates for extinction in the face of human-related threats. As a rough and ready rule, therefore, we might expect global climate change to see *r*-species do relatively better than *K*-species. It also seems likely that habitat generalists will be less vulnerable than habitat specialists in a changing climate scenario. This is because generalists are more likely to find their original (but climatically altered) habitat still inhabitable, or will be capable of spreading through marginal habitat to new habitable areas as these become available.

Chapter 4 made the distinction between traits related to dispersal (active or passive movement not in a specific direction or as a predictable population shift in the life cycle) and migration (predictable movement from one location to another as part of the life cycle). Migratory birds will be particularly challenged by global climate change because of altered circumstances at one or both ends of their migratory pathways, as well as in the staging posts that can be critical to their success.

Patterns of climate change at small scales have provided valuable pointers to the consequences of global change. Take the North Atlantic Oscillation (NAO) for example. This is a seesawing atmospheric pattern between the subtropical high- and polar low-pressure zones, each phase lasting for several years. A 'positive' NAO index occurs when there is a stronger than usual subtropical high-pressure center and a deeper than normal Icelandic low. The increased pressure difference produces more and stronger winter Atlantic storms and results in relatively warm and wet conditions in northwestern Europe and relatively cool conditions in North Africa and the Middle East. During periods with a positive NAO index, migrating barn swallows (*Hirundo rustica*) from Africa arrive earlier in the Danish springtime, while the diminutive wren (*Troglodytes troglodytes*), trapped by ornithologists in mid-migration, shows better body condition (Figure 11.7a). Global warming is expected to strengthen the NAO index and the success of many migratory birds seems certain to be affected, for better or for worse.

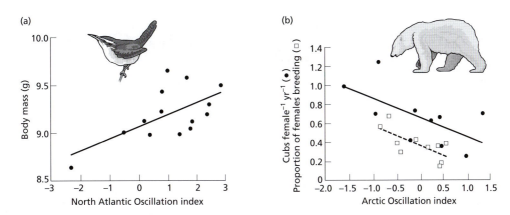

Fig. 11.7 (a) Average body mass of winter wrens trapped during their spring migration (on the island of Helgoland in the German North Sea) is greater in years with a strong positive North Atlantic Oscillation index. (After Barlein & Huppop, 2004.) (b) The birth rate and proportion of female polar bears breeding at Svalbard, Norway, are lower when the Arctic Oscillation index is positive. Global warming is expected to lead to changes in both these local climate indexes. (After Derocher, 2005.)

Most species do not migrate like swallows, but their powers of movement are equally important to future distribution. Note that the term 'power of movement' is a broad one that includes the ability of plants to progress through the landscape via dispersal of their seeds. A crucial question is whether species under the influence of global climate change will be able to keep up with the rate of movement of their habitable areas. History shows that tree species have changed and recovered distribution patterns with the waxing and waning of the ice ages. But is human-induced global warming going to move too fast? And in human-modified landscapes, will there be sufficient stepping-stones of habitable areas that are relatively undisturbed?

The question has been addressed for trees in North America including sweetgum (*Liquidambar styraciflua*), a species currently confined to the eastern half of the USA. Iverson et al. (2004) first used niche modeling to define the fundamental niche of the species and then ran a climate change model to predict habitable area in 2100. A considerable expansion of range is predicted to the north and east (Figure 11.8a,b). They then inserted another modeling step called 'SHIFT'. This involves calculating the probability of dispersal to each unoccupied 1 km^2 cell to the northeast of the current distribution. The probability of successful colonization is based on habitat quality in the unoccupied cell, abundance of sweetgum in each occupied cell, distances between occupied and unoccupied cells and a maximum migration rate of 50 km/century (estimated from historical records). The model was run for 100 years,

Fig. 11.8 (a) Current distribution of sweetgum trees across the eastern USA (darker colors show greater habitat quality – an index of density and size of trees) with the 1971 range outlined in black. (b) Predicted distribution of sweetgum in 2100, assuming that rate of dispersal can keep up with change in habitable area. (c) Actual occupied area in 2100 is predicted to be very much less than the potential area, because of dispersal restrictions. This panel, which focuses in on the current range boundary (indicated by the box in (b)), shows the presently occupied range in gray and the surprisingly small addition by 2100 of newly occupied habitat immediately to the northeast. According to the predicted distribution in (b), all of the area depicted in (c) could, in theory, be occupied by sweetgum if only they could shift fast enough. (From Iverson et al., 2004.) (This figure also reproduced as color plate 11.8.)

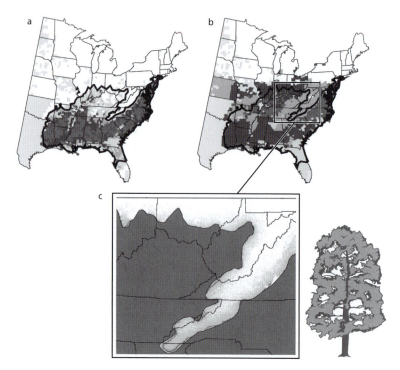

allowing the passage of four generations of sweetgum. Only when a colonist of an unoccupied cell reaches reproductive maturity (at 25 years old) can it start contributing to the colonization of other cells. You can see that only a very small percentage of the newly habitable area of sweetgum is likely to be actually occupied by 2100 (Figure 11.8c). Given the serious lag involved in species movement to newly habitable areas, managers may need to compensate by physically moving seeds or seedlings, especially of rare species unable to travel effectively through fragmented landscapes. However, it would be unreasonable to expect to be able to transplant whole communities.

11.3 Climate change predictions based on the theory of population dynamics

In Chapters 5–7 you saw how knowledge of the population dynamics of key species can be used when planning conservation, pest control and harvest operations. If we know how birth and death rates are likely to be modified as a result of climate change, this can be incorporated in plans with a timetable stretching through the twenty first century.

11.3.1 Species conservation – the bear essentials

A long-term study of polar bears (*Ursus maritimus*) at Svalbard in Norway has revealed a link between birth rate and an oscillating climate pattern where atmospheric pressure in the Arctic switches between high and low. This Arctic Oscillation is linked to the North Atlantic Oscillation discussed in Section 11.2.3. In years when the Arctic Oscillation is in negative mode (higher than normal pressure over the polar region), a larger proportion of female bears have cubs and also, on average, they have a greater number of cubs (Figure 11.7b). Improved breeding success might be due to the state of sea ice, and the access this provides to prey (seals). Alternatively, there may be a climate effect on the productivity of phytoplankton, which fuels the base of the polar bear food web (Derocher, 2005). What concerns us here, however, is the expectation from global climate models that the Arctic Oscillation is likely to change, with consequences for population viability of the bears. Polar bear populations have long been under pressure from hunting and high body pollutant levels. It will be important to quickly understand the nature of any further threat posed by climate change.

11.3.2 Pest control – more or less of a problem?

Just as polar bears may be affected by global climate change, so may animals that normally keep pest insects under control, whether naturally occurring spiders or deliberately introduced biological control agents such as parasitoid wasps (Chapter 5). But the picture is more complex than this, of course, because climate change might influence the success of the crops, their pests, or the pest's control agents, or indeed all three – and in different directions.

Take the case of the aphid *Rhopalosiphum padi* that attacks cereal crops in southern Britain. Newman (2005) begins with global climate change predictions. His analysis includes a range of scenarios differing in the extent to which atmospheric CO_2 increases during the twenty first century (baseline for 1961 to 1990 of 319 ppm; low scenario 525 ppm, medium low 562 ppm, medium high 715 ppm, and high scenario 810 ppm). He next couples these predictions with an ecological model that links local climatic conditions (particularly temperature and rainfall), the growth of cereal grasses (which depends on climate but also on soil water and nutrients) and the population dynamics of the aphi ds (which respond directly to climatic conditions, but also to plant drought stress and nitrogen available from the grasses). Figure 11.9 shows that pest aphid abundance is actually predicted to decline under global climate change, particularly for the medium to high CO_2 scenarios. It is important to recognize that climate change is likely to 'improve' circumstances for particular species (whether pests, invaders, harvested species or endangered ones) while it makes things more precarious for others.

11.3.3 Harvesting fish in future – cod willing

Mean annual temperature at the sea surface is predicted to increase in the North Sea by up to 1°C by 2040 and the population dynamics of fish may respond via changes to their rates of growth, birth and death. The situation is further complicated because the raised temperatures may change habitable areas, just as for terrestrial plants and animals. Kell et al. (2005) modeled the outcomes of climate change for cod (*Gadus morhua*), contrasting the effects of a likely change in juvenile survival as opposed to the expected reduction in habitable area. These two fishery models were evaluated for three climate change scenarios (Figure 11.10a): a best case scenario that assumes no increase in average temperature after 2001, a worst case scenario with a constant increase of 0.026°C per year (i.e. 1°C increase by 2040) and an intermediate scenario of a 0.2°C increase by 2040 (based on a model from the Hadley Center – like the one in Figure 11.2).

Recall, from Box 7.1, that a recruitment curve is a plot of net recruitment (births minus deaths) of a fish population versus the total size of the population (standing stock biomass). For any given stock biomass, a fishery can harvest all the associated net recruitment without depleting the stock – so that, in theory, the harvest is sustainable. The *maximum* sustainable yield (MSY), then, is the one at the highest point of the dome-shaped recruitment curve.

When the only effect of increased temperature is on fish survival (recruitment curves in Figure 11.10b), there is no change to the critical standing stock biomass at which the MSY is available – but the actual size of the yield is reduced. On the other hand, both the critical standing stock biomass and the size of the yield are reduced when increased temperature causes the habitable area of cod to shrink (Figure 11.10c). Setting aside the real difficulties of using recruitment curves to manage fisheries (Chapter 7), these results nevertheless provide valuable insights into the potential consequences of climate change for fisheries management decisions.

Fig. 11.9 Predicted decreases in mean abundance (mean ± standard error, compared to baseline in 1961–90) of aphid pests of cereal crops in southern England during the twenty first century assuming four increasingly extreme climate change scenarios – CO_2 concentrations increasing to 525, 562, 715 or 810 ppm, respectively. This is a case where climate change may bring a beneficial response. (After Newman, 2005.)

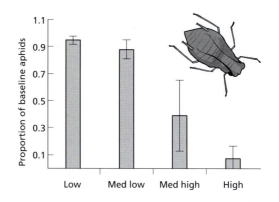

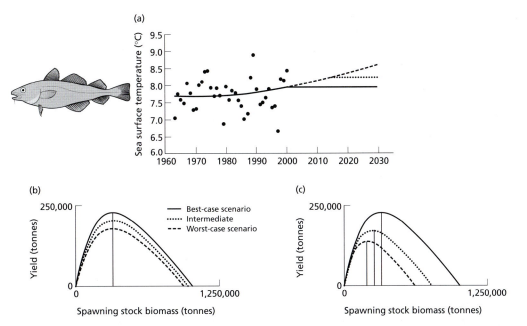

Fig. 11.10 (a) Historical sea surface temperatures in the North Sea and the scenarios assumed to affect the future of the cod fishery. The best-case scenario assumes no change to mean temperature beyond 2001, the worst case assumes a constant increase to a rise of 1°C and the intermediate case assumes a rise of 0.2°C by 2040. (b) Stock recruitment curves for each scenario, assuming that increased temperature affects cod population dynamics by reducing juvenile survival. (c) Stock recruitment curves that assume increased temperature reduces the size of the habitable area of cod. Vertical lines in (b) and (c) show, for each scenario, the standing stock biomass that provides the maximum sustainable yield. (From Kell et al., 2005.)

11.3.4 *Forestry – a boost for developing countries?*

Global climate change can also be expected to cause considerable change to forest harvests around the world. This will occur partly because of positive or negative effects on tree growth and forestry yields (both because of changes to temperature and moisture, but also through a direct fertilization effect of increased atmospheric CO_2). But in addition there will be potential losses as 'dieback', caused by increases to forest fires, storm damage or pest outbreaks. And finally, we can expect changes to the habitable areas of forest species. Table 11.1 shows the predicted responses to climate change in different parts of the world during the twenty first century.

Of course, it would be silly to imagine that foresters will fail to respond to climate change – in due course they will plant more appropriate trees. And, in fact, it is developing countries in tropical and subtropical regions that are expected to be best able to adapt, because their trees grow faster and their forest rotations are therefore shorter. Taken overall, forestry in many parts of the world is predicted to show long-term gains in production, particularly through increased growth rates, increases to forested areas and switches to more appropriate species. However, dieback can reduce these gains. North America, Russia and China seem particularly susceptible to the negative effects of dieback, at least for the first decade or two (Figure 11.11). South America, on the other hand, is predicted to perform strongly in this period because of low dieback and rapid adaptation by switching to more appropriate plantation trees. In the longer term, China and Russia catch up because of significant increases in forested land area and good increases in yield.

Table 11.1 Average equilibrium ecological effects on forestry of a global climate change model assuming a doubling of atmospheric CO_2. North America has the smallest predicted increase in forest productivity (cubic meters of wood per hectare per year) and, with Russia and China, a particularly large increase in the percentage of forest lost (dieback) because of fires, storms and pests. South America and China have the biggest predicted increases in forest habitable area under the climate change scenario. (From Sohngen & Sedjo, 2005.)

	% Increase in forest productivity	% Dieback	% Increase in forest area
North America	17	28	4
South America	23	10	27
Europe	34	9	7
Russia	52	21	14
China	38	20	20

11.4 Climate change predictions based on community and ecosystem interactions

Chapters 8 and 9 dealt with lessons for managers from the theories associated with community and ecosystem functioning. Global climate change can be expected to wreak effects on patterns of community succession (Section 11.4.1), via the complexities of food-web interactions (Section 11.4.2) and to alter the range and type of ecosystem services provided free to human society (Section 11.4.3).

11.4.1 Succession – new trajectories and end points

You saw in Chapter 8 how community composition is affected by disturbances, whether these are natural (such as hurricanes or volcanic eruptions) or manmade (such as forest clearing/burning for agriculture). The resulting community successions have characteristic trajectories and end points that are governed by local soils, topography and climate. Climate change, first, has the power to change disturbance frequency – by altering storm patterns and fires. Second, it can modify the interactions among species, particularly competition and facilitation (Box 8.1), which so influence the course of succession. And, third, as we have already seen, climate change is predicted to affect successional end points or, in other words, the characteristic biome of the area in question (Figure 11.3). Moreover, if species' ranges shift at different rates, then species' richness may also be affected, if only transiently. For

Fig. 11.11 Projected percentage increases in annual forestry harvest (in cubic meters per year), averaged for two climate change models that involve a doubling of atmospheric CO_2. The contrasting regional patterns reflect predicted differences in growth rate increase, dieback (due to storm, fire and pests) and changes to total tree habitable area. The models include the assumption that foresters will adapt to climate change by planting more appropriate trees. (From Sohngen & Sedjo, 2005.)

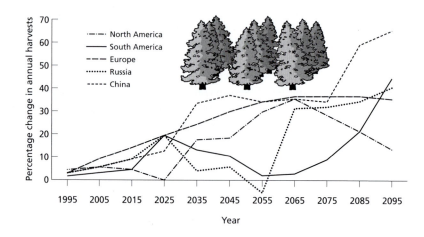

example, it has been suggested that new species may move in faster from lower elevations, and latitudes, than resident species will recede to higher elevations or the poles (Walther et al., 2002).

11.4.2 *Food-web interactions – Dengue downunder*

In Section 11.3.2 you saw how the effect of climate change on a target species (a pest aphid in this case) came about by altering a food-web interaction. Aphid density was modified partly by direct impacts of temperature and moisture, but also via changes to the performance of the cereal plant upon which the aphid feeds. Equally, changing patterns in the distribution and productivity of marine plankton (Figure 11.1) may exert effects on the productivity of the fish at the top of the food web, and thus change the sustainable yield of fisheries. And the Australian butterfly at most risk from climate change (*Hypochrysops halyetus* in Section 11.2.1) is unusually vulnerable not only because of its specialized requirement for a unique food plant, but also because it depends on the presence of a particular ant species in a mutualistic relationship. Climate change might therefore affect the butterfly directly, or indirectly via these food-web interactions.

I explained in Section 9.2 how the risk of certain diseases in humans depends on the way key species interact in food webs. Climate change can play a role here too. Take Dengue fever, for example. This is a potentially fatal viral disease currently limited to tropical and subtropical countries where its mosquito vectors occur. No mosquito species currently in New Zealand is capable of carrying the disease, but both of the world's most important vectors (*Aedes aegypti* and *A. albopictus*) have been intercepted at New Zealand's borders. If a vector mosquito population becomes established, it needs only a single virus-carrying human traveler to trigger an outbreak of the disease. de Wet et al. (2001) coupled knowledge of the mosquitos' fundamental niches (in terms of temperature and precipitation) with climate change scenarios, to predict areas of high risk of mosquito invasion and thus of establishment of the disease. Under present climatic conditions, *A. aegypti* is unlikely to be able to establish anywhere in New Zealand but *A. albopictus* could invade the northern, subtropical part of the North Island (Figure 11.12a). Under a climate change scenario, most of the North Island and some of the South Island would be at risk of invasion by *A. albopictus*. Under the same scenario, the greater Auckland area in the north of the North Island, where a large proportion of the human population lives, would become susceptible to invasion by the more efficient virus vector *A. aegypti* (Figure 11.12b). Vigilant border surveillance is the key, with emphasis on northern ports of entry and particularly Auckland, where most passengers and cargo arrive (including the imported tyres that provide a prime transport route for mosquito larvae) (Hearnden et al., 1999).

11.4.3 *Ecosystem services – you win some, you lose some*

People are part of ecosystems and obtain a variety of priceless ecosystem services free – provision of sufficient water, productive soils, flood control, recreational opportunities such as skiing and so on (Section 9.8). Schroter et al. (2005) evaluated for the whole of Europe the likely effect on selected ecosystem services of climate change. Predictions were based on an expected population size of 419 million people by 2080 (compared to 376 million in 1990), an atmospheric CO_2 concentration of 709 ppm and, according to four different model scenarios, average temperature rises of between 2.7 and 3.4°C and percentage changes in precipitation ranging from a reduction of 0.6% to an increase of 2.3%.

Fig. 11.12 Dengue fever risk maps for (a) *Aedes albopictus* for (i) present climatic conditions and (ii) for 2100 under a climate change scenario. (b) The fundamental niche of *Aedes aegypti* means that it cannot survive anywhere in New Zealand at present, but it could inhabit parts of the North Island as shown under the climate change scenario by 2100. (After de Wet et al., 2001.) (This figure also reproduced as color plate 11.12.)

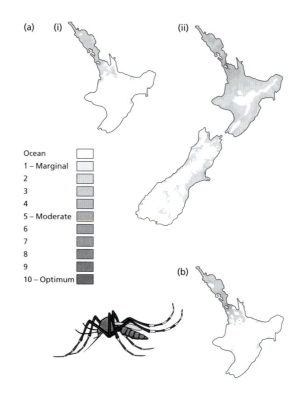

The models predicted large geographical divergences from these average values. Thus, when changes to water stress were modeled, parts of Spain, Portugal, southern France and Italy were worst affected (Figure 11.13). Changes in the provision of water can affect people profoundly. In 1995, about half of Europe's population were considered subject to water stress (living in catchment areas with less than $1700\,m^3$ of water per person per year). By 2080, between 20 and 38% of the Mediterranean population will be subject to even greater water stress.

Elsewhere, in the large river catchments of the Rhine, Rhone and Danube, changed flow patterns look set to reduce water supply at times of peak demand, and to increase the probability of winter floods. Navigation and hydropower will be affected. In the Alps, there will be an increase in the elevation of reliable snow (for skiing) from about 1300 m today to 1500–1750 m by 2080. On the other side of the ledger, European vegetation is predicted to make a positive contribution to reducing atmospheric CO_2 by locking up more carbon in biomass – because of a decreased area of agricultural land and an increase in forest.

11.5 A landscape perspective – nature reserves under climate change

A basic idea that derives from island biogeography theory (Box 10.1) is that smaller areas contain fewer species. One way to assess extinction risk of endemic species is to estimate, on the basis of climate change models, the loss in area of key habitats. Thus, for example, the South African fynbos, with its richness of *Protea* species (including the one shown in Figure 11.4), has been estimated to lose 65% of its area by 2050. On the basis of the general pattern relating species richness to area, this represents a reduction of 24% in number of species. Put another way, one quarter

Fig. 11.13 Stress status
of river catchment areas
throughout Europe by
2080 according to four
different climate change
models. In a few areas,
water stress is reduced
in comparison to no
climate change (blue),
in many places there is
no change (yellow), but
in some areas water
stress is set to increase
(orange, red). (From
Schroter et al., 2005.)
(This figure also
reproduced as color
plate 11.13.)

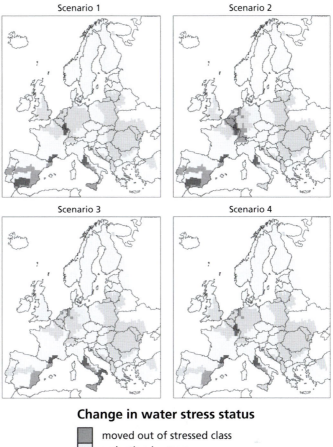

Fig. 11.13 Stress status
of river catchment areas
throughout Europe by
2080 according to four
different climate change
models. In a few areas,
water stress is reduced
in comparison to no
climate change (blue),
in many places there is
no change (yellow), but
in some areas water
stress is set to increase
(orange, red). (From
Schroter et al., 2005.)
(This figure also
reproduced as color
plate 11.13.)

Change in water stress status

moved out of stressed class
reduction in stress
basin never stressed
no significant change
increase in stress
moved into stressed class

of *Protea* species may be at risk of extinction due to climate change (Thomas et al.,
2004). And this conclusion is based on the optimistic assumption that all *Protea*
species are capable of dispersing to all currently uninhabited areas that become
inhabitable. If no dispersal is assumed, and future ranges are simply those reduced
parts of current ranges that remain inhabitable, 30–40% of species seem at risk of
extinction. Similar fates could await diverse animal and plant taxa around the world.
In many cases, though, suitable choice of protected areas can minimize the predicted
losses (Sections 11.5.1, 11.5.2).

11.5.1 *Mexican cacti
– reserves in the
wrong place*

Climate change poses a real conundrum for biodiversity managers. Will current
nature reserves, designed to protect particular elements of biodiversity, turn out to
be in the wrong places? Cacti are the dominant plant form in Mexico's Tehuacán-
Cuicatlán Biosphere Reserve. From knowledge of the biophysical basis of current
distributions and assuming one of three future climate scenarios, Téllez-Valdés and

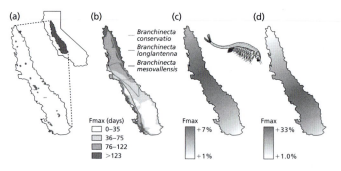

Fig. 11.14 (a) Distribution of biological reserves (black shading) in the Central Valley region of California, USA (inset). The current set of protected areas only includes a small proportion of locations where vernal pools are found. Thus, temporary pond dwellers, such as fairy shrimps (*Branchinecta* spp.), are at risk. (b) The current distribution of F_{max} – the annual average number of days of continuous inundation of ponds. F_{max} is a key factor determining the success of fairy shrimps. The current ranges of three fairy shrimp species are shown in outline. (c) Predicted changes to F_{max} under a cooler and drier scenario (–1°C, –10% precipitation). (d) Predicted changes to F_{max} under a warmer and wetter scenario (+3°C, +30% precipitation). The manager's objective will be to maintain shrimp biodiversity by adding new protected areas that incorporate the necessary future range of F_{max} values. (From Pyke, 2005.) (This figure also reproduced as color plate 11.14.)

Dávila-Aranda (2003) predicted future species distributions in relation to the location of the reserve.

Table 11.2 shows how the predicted ranges of species contract or expand under the various scenarios. Focusing on the most extreme scenario (an average temperature increase of 2.0°C and a 15% reduction in rainfall), you can see that more than half the species that are currently restricted to the reserve are expected to go extinct. Managers can use this information to make a case for setting aside new reserve areas. The second category of cacti, whose current ranges are almost equally inside and outside the reserve, will contract, but in such a way that their distributions become almost completely confined to the reserve. The final category, whose current distributions are much more widespread, also contract but are expected to be distributed both within and outside the reserve. In the case of these latter classes of cacti, then, the location of the reserve seems to cater adequately for potential range changes.

11.5.2 *Fairy shrimps – a temporary setback*

The inhabitants of temporary ponds, such as the ones known as vernal (springtime) pools, have a particular problem under climate change. Whereas some habitat types may move to new locations, vernal pools can only exist where there are topographic depressions underlain by dense, impermeable soil layers. And while the pools cannot shift, their properties can be altered dramatically by local climate. Thus, the length of the period during which these temporary ponds hold water will change with alterations to temperature and rainfall. Figure 11.14a shows the current distribution of protected areas in California's Central Valley – these only include a

Table 11.2 The core distributions (km²) of cacti in Mexico under current conditions and as predicted for three climate change scenarios. Species in the first category of cacti are currently completely restricted to the 10,000 km² Tehuacán-Cuicatlán Biosphere Reserve. Those in the second category have a current range more or less equally distributed inside and outside the reserve. The current ranges of species in the final category extend widely beyond the reserve boundaries. (After Téllez-Valdés & Dávila-Aranda, 2003.)

Species category	Current	+1.0°C −10% rain	+2.0°C −10% rain	+2.0°C −15% rain
Restricted to the reserve				
Cephalocereus columna-trajani	138	27	0	0
Ferocactus flavovirens	317	532	100	55
Mammillaria huitzilopochtli	68	21	0	0
Mammillaria pectinifera	5,130	1,124	486	69
Pachycereus hollianus	175	87	0	0
Polaskia chende	157	83	76	41
Polaskia chichipe	387	106	10	0
Intermediate distribution				
Coryphantha pycnantha	1,367	2,881	1,088	807
Echinocactus platyacanthus f. grandis	1,285	1,046	230	1,148
Ferocactus haematacanthus	340	1,979	1,220	170
Pachycereus weberi	2,709	3,492	1,468	1,012
Widespread distribution				
Coryphantha pallida	10,237	5,887	3,459	2,920
Ferocactus recurvus	3,220	3,638	1,651	151
Mammillaria dixanthocentron	9,934	7,126	5,177	3,162
Mammillaria polyedra	10,118	5,512	3,473	2,611
Mammillaria sphacelata	3,956	5,440	2,803	2,580
Neobuxbaumia macrocephala	2,846	4,943	3,378	1,964
Neobuxbaumia tetetzo	2,964	1,357	519	395
Pachycereus chrysacanthus	1,395	1,929	872	382
Pachycereus fulviceps	3,306	5,405	2,818	1,071

small proportion of the locations where vernal pools occur. The protection of vernal pool dwellers, such as fairy shrimps (*Branchinecta* spp.), is thus far from assured even under current climatic conditions.

One of the key physical parameters that determine the success of fairy shrimps is F_{max} – the annual average number of days of continuous inundation of a temporary pond. The fairy shrimps are restricted to temporary ponds that remain inundated long enough for the shrimps to mature and breed, but dry up soon enough to prevent the build up of their predators. Figure 11.14b shows the geographical variation in F_{max} under current climatic conditions, while Figure 11.14c and d show changes to F_{max} under two climate change scenarios. Both global climate models predict California will become wetter, but there is less certainty about whether local temperatures will decrease (Figure 11.14c) or increase (Figure 11.14d).

Species of fairy shrimp at risk of extinction have different niche requirements (in terms of F_{max}, etc.) and thus different distributions across California. The current ranges of three of the species are superimposed on Figure 11.14b. The predicted effects of changed inundation period on each species can be used to identify those that will be adequately protected by current reserves and to assist managers to select new protected areas to provide for the needs of fairy shrimp biodiversity overall.

Summary

Predicting the ecological effects of global climate change

Evidence continues to accumulate of increases in greenhouse gases, changes to climate and biological consequences. Climate scientists have developed models that predict the magnitudes of future change in temperature and precipitation. The concentration of CO_2 is expected to rise from 368 ppm to between 540 and 970 ppm by 2100, with a concomitant rise in average global surface temperature of between 1.8 and 4.0°C, but with considerable variation in temperature and precipitation from place to place. Climate models are used by hydrologists to forecast patterns of flooding and runoff from land, and by ocean scientists to forecast changes in sea level and ocean currents. The models of the physical scientists are, of course, simplifications and it is important to consider a variety of models to provide a range of change that encompasses the most probable outcome. The role of ecologists is to take the physical predictions and envision the consequences for species and ecosystems.

Climate change predictions based on the ecology of individual organisms

Future species' distributions can be expected to be determined in line with their niche requirements, life-history features and, in particular, their ability to move from where they are now to where their optimal niche conditions will be in future. Bioclimatic modeling of current distributions can be used to produce climate envelopes for individual species. These can be superimposed onto regional templates of predicted climate to indicate where species may occur in future.

Different threatened species might be expected to make gains or losses in habitable area and, in theory, the effect of climate change might be neutral. However, the expected shifts in range towards the poles and higher in elevation will often involve contractions in species ranges and greater risks of extinction. We might expect global climate change to see r-species do relatively better than K-species. It seems likely, too, that habitat generalists will be less vulnerable than habitat specialists in a changing climate scenario. The potential ranges of invasive species can also be expected to shift; managers can benefit from predictions about where these might pose problems in future.

The powers of movement of species are critically important to future distributions. Not only can we expect a general loss of biodiversity because of reductions in habitable area. Recognizing that many species will not be able to get to newly habitable areas, the risk to biodiversity is even greater.

Climate change predictions based on the ecology of populations

Knowledge of population dynamics, and the way birth and death rates will be modified as a result of climate change, can be incorporated in future management plans. This is true for species deserving of conservation effort as well as pests (and their natural enemies) and potential invaders that pose a biosecurity risk. Equally important is a thorough understanding of the population consequences of climate change for the sustainability of fish harvests and forestry practices.

Climate change predictions based on the ecology of communities and ecosystems

Community composition is affected by disturbances that set in train community successions with characteristic trajectories and end points. Climate change may change disturbance frequency (by altering storm and fire patterns and fires). It may also modify the interactions among species that influence the course of succession. And finally, climate change is predicted to affect successional end points or, in other words, the characteristic biome of the area in question.

Climate change may affect particular species and these, because of their community role, can produce ramifications that spread through the food web. Some species are more vulnerable because they are affected both directly by climate change and indirectly through interactions with their prey, predators or mutualists (which themselves will be influenced in diverse ways by climate change). The predicted spread of human disease will sometimes depend on the consequences of climate change for the distribution of insects that act as disease vectors.

Climate change can also be expected to impact on a variety of ecosystem services, including the provision of sufficient water, productive soils, flood control and recreational opportunities such as skiing.

Climate change predictions based on landscape ecology

When considering the effects of climate change, a landscape perspective is generally appropriate because managers need to envision how global climate change will reorganize regional landscapes. A prediction based on island biogeography theory is that smaller areas contain fewer species. Extinction risk of endemic species due to climate change can be estimated in terms of loss in area of key habitats. A large proportion of the world's biota seems at risk but suitable choice of protected areas can minimize the predicted losses. In some cases, current nature reserves will prove to be in the wrong places – climate change models can assist managers to plan future reserve networks.

The final word

Valerie the anthropologist can see the parallels between human populations losing their habitat on low-lying Pacific islands and the many examples where the distributions of plants and animals are predicted to change. *'It occurred to me that the average effects of global climate change might, in theory, be neutral – with some species benefiting and others losing out. But this isn't the case for biodiversity, is it? With global warming the ranges of species must generally shift towards the poles and up in elevation. And doesn't this mean that species ranges will tend to contract into smaller areas, with consequent losses to biodiversity?'*

Imagine the world is faced by global cooling rather than global warming. Outline the patterns of change that you would expect in species ranges, biodiversity, harvests, human disease and pest control under a global cooling, as opposed to global warming, scenario.

References

Barlein, F. & Huppop, O. (2004) Migratory fuelling and global climate change. *Advances in Ecological Research* 35, 33–47.

Beaumont, L.J. & Hughes, L. (2002) Potential changes in the distributions of latitudinally restricted Australian butterfly species in response to climate change. *Global Change Biology* 8, 954–971.

Begon, M., Townsend, C.R. & Harper, J.L. (2006) *Ecology: from individuals to ecosystems*, 4th edn. Blackwell Publishing, Oxford.

Bryden, H.L., Longworth, H.R. & Cunningham, S.A. (2005) Slowing of the Atlantic meridional overturning circulation at 25°N. *Nature* 438, 655–657.

Derocher, A.E. (2005) Population ecology of polar bears at Svalbard, Norway. *Population Ecology* 47, 267–275.

de Wet, N., Ye, W., Hales, S., Warrick, R., Woodward, A. & Weinstein, P. (2001) Use of a computer model to identify potential hotspots for dengue fever in New Zealand. *New Zealand Medical Journal* 114, 420–422.

Hannah, L., Midgley, G., Hughes, G. & Bomhard, B. (2005) The view from the Cape: extinction risk, protected areas, and climate change. *BioScience* 55, 231–242.

Hays, G.C., Richardson, A.J. & Robinson, C. (2005) Climate change and marine plankton. *Trends in Ecology and Evolution* 20, 337–344.

Hearnden, M., Skelly, C. & Weinstein, P. (1999) Improving the surveillance of mosquitoes with disease-vector potential in New Zealand. *New Zealand Public Health Report* 6, 25–28.

IPCC (2007) *Fourth Assessment Report of the Intergovernmental Panel on Climate Change.* Working Group 1, Intergovernmental Panel on Climate Change, Geneva.

Iverson, L.R., Schwartz, M.W. & Prasad, A.M. (2004) Potential colonization of newly available tree-species habitat under climate change: an analysis for five eastern US species. *Landscape Ecology* 19, 787–799.

Kell, L.T., Pilling, G.M. & O'Brien, C.M. (2005) Implications of climate change for the management of North Sea cod (*Gadus morhua*). *ICES Journal of Marine Science* 62, 1483–1491.

Kriticos, D.J., Sutherst, R.W., Brown, J.R., Adkins, S.W. & Maywald, G.F. (2003) Climate change and the potential distribution of an invasive alien plant: *Acacia nilotica* spp. *indica* in Australia. *Journal of Applied Ecology* 40, 111–124.

Neilson, R.P. & Drapek, R.J. (1998) Potentially complex biosphere responses to transient global warming. *Global Change Biology* 4, 505–521.

Newman, J.A. (2005) Climate change and the fate of cereal aphids in Southern Britain. *Global Change Biology* 11, 940–944.

Pyke, C.R. (2005) Interactions between habitat loss and climate change: implications for fairy shrimp in the Central Valley Ecogregion of California, USA. *Climatic Change* 68, 199–218.

Roura-Pascual, N., Suarez, A.V., Gomez, C. et al. (2004) Geographical potential of Argentine ants (*Linepithema humile* Mayr) in the face of global climate change. *Proceedings of the Royal Society London B* 271, 2527–2534.

Schroter, D., Cramer, W., Leemans, R. and 32 others (2005) Ecosystem service supply and vulnerability to global change in Europe. *Science* 310, 1333–1337.

Sohngen, B. & Sedjo, R. (2005) Impacts of climate change on forest product markets: implications for North American producers. *Forestry Chronicle* 81, 669–674.

Téllez-Valdés, O. & Dávila-Aranda, P. (2003) Protected areas and climate change: a case study of the cacti in the Tehuacán-Cuicatlán Biosphere Reserve, Mexico. *Conservation Biology* 17, 846–853.

Thomas, C.D., Cameron, A., Green, R.E. and 16 others (2004) Extinction risk from climate change. *Nature* 427, 145–148.

Volk, M., Niklaus, P.A. & Korner, C. (2000) Soil moisture effects determine CO_2 responses of grassland species. *Oecologia* 125, 380–388.

Walther, G-R., Post, E., Convey, P. et al. (2002) Ecological responses to recent climate change. *Nature* 416, 389–395.

Index

Figures in *Italic*; Tables in **Bold**; information in Boxes indicated by B

abalone food web   236–7
  abalone no-take areas   237, *237*
  effects of resurgent sea otters   236–7
  harvests prone to collapse   236
Acacia nilotica subspecies *indica* (spiny
      acacia)   298, *299*
Acacia spp.   241
Acaulospora   213
Acer opalus   211
Achillea mellefolium   62
acid rain   2
  effects of   17
acorn barnacle   *231*B
acorns, 'mast years' followed by increase in
      Lyme bacterium infected ticks   235
adaptive management   186
Aedes aegypti
  transmission of yellow fever   235
  vector for dengue fever   307, *308*
Aedes albopictus
  failure to replace *O. triseriatus*   236
  transmitter of dengue fever in
      Hawaii   235
  vector for dengue fever   307, *308*
African witchweed (*Striga asiatica*),
      eradication campaign, eastern
      USA   142–3
Agent Orange   150
agricultural intensification   17–18
  places pressure on agrosystems   253
  risks to biodiversity   51–2, 283–4
agricultural land
  marginal, available for 're-wilding'   13
  tropical, abandoned in Panama   62–3,
      64–5, 209
agricultural landscape, management a
      multidisciplinary endeavor
      283–5

agricultural weeds
  early-successional, *r*-selected   219
  greater richness in organically farmed
      wheat fields   274
agriculture
  arable farming   274, *274*
  change in practices can turn native
      species into pests   141
  and ecosystem health   253
  effects of increasing intensity   32
  and habitat degradation   17–18, *18*, *19*
  intensive   2, 245–7, 257, 276, 284
  minimizing fertilizer input and nutrient
      loss   245–7, 257
agroecosystems, sustainable   245–9
  constructing wetlands to manage water
      quality   247–8
  managing lake eutrophication   248–9,
      250
  stopping caterpillars eating broccoli
      245
agroenvironmental subsidies   53
Agrostis capillaris
  cultivar 'Goginan' (acidic lead/zinc
      wastes)   49
  cultivar 'Parys' (acidic copper
      wastes)   49
Aira caryophyllea, restricted to dry,
      nutrient-poor soil   *50–1*, 52
Alabama leafworm (*Alabama argillacea*)
  development of resistance to chemical
      pesticides   162
  primary cotton pest   151
Alaska, *Exxon Valdez* incident   17
alder (*Alnus*)   97
algae   *231*B, 240
  control of nuisance blooms in lakes
      256

algal biomass, highest in streams with
    brown trout   240–1, *241*
aliens, unwanted   41–6, 55, 141
    alien plants, British Isles   41, *41*, 42
    alien plants, Czech Republic   68–9, *69*
    disrupted ecosystems make invasion
        easy   44–6
    ecological niche modeling   42–3
    establishment largely dependent on niche
        requirements   42
    see also invaders; invasions
alleles, dominant or recessive   114B
allelochemicals   3
Alliaria petiolata   221
Allium spp., pest and food   141
allogenic succession   204, 208, 224
Alnus japonica   207
Alopecurus aequalis   207
alpine marmots (*Marmota marmota*),
        benefits of constant escapement
        harvesting   183
Alyssum bertolinii, heavy metal tolerant   49
American eels (*Anguilla rostrate*)   255
amphibians, signs of population decline
        6, 7
Amphiprion percula, a mutualist   40B
Aneilema keisak   207
angling   37
animals, restoration timetable for   208–9
    coastal sand dunes, South Africa   208,
        208
    tidal salt marshes   208–9, *209*
Anopheles gambiae, transmission of malaria
        in Brazil   235
aphid alarm pheromone   150–1
aphid pest (*Rhopalosiphum padi*)   307
    decline in abundance under global
        climate change   303–4, *304*
apples, blemished   141
Arabis hirsute   277
Arctic Oscillation
    likely to change, consequences for
        viability of polar bears   303
    linked to North Atlantic Oscillation   303
Arctic sea ice, thinning and reduction
        in   293B
arctic tern (*Sterna paradisaea*), migration
        from North to South poles   82, 83B,
        105
Arcto-Norwegian cod fishery   187–8
    low effort and large mesh size gave best
        results   188

mesh sizes eventually increased   188
    stocks seriously depleted through
        overfishing   189
Ardisia crenata, experimental inoculation
        with mycorrhizal fungi   44
Argentine ants
    development of niche model, future
        predictions of distribution   27–8B,
        300, *300*
    found on every continent except
        Antarctica   300
    unpleasant consequences for biodiversity
        and domestic life   300
armadillo (*Priodontes maximus*)   271
Arnhem Land, aboriginal fire regime
        contrasted with typical modern
        pattern   218–19, *219*
    Dukaladjarranj vs. Kakadu National Park
        species diversity   218
    fire-sensitive vegetation remained well
        represented   218
    planned burning to provide green forage
        for game animals   217, *219*
Artemesia gmelinii   296B
Artemesia scoparia   296B
aspen (*Populus tremula*)   97
Aster oblongifolius   216
atmospheric composition and climate
    increase in atmospheric CO_2 18, *19*, 20,
        32
    observed changes   293B
    observed ecological consequences   293B
    observed physical environmental
        consequences   293B
    see also greenhouse gases
Atriplex prostrata, halophytic   50
augmentation biological control
    inoculation   154–5, 158–9
    inundation   154–5, 159–60
Australia
    Arnhem Land, aboriginal fire regime
        contrasted with typical modern
        pattern   218–19, *219*
    Bongil Peninsula, invader-dominated
        sand-mined sites   210
    effect of climate change on some butterfly
        distributions   298
    IPM used against an invasive weed
        166–7
    koalas at risk   120–2
    largest reserves on unwanted land
        13, *13*

marsupials, extinct or endangered   73, 73

New South Wales, effects of logging in *Eucalyptus* forest   97–100

South, southern emu-wren conservation   267–8

southwestern, rufous bristlebird threatened by habitat loss   46–7

Western, minimum number of sites to include all fish biodiversity   269, *280*

Austria, grassland study in meadows   52–3

autogenic succession   204B, 224–5

primary and secondary   204B, 225

Bacillus thuringiensis (Bt)

genetically engineered to produce novel combinations of endotoxins   164

in inundation biological control   159

two strains effective against malaria, *israelensis* and *sphaericus*   164

widespread concern over insertion into GM crops   159

Balanus balanoides   40B

baldy cypress swamp (*Taxodium distichum*)   92

seeds short lived   92

baleen whales, migration to Antarctic   83B

ballast water, source of many alien species   92–3

compulsory to dump in open ocean (Great Lakes)   93

other possible cleaning methods   93

banana poka (*Passiflora tarminiana*), serious threat to Hawaiian high elevation forests   156

Barents Sea study, showing biomagnification of chlordane   *148*

barn swallows (*Hirundo rustica*), and a positive NAO index   301, 302

barnacles, coexistence of   40B

barndoor skate (*Dipturus laevis*)   75

Bashania fargesii, panda food   24B, 85, 86

bats   97–100, 253

activity and species richness assessed, southern England   253

community recovers well within 15 years of logging   99–100

comparison of activity, logged and unlogged forest patches   98–100, *99*

as ecosystem health indicators in agricultural soils   253

essential maternity roosts   100

importance of trackways as dispersal pathways for feeding   98

may be favored by some forestry practices   97

species richness highest on forest tracks   98

bearded vultures (*Gypaetus barbatus*)

important variable, biomass of ibex   275

reintroduction program for   275–6

Beauveria bassiana   159

bees, native, pollination services of   103–4

maximum foraging distance   103–4

beet army worm (*Spodoptera exigua*)   152

behavior management   89

Bellis perennis   *50, 52*

benzoin gardening in Sumatra   216–17, 225

indigenous people know range of uses for forest plants   217, *218*

tapping trees of genus *Styrax*   216–17

three categories of garden and tree species richness   217, *218*

Berteroa incana   219

big-horn sheep (*Ovis canadensis*)

long-term population records in desert areas   116

smaller the population greater risk of extinction   116, *116*

bigeye tuna   *238*

bioclimatic niche modeling, *Protea* species in Cape Floristic Region   297–8, *297*

biodiversity   4, 287

in Berlin   277

can be a matter of economics   52–3, *53*

compromised by human actions   61

consequences for in regions fulfilling economic potential   6–7

ecosystem function and ecosystem services   7–10

regional, depleted by intensive agriculture   284

selecting new reserves for   279

unwanted   13–14

biodiversity change, principal drivers   11–12, *12*

biodiversity conservation

designing reserve networks for   277–80, 284B, 287

possible in big cities   277

relevance of Island Biogeography Theory   264B

biodiversity crisis   4–20, 32
    scale of problem   6–7
biodiversity hotspots   265B, 277, *278*
biodiversity loss   2, 312
    awareness of   8, *9*
    causes   32
    drivers of, the extinction vortex   11–12
    through agricultural intensification
        51–2, 283–4
    tropical forest in Panama   62–3
biodiversity managers, climate change a
        problem for   309–10
biological control agents   25B, 162–5
    occasional use of vertebrates   157–8
    unwanted outcomes   160–1
biological pest control   143, 154–61, 167,
        168
    augmentation   154–5, 158–60
    conservation biological control   154,
        156–8
    importation of a natural enemy   154,
        155–6
    introduction/augmentation of enemies of
        target pest   *145*B, 146B
    when it goes wrong   160–1
biological pesticides   154–5, 159–60
biological pollution, invasions as   96
biomagnification   147
    of DDT and related products   147–8,
        148
biomanipulation, managing
        eutrophication   248–9, *250*
biosecurity managers, animal
        invaders   241, 256
biosecurity precautions   14, 28B
    routine in many parts of the world   92
biosecurity strategies, national   96
birch (*Betula* spp.)   97
bird biodiversity
    recovery after logging   99–100
    relative importance of extinction
        drivers   11, *11*
bird populations
    dispersal, migration and wind farms   82
    uptake of GM technology impacts on
        weed abundance   133, *133*
birds
    access to, consequences for
        broccoli   245, **246**
    British, species in decline   6, 7
    extinctions   11, *11*
    invasion success   66–7

migrating and dispersing, threatened by
        wind farms   100–1
migratory, challenged by global climate
        change   301
non-migrating, powers of movement
        important for future
        distribution   302
vulnerable to predation of eggs and
        young and nest parasitism   268–9
bison (*Bos bison*), grazing in shifting prairie
        mosaic   207
black fly (*Simulium damnosum*)
    larva control program   162–3
    strategy, rotating range of
        pesticides   163, **163**
black guillemots (*Cepphus grylle*), evidence
        of biomagnification   *148*
black rockfish (*Sebastes melanops*), larvae
        from older females   189, *189*
black-legged kittiwake (*Rissa
        tridactyla*)   102
black-throated diver (*Gavia arctica*),
        nonmaneuverable and
        vulnerable   102
bog restoration   89–91
boll weevil (*Anthonomus grandis*), primary
        cotton pest   151
Bombycilaena erecta, tolerant of
        grazing   *51*, 52
Borrelia burgdorferi   236
    bacterium causing Lyme disease   26B,
        234
    white-footed mouse, efficient transmitter
        to ticks   235
botanicals   147, 167
Bothriochloa ischaemun   296B
Brachypodium pinnatum, tolerates only
        minimal grazing pressure   *51*, 52
bracken fern (*Pteridium aquilinum*), spreads
        by underground rhizomes   83B
British Isles, alien plants   41, *41*, 42
    many arrive from similar climatic
        locations   41, 42
    more found in disturbed habitats close to
        transport centers   41
broccoli (*Brassica oleracea*), predation by
        caterpillars   245
brown trout (*Salmo trutta*)
    alter energy and nutrient flux   240–1
    change behavior of grazing herbivorous
        algae   240
    fare better in floods than native fish   45

an invasive species 37
predation by 40B
brown-headed cowbird (*Molothrus*
    ater) 268
bushmeat harvest, tropical forests
    significant threat to biodiversity 196
    vital food resource for local people 197
butterflies
    Australian, effect of climate change on
        distributions 298
    first aid for 222–3
by-catch
    and extinction risk 185
    monitoring of 185–6

C:N ratio 234B, 244
cabbage loopers (*Trichoplusia ni*) 152
Cache River floodplain, Illinois, possible
    restoration of baldy cypress
    swamp 92
cacti, Mexican, reserves in wrong
    place 309–10
Caesalpina violacea 214
California
    abalone population, human harvest and
        sea otter predation 236–7, 237
    biological reserves, Central Valley 310,
        310–11
    Channel Islands, outfoxing the
        foxes 130, 130
    forest fires 203, 226
    invasions and failed introductions of fish
        in catchment areas 70–1
    northern, role of native bees in
        watermelon farms 103–4, 103, 104
    successful eradication of *Terebrasabella
        heterouncinata* 142
Californian red scale insect (*Aonidiella
    aurantii*) 158
Callitris intratropica woodlands 218
Callophrys affinis 223
Calophyllum longifolium, survives well in
    Saccharum grassland 63–4
Canada lynx (*Lynx canadensis*)
    clear-cutting beneficial in long term 222
    occurs in areas with regenerating
        forest 222
    specialist predator of snowshoe
        hares 221
Canis simensis (Ethiopean wolf) 123–6
canonical correspondence analysis 45, 45
Cape Floristic Region, South Africa 297B

conservation planning for 279–80
'fynbos' 297–8, 297
irreplaceability analysis 279, 281
Carapa guianensis, survives well in
    Saccharum grassland 63–4
carbamates 149, **149**
carbon addition, lowers soil nitrate
    levels 244, 245
carbon dioxide (CO$_2$) 18–20
    indirect and direct effects on plant
        performance and distribution 298
    sources and sinks 19–20, 20
Caribbean black-striped mussel (*Mytilopsis
    sallei*), blunderbuss approach to
    eradication 142
carrying capacity 112B, 123–6, 175B,
    176B
Carteret Islands, disappearing 291
Chalinobus motia 99
chalk grassland species, plotting mean
    locations and variance 50–1, 51–2
Chamaecytisus supinus, tolerates only
    minimal grazing pressure 51, 52
Channel Islands (Californian coast)
    long-term records of birds indicate
        minimum viable population 116
    outfoxing the foxes 130, 130
chemical pesticides 146–54, 167
    undesirable outcomes argue for
        precautionary approach 154
    use in poorer countries 154
China, giant pandas 83B, 85–6, 86
    design of nature reserves for 24B
Chinese bush clover (*Sericea lespedeza*),
    native and potentially invasive 42,
    43, 43
chitons 231B
chlordane, biomagnification of 148
chlorinated hydrocarbons 147–8, 148, **149**,
    151
Chrysanthemum spp., pyrethrum from 147
Chthamalus stellatus 40B
cityscape ecology 277, 287
classical biological control (importation of
    natural enemy) 155–6
climate change 4
    caused by CO$_2$ 17
    caused by greenhouse gases 293B
    future, ecological consequences
        of 293B, 296B, 297B
    likely effect on European ecosystem
        services 307–8

climate change (cont'd)
    modification of interactions among
        species 306
    power to change disturbance
        frequency 306
    predicted to affect successional end
        points 296B, 306
    predictions for twenty first
        century 293B, 294–5B
    sociopolitical scenarios 30, **31**
    see also global climate change
climate change modeling 291
    changes to water stress in Europe 308,
        309
    including SHIFT, of sweetgum 302–3,
        302
    large river catchments, effects in 308
    positive effect on European
        vegetation 308
    predicted twenty first century
        changes 293, 294–5B
    simulated world distribution of
        biomes 296B
climate change predictions 297–303
    based on community–ecosystem
        interactions 306–8, 312–13
    based on ecology of individual
        organisms 297–303, 312
    based on landscape ecology 313
    based on theory of population
        dynamics 303–5, 312
    life history traits and the fate of
        species 300–3
    niche theory and conservation 297–8,
        297B
    niche theory and invasion risk 298, 299,
        300, 300
climate envelope models 40B
    used to indicate future species
        occurrences 293B, 297B, 312
climate matching 40B, 44
climate models 40B, 44, 312
    climate change models 291, 293,
        294–5B, 296B, 302–3, 302, 304, 308
climatic conditions, year-to-year vagaries
        in 179–80, 184
coastal sand dunes, South Africa
    millipedes colonize spontaneously
        regenerating dunes 208, 208
    recovery from opencast mining 208
    successional pattern in dung beetle
        community 210

Cochlospermum vitifolium 214
cod (Gadus morhua)
    Arcto-Norwegian cod fishery 187–8
    North Atlantic, decline in maturity
        size 26B, 192, 193
    outcomes of climate change
        models 304, 305
    shows evidence of biomagnification 148
    see also Newfoundland cod fishery; polar
        cod
Colias eurytheme 223
Colorado pikeminnow (Ptychocheilus lucius)
    present distribution 53, 54
    rarity due to river bed accumulation of
        fine sediment 53, 54
coltsfoot (Tussilago farfara), an invader 60
Columbia, newly discovered rice virus 141
commercial whaling, strictly
        regulated 198
common skate (Dipturus batis) 75
    locally extinct 185
communities
    composition affected by
        disturbances 303, 312–13
    patch dynamics concept 263B
    some richer in species than others 265B
community ecology 22, 26–7B, 33
    and concept of ecological
        succession 203
    food web theory 22, 26–7B
    succession theory 22, 26B
community groups, confronting a
        sustainability issue 27B, 29–30
competition, role in reducing fundamental
        niche size 40B
competition–colonization trade-off 204B,
        225
    in Panama and Costa Rica 209
competitiveness 77
complementarity 279, 280, 288
    complementarity analysis in Western
        Australia 279, 280
conifers, invasive, in the USA 23B, 66, 67
conservation arguments, in cost–benefit
        terms 28
conservation biological control 154,
        156–8
conservation genetics 127–30
    genetic rescue of the Florida
        panther 128
    outfoxing the foxes 130, 131
    the pink pigeon 128–9, 129

reintroduction of a 'red list' plant
129–30, *130*
conservation management
and food webs   239, *240*
information used to define adequate
management   *50, 51*, 52
knowledge of intricacies of food
webs   239
lessons from emu-wren problem   268
one large or several small reserves
264B
painful decisions about priorities   130–1
conservation planning
for Cape Floristic Province, South
Africa   279–80
regional or national scale   279
constant escapement, in space   183–4
spatial controls on fishing and
hunting   183
constant escapement, in time   182–3
appropriate for annual species   183
strategy used for the *Loligo* squid   183,
183
coral reefs
cold water, destruction by bottom
trawling   17, 28
protection against tsunamis   230–1
use of dynamite   17
corncrake (*Crex crex*), declining   113
cost–benefit economic analyses, for
estimates of EIL and ET   143–4B
Costa Rica, migratory behavior of
three-wattled bellbird   84–5
complex behavior poses big challenges
for managers   84–5
Costelytra zealandica, became a pest   141
costs, fixed and variable   194
Cotesia plutellae, parasitoid   165, *165*
cotton
emergence of secondary pests   151
example of pesticides gone wrong   151,
152
mass use of organic insecticides   151
cotton boll worm (*Heliothis zea*)   *152*
cottony cushion scale insect (*Icerya
puchasi*)
and first use of DDT in California   165
importation of two candidate
enemies   155
long-term outcome, both enemies are
necessary   155
Crataegus monogyna   211

crop plant pests
the most costly   14, *15*
see also pests
Cryptochaetum sp.   155
CSR concept   61B, 71, 77
competitive strategy (C)   61B
ruderal strategy (R)   61B
stress-tolerant strategy (S)   61B
Cuba, spiny lobster fishery
determination of long-term maximum
profit   194
effort (boat fishing days)   *195*
previously managed according to biomass
yield   194
cultural eutrophication   247
cultures, identity can be dependent on
native flora and fauna   224
cyclamen mites (*Steneotarsonemus pallidus*),
target pest resurgence   151
Czech Republic, transport, release and
establishment of alien plants   68–9,
69

data-less fisheries management   186
DDT   163
and related products, problems of toxicity
and environmental long life   147, **149**
resistance to   162
decomposer system   232B, 256
secondary production and respiratory
heat loss   232B
deer mouse (*Peromyscus maniculatus*),
primary vector of Sin Nombre
hantavirus   160, *161*
deforestation, fractures within-forest
nutrient cycling   234B
degradation
of habitat   4, 17–18
of provisioning services   9
of regulating services   9–10
demographic uncertainty   135
in small population   113B
dengue fever   235, 307, *308*
developed nations, responsible for climate
change   291
diamondbacked moths, parasitoid attack,
application of botanical
pesticide   165, *165*
Dianthus monspessulanus   51, 52
diclofenac, used for sick cattle   133–5
and associated vulture deaths   134–5,
134

dieldrin   153–4
Dipteryx panamensis   209
  survives well in *Saccharum*
        grassland   63, 64
dispersal   83B, 301
  active or passive   83B
  by clonal growth   83B
dispersal behavior   92–6, 104
  thorough understanding of necessity to
        managers   96
dispersal corridors   207–8, 264B, 270
  novel dispersal pathways   92
dispersal and migration   83–4B, 105, 263B
  dispersal   83B
  dormancy   84B
  migration   83B
Distichlis spicata   208
disturbances   203
  affecting community composition   306
  and concept of the successional
        mosaic   205–6B
  in grasslands   204B
  open up forest gaps   204B
  on rocky shores   204B
  variation in relative importance of
        species on abandoned plots   206B
DNA analysis, can identify closeness of
        relationships   130
dormancy   60B, 64, 77, 84B, 105
  see also seed-banks
dynamic pool models in fisheries
        management   187–90

early-successional species   204B, 219, 225
Eastern red cedar (*Juniperus
        virginiana*)   215
Echinacea paradoxa   216
ecological communities   203
ecological consequences
  of freshwater invaders   240–1
  observed, of global climate change   293B
ecological engineers, changing habitats   3
ecological factors, conditions and
        resources   37–8B
ecological niche modeling   39–40B, 39B,
        42–3
  climate envelope models   40B
  climate matching   40B, 44
  modeling fundamental or realized
        niches?   44
  studies of four invasive plants in North
        America   42–3, 43

of sweetgum fundamental niche   302–3,
        302
ecological problems, may have social
        aspect   56
ecological restoration   242–4, *245*
ecological specialists, vulnerable   72
ecological succession   86
ecological succession theory   203, 204–6B
  allogenic succession   204B
  autogenic succession   204B
  primary or secondary successions   204B,
        *205*B, 206B
  species traits determine course of
        succession   204–5B
  successional mosaic and central role of
        disturbance, concept   205–6B
ecological theory, hierarchical
        organization   22, 23–8B
ecologists, as advocates or advisers   xv
ecology, meets economics and social
        sciences   xiv–xv
ecology and economics, and the
        sociopolitical dimension   131–3,
        195–8, 200
Economic Injury Level (EIL)   143–4B
  pest control and economic gain   143–4B,
        *144*B
economic optimum yield (EOY), and MSY,
        difference between   194, *194*
Economic Threshold (ET)   144B
economics, traditional, environmental costs
        not taken into account   28
ecosystem ecology   26–7B, 33
ecosystem effects
  of freshwater invaders   240–1
  of invasive plants   241–2
ecosystem health   231, 257
  in an agricultural landscape   253
  of forests   252–3
  of a marine environment   255–6
  often or partly a social construct
        154–5
  of rivers   254–5
ecosystem management, pressure, state and
        management response   252, 253
ecosystem processes, possibly responding
        to biodiversity change   7–8
ecosystem properties   8
ecosystem provisioning services   9, 230–1,
        249, 257
ecosystem regulating services   9–10, 231,
        249, 257

ecosystem services   *10*, 28, **31**, 136–7,
        207–8, 230–1, 257
    contributions of species to   251–2
    cultural and supporting services   9
    from intact wetland   251
    losses to   32
    lost, determination of real costs
        possible   29
    lost through pesticide use   284
    many degraded by humans   9–10, 252,
        257
    provided by native plants and
        animals   104
    provided by vultures   109, 135
    value of   231, 249, 251–2
    of the world, estimated total value   29
ecosystem theory   22
ecosystems   7, *10*, 230
    abiotic compartments   233B
    disrupted, make invasion easy   44–6
    energy and nutrient dynamics in   256
    natural, destroyed by humans   2
    unhealthy   252, 257
eels (*Anguilla* spp.), long-distance one-way
        migrant   83B
El Niño events   *180*, 185
Elytrigia repens   219
emerging infectious diseases   141
emu-wren (*Stipiturus malachurus
        intermedius*)   267–8
    metapopulation in patches of dense
        swamp   267, *268*
    need for interpatch corridors   207–8
    optimal management decisions   268
Encarsia formosa, used for inoculation   158
endangered species   239, 286, 287
    dealing with   109, 110, 110–13B
    plants, found in city lawns and
        gardens   277
    population of   262
    relevance of successional theory
        to   221–4
    restored to locations where extinction
        occurred   89
endangered species conservation   46–9, 55,
        108–38
    historical reconstructions help identify
        best reserve sites   49
    Monarch butterfly   46–7
    setting aside areas where exploitation is
        restricted/prohibited   46
    translocation of the takahe   23B, 48–9

endemism hotspots   277
enemy-interaction theory   215–16, 225
energy flow
    general patterns of   233B
    ocean food web, central north
        Pacific   *238*
English pasture, recovery to species-rich
        meadow   208
entomopathogenic nematode worms, as
        biological control agents   159–60
environmental factors, influencing
        birth, death, migration rates
        and λ   110–11B
environmental impact, of human
        population density and
        technology   3–4
environmental issues
    nearly all have a sociopolitical
        angle   29–30, **31**, 33
environmental problems, risky to assume
        availability of technical fixes
        21
environmental uncertainty   135
    in small population   113B
epidemiology theory, size of epidemics and
        intervals between   123–4
Equilibrium Population   144B, *144*B
Eryngium campestre, defence against
        grazing   *51*, *52*
Ethiopean wolves, dogged by
        disease   123–6
    carrying capacity of each occupied
        patch   123–6
    critically endangered   124
    domestic dogs reservoir of rabies   124
    female dispersion   124
    live in close-knit, male biased
        packs   124
    population viability sensitive to female
        recruitment to packs   123–4
    potential of vaccination intervention to
        limit epidemics   125
Euglandina rosea, biological control agent
    disaster in Polynesian islands   160
    unsuitable, generalist diet   160
Euphorbia maculata   219
European hare (*Lepus europaeus*)   276
European rabbit (*Oryctolagus
        cuniculus*)   140
European stem borer (*Dioryctria
        sylvestrella*), moth caterpillar
        272

eutrophication
  cultural 247
  and excess nutrients 233B
  see also lake eutrophication
exotic species
  abundance and richness declines during
      succession 221, *221*
  lost in forests due to canopy
      closure 221
  newly arrived, eradication of 142
exploiter mediated coexistence 242–3,
      244
extinction 4
  of birds 11, *11*
  chances of 110
  humans adding to causes 109, 110
  local 264B, 270
  uncertainty and risk of 113B
  vulnerability to 71–2
extinction drivers 11, 32
extinction hotspots 277
extinction rates
  difficult to establish 6
  natural 6
extinction risk 185
  assessment from correlational data 113,
      115B, 116
  categories of (IUCN Red List) 5B
  classification 5B
  economic extinction occurs before
      ecological extinction 185
  of endemic species estimated 308–9
  fairy shrimps, California *310*, 311
  high, Cape Floristic Region 297
  highest for large-bodied, *K*-selected
      species 73–4, *73*, 116
  increase in 298
  skates and rays 74, *75*
  species traits as predictors of 71–7
extinction vortex/vortices 11, 136
  progressively lower population size to
      extinction 115B, *115B*
extinctions
  modern 16
  prehistoric 14, 16
Exxon Valdez (oil tanker), 1989
      incident 17

facilitation, by early species 205B
facilitation theory 210–15, 225
  nurse plants 210–13
fairy shrimp (*Branchinecta* spp.) *310*, 311

fairy shrimp (*Streptocephalus vitreus*),
      dormant phase 60B
fairy shrimps, inhabitants of temporary
      ponds *310*, 311
  current distribution of protected areas,
      California *310*, 311
  different niche requirements for different
      species 311
  variation in F_{max}, current climate
      conditions *310*, 311
Falkland Islands, constant escapement
      strategy for *Loligo* squid *183*, 193
Falsistrellus tasmaniensis 99
Fargesia spathacea, panda food 24B, 85, 86
farm animals, nitrogen-rich waste
      from 247
farmed habitat, restoration for hares 276,
      277
farmers, use of honey bees 103
fat hen (*Chenopodium album*), suitable to
      show GM sugar beet effects on
      weed–bird interaction 132–3
Fender's blue butterfly 117
  probability of population
      persistence 117, *118*
ferrets (*Mustela furo*), imported into New
      Zealand to control rabbits 140–1
fertilizers 253
  heavy use 2, 17–18, *19*
  indiscriminate use should cease 21
  runoff from fuelling ocean dead
      zones 247
  and sustainability 21
  tools to minimize loss from the
      land 246
Festuca ovina *50*, 52
Festuca rubra 62
  cultivar 'Merlin' 49
fires 64, 203, 225
  burning brush piles in Missouri 215–16
  creating successional mosaic benefiting
      hunters 217–19, *219*
  effects on tropical dry forests 213
  now less frequent but more intense 206
  see also forest fires
fish
  effect of size on quality of
      offspring 189–90
  food web consequences of
      harvesting 238–9
  harvesting in the future 304
  influenced by high river discharges 53

management by fixed quota   178–81
survival affected by increased sea
    temperature   304, *305*
see also shark species; skates and rays
fisheries collapses
    from overfishing   184
    unfavorable climatic conditions   179–80,
        180, 184–5
fisheries management, dynamic pool
    models   187–90
    Arcto-Norwegian cod fishery
        example   187–8
    management measures, input controls or
        output controls   190
fisheries managers   304
    use of ecosystem-based fisheries
        models   238, *238*
'fishing down marine food webs'   239, 257
fishing industry, a keystone predator   238
fixed effort, management by   177–8B, *178*B,
        181–2, 199
    Pacific whiting   181
    Saiga antelope   181–2
fixed quota, management by   176–7B,
        *177*B, 178–81, 199
    of fish   178–80
    of moose   180–1
    setting of quota for the year   178–9,
        179
flood protection   284
Florida panther (*Puma concolor coryi*),
        genetic rescue of   24–5B, 128
    cougar ancestry, dramatic reversals of
        undesirable traits   128
    suite of undesirable traits   128
    translocation of individuals from another
        subspecies   128
flower-visitation webs   *214*
fluoracetate (basis for pesticide 1080)
        163
flying squirrel (*Pteromys volans*)   97
    core breeding habitat   97
    dramatic decline in Finland   97
    favors spruce-dominated forest   97
    optimal breeding habitat
        recommendations   97
    typical breeding and dispersal
        landscape   *98*
food or feeding techniques, novel, adoption
        by birds   210
food web theory   22
    and human disease risk   26–7B, 234–6

food webs   230, 256
    aquatic, affected by leaching
        fertilizers   245–6
    and climate change   313
    and conservation management   239, *240*
    direct and indirect effects in   231B, 307
    and harvest management   234–9
    ocean, central North Pacific   238–9, *238*
    top-down and bottom-up control   232B,
        256
food webs and pathways of energy and
        nutrients   231–4B
    direct and indirect effects in food
        webs   231B, 307
    ecological energetics   232B
    ecological stoichiometry   233–4B, 241,
        256
    nutrient dynamics   232–4B
    strong interactors and keystone
        species   231–2B
food-web interactions   307
forbs   242
    in grassland restoration   62, **63**
    morphological defences to grazing   *51*, 52
forest fires
    California   203, 226
    Indonesia   *19*, 20
forest fragmentation, variation in wood
        thrush fundamental rate of increase
        λ   268–9, *269*
forest health, potential value of predatory
        soil mites as indicators of   253, *254*
forest improvement, willingness to pay
        for   276
forest patches
    different harvest values, Peruvian
        forest   271, **272**
    linking forest fragments   213
    logged and unlogged, bat activity
        in   98–100, *99*
    vulnerability of birds living on edges
        of   268–9
forested wetlands, much lost to
        agriculture   91
forestry   190–1
    a boost for developing countries   305,
        306, *306*
    plantation forestry in the
        landscape   272–3
    planting of more appropriate trees   305
    sustainable harvesting of New Zealand
        beech forest   190–1

forestry management 105
forests
  community forest concessions 173, 174
  dieback and global warming 305, **306**
  ecosystem health of 252–3
  gaps caused by disturbances 204B
  landscape management in 271, **272**, 287
  natural, protection for future
      generations 173–4
  regenerating, home to lynx and snowshoe
      rabbit 221–2
  using succession to control
      invasions 220–1, 226
  see also Mediterranean forest; tropical dry
      forests; tropical forests
fossil fuels
  causing atmospheric pollution 17
  increase in atmospheric CO_2 18, *19*, 32
  indiscriminate use of threatens
      sustainability 21
  oil pollution 17
French pine plantations, attacks by
      pests 272–3, *273*
fundamental niche 53, 55
  of hares 276
  many dimensions 38–9B, *38B*
  sand shrimp (*Crangon septemspinosa*),
      two-dimension niche *38B*, *39B*
  summary of organism tolerance and
      requirements 38B
  of sweetgum, modeled 302–3, *302*
fur seals (*Callorhinus ursinus*), anticlockwise
      predator–prey spiral 195, *196*
fynbos vegetation, South Africa
  assessment of extinction risk 308–9
  extinction risk as result of habitat
      loss 297
  Protea species, possible gains and
      losses 297–8, *297*

Galaxias anomalus, fundamental
      niche 45–6
Galaxias depressiceps, fundamental niche
      invaded 40B
Galaxias spp. 240
gallflies (*Urophora* spp.) 160, *161*
garlic mustard (*Alliaria petiolata*) 43, *43*
  adversely impacts rare native insects 42
genetic diversity
  loss of in small populations 114B
  lower in small populations 127
  and population dynamics 270

genetic drift, in small populations 114B,
      136
genetic mixing, reduces inbreeding
      effects 127, 128–30
genetic modification 131–3
  control action of *Bacillus thuringiensis*
      (Bt) 159
  use of glyphosate on crops 149
genetic variation 114B, 136
  determined by natural selection and
      genetic drift 114B
  loss of in small populations 114B
genetics of small populations 114–15B,
      136
  inbreeding depression 114–15B, 136
  loss of genetic variation 114B
  relative importance of genetic and
      demographic risks 115B
Genipa americana 209
Germany
  development of species sensitivity index
      for some seabirds 101–2
  North Sea, areas of concern about wind
      farms 102, *102*
  wind farm sensitivity index (WSI) 102
Ghana, bushmeat consumption vs. fish
      consumption 197, *197*
  harvests inextricably linked, risk of
      overexploitation 197
giant pandas (*Ailuropoda melanoleuca*),
      conservation of 85–6
  core panda habitats 86, *86*
  elevational migrants 85
  extreme dietary specialists 85–6
  migratory behavior and design of nature
      reserves for 24B
  nature reserves are insufficient 86
  seasonal variation in feeding habits
      83B
Gibraltar, installation of two land-based
      wind farms 100–1, *100*
  important migration bottleneck between
      Africa and Europe 100
  migrating soaring birds not badly
      affected 101
  mortality rates 101
Gigaspora 213
glaciers and icecaps, melting 18–19
glasshouses, inoculation biological control
      in 158–9
glaucus gull (*Larus hyperboreus*), shows
      evidence of biomagnification *148*

Gliricidia sepium   214
global climate change   8, 11, 18–20, 22,
        290–314
    effects on tree growth and forestry
        yields   305, **306**, *306*
    observed changes to atmospheric
        composition and climate   293B
    observed ecological consequences
        293B
    observed physical consequences   293B
    predicting ecological effects of   292–7B,
        312
    r-species and habitat generalists less
        vulnerable   301
    see also climate change
global extinction   113B
global hotspots   287
    Cape Floristic Province, South
        Africa   279
    identification to guide international
        conservation efforts   277, 279
global mean sea level, rise in   293B
globalization, how to manage invasions
        under   96
Glomus spp.   213
glyphosphate, genetic modification in
        certain crops   149
golden whistlers (*Pachycephala pectoralis*),
        declined after logging   72
goose barnacle (*Pollicipes* sp.)   231B
gorse (*Ulex europaeus*)   223–4
grass genotypes, metal tolerant   49
grassland
    Californian, change in grass types
        242
    disturbances in   204B
    using succession to control
        invasions   219–20
grassland experiments, Europe   8, 9
grassland restoration   62, **63**
    forbs, good establishment   62, **63**
    grasses   62
    increasing restoration efficiency   62
    relating species performance to
        life-history traits   62
    to perennial or annual grassland   242
    see also prairie grassland restoration
grazer system   232B, 256
grazing, by zooplankton, fish-induced
        reduction in   240–1
grazing intensity, can affect likelihood of
        invader success   44

grazing pressure
    'decreasers' and 'increasers'   77
    and perennial species   77
grazing trials, prediction of grazing
        vulnerability   77
great auk (*Alca impennis*), globally
        extinct   185
Great Lakes of North America,
        invasions   92–3
    invaders arrived in ballast water   92–3
    sea lampreys, treated with lampricide
        TFM   150
    zebra mussels, devastating effect of   14,
        15, 93–4, 241
Great Lakes–Baltic Sea trade link, biotic
        homogenization at each end   14
Great Lakes–Caspian Sea trade route,
        arrival of the zebra mussel   93
great tits (*Parus major*), as biological
        control agents in orchards
        157–8
greater prairie-chickens (*Tympanuchus
        cupido pinnatus*)
    current populations isolated   114B
    relationship between population size and
        genetic diversity   114B
greenhouse gases   293B, 312
    further rises in and continuing
        temperature rise   18–19
    future emissions, varying estimates   30,
        31, 33
    principal causes of increase   19
griffon vultures (*Gyps fulvus*), deaths in
        Gibraltar   101
grizzly bears (*Ursus arctos*), probability of
        population persistence   270
growing season, lengthened   293B
Guazuma ulmifolia   214

habitat degradation   2, 32, 71
    by agricultural development   17–18, *18*,
        19
    by fossil fuels   17
habitat fragmentation   12, 71, 73
    vulnerability of shade-tolerant
        species   77
    vulnerability of the southern
        damselfly   87
habitat loss   4, 12–13, 32, 71
    biggest problem facing threatened bird
        species   11, *11*
    management response to   13

habitat restoration   49–54, 55–6
  agricultural intensification   51–2, 55
  cost of restoring a species   52–3
  land reclamation   49–51
  mined sites   49–51, 55
  river restoration   53–4, 55–6
habitat specialists, at higher risk of
    extinction   73
habitats
  continuous, diffusive spread of
      invaders   96
  discontinuous, invasion hubs   95–6
  K-selecting   61B
  r-selecting   61B
hairy vetch (Victa villosa), as a nurse plant
    to sweetgrass   224
halibut (Hippoglossus stenolepis), alternative
    to Pacific fur seal fishery   195
halophytic plants   50
hare barley (Hordeum murinum)   242
hares (Lepus europaeus), restoration of
    farmed habitat for   276, 277
  agriculture now too intensive   276
  habitat availability   276, 277
  resource availability   276, 277
harvest management   172–201, 257
  adaptive management approach   186
  avoiding the tragedy of the
      commons   173–4, 198
  counteracting evolution towards small
      size   25–6B
  difficulties in using MSY as basis
      for   176–7B, 177B, 178–86
  early maturation a threat to   192
  ecologist's role is stock assessment   186,
      199
  and food webs   236–9
  killing just enough   174, 178
  MSY concept central to   174, 178–86
  role for nature reserves   183
  setting aside nonharvested areas for
      larger fish   192
  socioeconomic view of   195–8, 200
harvest management in practice   178–86
  assessment of MSY, ecologist's role
      186
  evaluation of the MSY approach – role of
      climate   184–5
  management by constant
      escapement   199
  management by fixed effort   181–2
  management by fixed quota   178–81

species especially vulnerable when
    rare   185–6
harvest models recognizing population
    structure   186–91, 199
  dynamic pool models in fisheries
      management   187–90
  a forest bird of cultural importance   191
  forestry   190–1
  Saiga antelope, male worth more than
      female   181–2, 186–7
harvested populations, evolution of   191–3,
    199
  setting aside nonharvested areas   192–3
  size-selective harvesting, effects
      of   191–2
  trophy big-horn rams, decrease in body
      size and horns   192, 193
harvesters
  exert strong selection pressures on
      exploited populations   175
  indigenous, regulations reduce chance of
      overexploitation   189
harvesting
  involves predator–prey interaction   195
  managing succession for   216–19
  reduces population below carrying
      capacity   176B
harvesting efficiency, and small
    populations   185
harvesting effort, effective regulation
    difficult   16
harvesting, managing succession for
    216–19, 225–6
  aboriginal burning enhances
      harvests   217–19, 219
  benzoin gardening in Sumatra   216–25
Hawaii
  customary fishers take only
      medium-sized fish   189
  extinction of many endemic birds   40B
health hazards, excess nitrates in drinking
    water   245
hedgehog (Erinaceus europaeus)   223
hemiparasites, used to enhance meadow
    species richness   243, 244
herbicides   149, 150, 253
herbivores
  influence on succession   215
  vertebrates, as influential as insects
      215
Heteractis magnifica, a mutualist   40B
Heteranthera dubia   92

Hieracium pilosella, aggressive invader in New Zealand   50, 52
historical data sets, long-term, provide valuable insights   116
Homo sapiens   2–4
  (not) just another species   3, 32
honey bees (*Apis mellifera*)   103
horsechestnut leaf-mining moth (*Cameraria ohridella*)   158–9
horticulture in the landscape   273
  brussels sprout study   273
  natural enemies depress pest populations   273
host-specificity testing, expansion of recommended   161
house sparrow (*Passer domesticus*), successful invader, more novel behavior   67
houseflies (*Musca domestica*), resistance to DDT   162
human actions, reduced abundance and range of many species   110
human behavior   195–7
  behaving as classic predators   195, *196*
  poaching   196, *196*
  problem of bushmeat harvest   196–7, *197*
human disease organisms, imported   14
human disease risk   230, 234–6, 257, 313
  dengue fever   235, 307, *308*
  Lyme disease   26–7B, 230, 234–5, 257
  minimizing of   26–7B
  understanding transmission from wild animal populations   235–6
human health
  and emerging infectious diseases   141
  fertilizers pose risk to   245
  may be compromised by indirect food-web effects   160, *161*
  pesticides pose risk to   154
  reflects aspects of ecosystem health   252
human impact   5, 14
  excreta rarely placed on land of origin   247
  felt by different species in different ways   76
human migration, mirrored by megafaunal extinction   14, 16, *16*
human population
  expansion causing environmental problems   3–4

greater growth expected in species-rich tropical areas   18
human pressure
  and extinction risk   73
  and fish extinction   72
Huron, Lake, dispersal of waterfleas via regular boat traffic from   95, *95*
Hydrilla verticillata, native and potential invaded distributional areas   42, 43, *43*
Hyperaspis pantherina, ladybird beetle   156
  biological control agent   25B
Hypochrysops halyetus, butterfly at risk   298

ibex (*Capra ibex*)   275
Icaricia icarioides (Fender's blue butterfly)   117, *118*
'immigration potential by wind'   90–1
Imperata cylindrica, invasive perennial grass   210
inbreeding depression   11, 114–15B, 127, 136
inbreeding, frequent, purges more lethal recessive alleles   115B
Indonesia
  economic cost of deliberate vegetation burning   29
  forest fires and atmospheric CO_2 *19, 20*
inoculation biological control
  in glasshouses   158–9
  under field conditions   158
inorganics   167
  early natural chemical weapons   147
insect growth regulators   **149**, 150
insect pollinators   103
insecticides
  organic, application to cotton crops   151
  third-generation   150–1
  toxicity to nontarget organisms and persistence of   **149**
insects for inoculation, from local plant material   158–9
integrated pest management (IPM)   143, 164–7, 168
  accepts presence of pest not always a problem   164
  against an invasive weed in Australia   166–7
  against potato tuber moths in New Zealand   165–6

integrated pest management (IPM) (cont'd)
 integrates control measures in a
  compatible manner 165
 involves use of computer-based expert
  systems 165
 relies on natural mortality factors 165
intermediate disturbance hypothesis 206B
International Commission for the
  Conservation of Atlantic Tunas
  (ICCAT) 179
International Whaling Commission
  (IWC) 198
interspecific competition 231B
inundation biological control 159–60
 bacterium Bacillus thuringiensis (Bt)
  159
 fungal agents 159
 viruses 159
invader traits, what we know and don't
  know 71
invader-dominated community, controlling
  succession in 223–4
invaders 3, 11–12, 60
 favored by present combinations of
  conditions 45
 freshwater, consequences of 240–1
 loss of exotics in forests due to canopy
  closure 221
 more in disturbed habitats, fewer in
  remote areas 41, 41
 rapid response to, blunderbuss rather
  than surgical strike 142
 shade-tolerant, require more active
  management 221
 unwanted biodiversity 2, 13–14
 what we know and don't know about
  invader traits 71
 see also aliens, unwanted
invaders, pests and diseases, new set
  of 292
invaders, predicting arrival and spread
  of 92–6, 105
 Great Lakes 92–4
 invasion hubs or diffusive spread 95–6
 lakes as infectious agents 94–5
 managing invasions under
  globalization 96
invasion hubs 94–5
 or diffusive spread 95–6
 identified for special management
  treatment 95
invasion risk-related tariffs 96

invasion success, predictors of 65–71
 importance of flexibility 66–8
 invasive conifers, success for some
  predictable 66, 67
 separating invasions into sequential
  stages 68–71
invasions
 Bongil Peninsula, Australia, of aggressive
  perennial grass 210
 ecosystem effects of invasive
  plants 241–2
 ecosystems consequences of 240–2
 how to manage under globalization 96
 spiny acacia in Australia 298, 299
 successful, and niche theory 37
 using succession for control 219–21,
  226
IPM see integrated pest management
irreplaceability 279–80, 281, 288
 irreplaceability analysis, for Cape
  Floristic Province 279, 281
Island Biogeography Theory 263–4B,
  264B, 286, 313
 applicable to oceanic islands and habitat
  islands 263B
 island size and isolation important
  263B
 relevance to biodiversity
  conservation 264B
 and wildlife reserves 269–70
Italy, marine zoning plan 280, 281,
  282–3
 final plan, for sustainable desires of all
  interest groups 282, 283
 maps, factors important to interest
  groups 281–3, 282
 permitted or prohibited activities 280,
  281
IUCN Red List of Threatened Species 5B,
  126–7
Ixodes scapularis (black-legged tick) 235,
  236
Ixodes (ticks)
 carrier of Lyme disease bacterium
  26–7B, 234–5
 life cycle 234–5
 most abundant host, white-footed
  mouse 235, 236

Jacaranda copaia 63
Japanese grass (Microstegium vimineum),
  weed of Asian origin 67

Japanese quail (*Coturnix coturnix japonica*), evolved resistance to DDT   163
jarrah (*Eucalyptus marginata*) forest
  passerine species and selective logging   72
Juncus effusus   207

K-selected traits   142
  in endangered pine species   61B, 66
K-selection, skates and sharks   75
K-strategists, most vulnerable to overexploitation   75
Kalahari bushmen
  hunting techniques (film)   262
  problem of wildlife reserves   262
Kenya, Tana River crested mangabey study   118
kereru (*Hemiphaga novaeseelandiae*)   191
  Maori plans, reduce predation and protect habitat   191
  population viability analysis   191, *192*
  shrinking population   191
kestrels (*Falco tinnunculus*), juvenile deaths in Gibraltar   101
key concepts   xv
keystone species   231–2B, 256
  sea otter (*Enhydra lutris*)   236–7
Kincaid's lupin (*Lupinus sulphureus kincaidii*)   117
koalas (*Phascolarctos cinerreus*), an icon at risk   120–2
  modeling of two populations in Queensland   120–2, *120, 121*
  potentially threatened nationally   120
  sensitivity analysis   121, **121**
kokako (*Callaeas cinerea*)
  advice to managers   239, *240*
  diet shared by rats and possums   239, *240*
  nesting success poor due to predation   239
Korea, central, abandoned rice fields to species-rich alder woodland   207

lake eutrophication management   248–9, *250*
  biomanipulation, top-down control   248–9
  bottom-up control of nutrient availability   248
  reversed by chemical or biological means   248

lakes
  cultural eutrophication   247
  examples of discontinuous habitat   95–6
  as infectious agents   94–5
  invasion hubs or dead ends   94–5
  managing eutrophication   248–9, *250*
land reclamation, mined sites   49–51
  candidate plants for restoration, fundamental niches   49
  ecotypes having evolved resistance in mined areas   49
  phytoremediation   49–50
landscape ecology   22, 27–8B, 33, 263B, 313
  important general principles   263–7B
  landscape harvest management   270–1, **272**, 287
  marine zoning plan for sustainability   27B
  predicting future invasions due to climate change   27–8B
landscape perspective   308–11
  fairy shrimps   311
  Mexican cacti   309–10
landscape theory   263–7B
  determination of species richness   264–5B
  metacommunities and Island Biogeography Theory   263–4B
  metapopulations   263B
  partitioning of species richness, local and landscape scales   265–6B, 267B
  patches, patch dynamics and landscapes   263B
landscapes   285–6
  agricultural, management of   283–5
  arable farming in   274
  patchy, oceanic and terrestrial   270
  regional, envisioning future reorganization   292, 313
  species richness similar in organic and conventional fields   274, *274*
Lantana camara, successful invader   65, *65*
late-successional species   204B, 225
latex   173
leguminous plants   3, 223
  as nurse plants   211–12, *213*
Lemna minor   92
Lespedeza davurica   296B

lesser white-fronted goose (*Anser erythropus*)
  alteration of migration route achieved 89
  knowledge of migratory behavior useful for management 89
Leucaena leucocephala 214
Leucanthemum vulgare 62
life-history theory 60–1B
  the CSR concept 61B
  life cycles 60B
  the r/K concept 61B
  species traits 60–1B
life-history traits 62
  and the fate of species 300–3
  used to make management decisions 60
limpets 231B
Linepitheme humile (Argentine ant) 300, *300*
Loess Plateau, China, secondary succession on 206B
logging
  effects of in a *Eucalyptus* forest 97–100
  recovery of bat communities after 99
  selective, and passerine species in jarrah forest 72
logistic equation 112B
Loligo squid, constant escapement strategy for 183, *183*
Lolium perenne, high tolerance of trampling *50, 51, 52*
long-nosed skate (*Dipturus oxyrhincus*) 74
Lonicera maackii 221
Lonicera tartarica 221
Lotus wrightii 223
lupin (*Lupinus polyphyllus*)
  an invader 60
  see also Kincaid's lupin (*Lupinus sulphureus kincaidii*)
Lupinus arboreus 241
lygus bug (*Lygus hesperus*) 152
Lymantria dispar, nuclear polyhedrosis virus 159
Lyme disease 234–5, 257
  minimizing risk to humans 26–7B
  spread by ticks 230, 234–5

Maculina arion, large blue butterfly, extinction of 160
malaria 164
  transmission of in Brazil 235
  and use of DDT 147

mallard duck (*Anas platyrhynchos*), successful invader 67
Mamestra brassicae, caterpillars pests of brussels sprouts 273
management plan for Fiordland region, New Zealand 283
  bottom-up effort by local community 283
  concept of gifts and gains 283
  sustainable middle ground identified 283
marine environment, ecosystem health 255–6
marine fisheries 257
  management for MSY difficult to achieve 186
  truly difficult to manage 174
marine protected areas 287
  benefits for harvest management 184, 190, 270
  fish density increase may lead to migration 270
  increase in target species density and size 184
  modeling movement of 'virtual fish larvae' 270, *271*
  sustainability of abalone harvest 236–7, *237*
marine reserve design 88
  bimodal distribution of dispersal distances 88, *88*
  duration of propagules in floating phase 88
  self-replenishment possible 88
marine zoning plans 27B, 280, **281**, 282–3
marri trees (*Corymba calophylla*) 72
marsh wren (*Cistothorus palustris*) 208
maximum sustainable yield (MSY)
  approaches to management 174–86, 199
  difficulties in use as basis for harvest management 176–7B, 178–86
  evaluation of and the role of climate 184–5
  finding the top of the curve *177B, 180,* 184
  putting theory into practice 176B
  simplification of the approach 174
  see also constant escapement; fixed effort; fixed quota
meadow goldenrod (*Solidago altissima*) 215

meadow restoration   *50–1, 51–2*
    manipulation of the food web   242–3,
        244
    plant–pollinator interactions
        compared   213, *214*, 215
meadow species   *50–1, 51–2*
    cannot use higher nitrogen inputs
        efficiently   44
    European springtime flowers, peasant
        farming areas   60, 153–4
meadows, species rich, uncommon in
        European agricultural
        landscapes   242
Mediterranean forest
    restoration of   26B
    Spain, planting of seedlings under nurse
        plants, high success rate   210–12
megafaunal diversity, disappearance of   14,
        16, *16*
Melampus bidentatus   208
Mendota, Lake, Wisconsin, attempted
        biomanipulation of   248–9, *250*
    biomass of zooplanktivorous fish
        declined   249, *250*
    density of two species of piscivorous fish
        increased   249, *250*
metacommunities   263–4B, 286
    metacommunity concept   263B
metapopulation conservation   267–70
    the emu-wren   267–8
    endangered species exist as
        subpopulations   267
    problem with large carnivores   269–70
    the woodthrush   268–9
metapopulation dynamics
    importance of dispersal and
        migration   263B
    may counter local extinction and
        enhance genetic diversity   270
metapopulations   263B, 286
methane   293B
Mexico
    overwintering habitat for monarch
        butterfly   46, 47
    protected rainforest, carbon storage by
        one key species   252, *252*
    target pest resurgence   151
migrants
    elevational (giant panda)   85
    one-way (Pacific salmon)   83B
migration   83B
    and global climate change   301

    long distance   83B
    mass directional movement from one
        location to another   83B
    movement on different timescales   83B
migration and dispersal, lessons for
        conservation   84–8
    designing marine reserves   88
    dispersal of a vulnerable aquatic
        insect   86–7
    panda conservation   85–6
    the three-wattled bellbird   84–5
    see also dispersal and migration
migratory species, attempts to restore
        populations   89
Millennium Ecosystem Assessment   22
millipedes, important role in ecosystem
        recovery, sand dunes   208
Mimosa pigra, invasive weed in
        Australia   156
    IPM used against   166–7
minimum viable population   110
Mohawk Nation, value sweetgrass for use in
        basketry   224
    nearest gathering site too far   224
    planting sweetgrass with hairy
        vetch   224
moi (*Polydactylus sexfilis*), sustainably
        harvested in Hawaii   189
molecular (DNA) technology, use of   127,
        130, 136
monarch butterfly (*Daneus plexippus*)
        83B
    critical dimensions of overwintering
        niche in Mexico   46, 47
    no-constraint scenario preferred and
        accepted   46, 47
Monardia fistulosa   244
monocultures
    and nitrate leaching   246
    and pest attacks   272–3
moose (*Alces alces*), realistic quota
        set   180–1
mosquitos   257
    carry many transmittable
        diseases   235–6
moth populations, controlled by sex-
        attractant pheromones   150
multipurpose reserve design   262–3,
        280–5
    marine zoning   280, **281**, 282–3
Mus musculus, development of evolved
        resistance to bromadiolone   163

Muskoka, Lake
    development as an invasion hub   95
    first inland lake invaded by
        waterflea   95, *95*
mussel (*Mytilus edulis*)   231B
mutualisms/mutualists   213
    pollination   213, *214*, 215
    positive effects of   40B
mycorrhizal fungi, enhance seedling
        growth   213, *214*
mycorrhizas   *41*, 213
Myrica faya   241
myxoma virus
    biocontrol of UK rabbits led to extinction
        of *Maculina arion*   160
    evolution of resistance to in
        Australia   163–4
myxomatosis   140

natural enemies   145B, 152–3
    adequate distribution of essential   158
    importation of   154, 155–6
    parasitism by   272, 273, *273*
natural habitats, economic value of
        retention or conversion   249, 251,
        251
    coral reefs, sustainable vs. destructive
        fishing, Philippines   251, *251*
    draining of freshwater marshes for
        agriculture   251
    forest conversion to plantations,
        Cameroon   249, *251*
    logging in Malaysia   249, *251*
    mangrove swamps vs. shrimp farming,
        Thailand   249, 251, *251*
natural mortality
    density-dependence in   191
    and harvest mortality   190–1
natural selection, and genetic
        variation   114B, 136
nature, economic valuation of   28–9
nature reserves   292
    and future fynbos habitat   297
    for giant pandas   24B, 86
    a role in harvest management   183
    in the wrong place?   298
    see also dispersal behavior
nest parasitism   268
net primary production (NPP)   232B
New Zealand
    and the brown trout (*Salmo trutta*)   37,
        40B, 45, 240–1

conservation of the kokako   239, *240*
Fiordland, ecosystem management
        plan   27B, 283
    forest bird of cultural importance   191,
        192
Hieracium pilosella an aggressive
        invader   *50*, 52
IPM against potato tuber moths   165–6
Kakaunui River, use of Maori Cultural
        Stream Health Measure
        (CSHM)   255, *255*
marine zoning plan for   27B, 283
offshore islands, brute-force methods to
        destroy small weed population   142
possibility of dengue fever if vector
        mosquitos become established   307,
        308
problems of imported ferrets (*Mustela
        furo*)   140–1
proposal for sustainably harvesting
        mixed beech forest   190–1
rabbits, myxomatosis and rabbit
        calcivirus disease   140
rediscovery of giant weta (*Deinacrida
        mahoenuiensis*)   223–4
river managers use macroinvertebrate
        community index   254
success and failures of bird
        introductions   66–7
translocation of the takahe   23B, 48–9
New Zealand woodpigeons   191, *192*
Newfoundland cod fishery, an ecosystem
        lost   29
niche analyses, vulture sightings in Valais
        region, Switzerland   275
niche breadth   52, 265B
    and behavioral flexibility   73
    broad   68
niche matching, for species success   60
niche opportunity   44
niche theory   33, 36–58
    conditions and resources   37–8B
    and conservation   297–8, 297B
    the fundamental niche   38–40B
    and invasion risk   298, 299, 300, *300*
    niche an n-dimensional
        hypervolume   39B
    the realized niche   40B
    restoration ecology   49–54, 55–6
    and translocation of New Zealand's
        takahe   23B, 48–9
    unwanted aliens   41–6, 55

Nile perch (*Lates nilotica*), Lake Victoria 2, 3, 13
nitrogen (N)
  minimizing unintended loss from the land 246
  reasons for ready nitrate losses from land 246
  in runoff, fueling ocean dead zones 247
nitrogen oxides NO$_x$ 17
nonharvested areas, set-aside, for and against 192–3
North Atlantic, decline in cod maturity size 26B, 192, *193*
North Atlantic Oscillation (NAO), NAO index would be strengthened by global warming 301, *301*
North Pacific fur seal fishery, fleet size response to seal herd size 195, *196*
northern cardinal (*Cardinalis cardinalis*) 245
northern corn root worm (*Diabrotica barberi*), evolved resistance to cultural control method 164
Northern fence lizard (*Sceloporus undulatus*) 216
nurse plants 210–13
  in restoration of Spanish forest 210–12, 213
nutrient cycling 233B
nutrient dynamics 232–4B
  nutrient elements, availability to plants 233B
nutrients
  can cause oligotrophic lakes to become eutrophic 247
  enhancement of availability 40B, 44
  see also nitrogen (N)

oak tree (*Quercus robur*) 82
oak trees (*Quercus* spp.) 211
ocean dead zones 247
oceans
  management to alleviate runoff problems 247
  profound changes in 293B, 294B
  and sea level rise 291
Ochlerotata triseriatus (North American mosquito) 236
Oenothera parviflora 277

oil pollution 17
Onionis spinosa, defence against grazing *51, 52*
optimal economic yield, not always possible to achieve 198
ordination 44–5, *45*
organic farming 253, 274
organisms, individual
  ecology of 22, 23–4B
  species traits, importance of 23B
organophosphates 148–9, **149**, 151
Orthezia insignis, killing St Helena gumwoods 25B
overexploitation 3, 12, 14, 16–17, 32, 109, 110, 195
  of bushmeat, a risk 197
  by world's industrial fishing industries 16
  chance of reduced by indigenous harvesters 189
  collapse of Newfoundland cod fishery 29
  in forests 174
  impacts, coupled to destructive harvesting techniques 17
  K-strategists most vulnerable 75, *75*
  and local or global extinctions 185
overharvesting, of wild populations 71
overstory reduction and slash mulching (ORSM) 222–3
Ozark glades, intrusion of fire-intolerant species 215–16
  burning of brush piles bad for Northern fence lizard 215–16

pacific mites (*Tetranychus pacificus*), target pest resurgence 151
Pacific Ocean, islands disappearing with sea level rise 291
pacific salmon (*Oncorhynchus nerka*), long-distance one-way migrant 83B
Pacific whiting (*Meluccius productus*)
  recruitment and spawning stock estimates 178–9, *179*
  total allowable catch shared 181
Pakistan, diclofenac, use banned for sick cattle 135
pampas grass (*Cortaderia selloana*) 142
Panache, Lake (northern Ontario), future invasion hub 95

Panama, tropical forest restoration   62–5
  invasion by *Saccharum spontaneum*   63
  lack of naturally arriving tree seeds   63
  large-seeded late successional trees
    planted by hand   65, 209
  loss of biodiverse tropical forest   62–3
  medium or large seed species do
    best   63–4, *64*, 209
pandas *see* giant pandas (*Ailuropoda
    melanoleuca*), conservation of
parasite wasp *Aphytis melinus*, an inoculant
    in lemon orchards   158
parasites
  Giardia   284
  as good ecosystem health indicators
    255
parasitism
  by natural enemies of pests   272, 273,
    273
  rates increase with area of pasture
    273
parasitoid wasp (*Macrocentrus
    sylvestellae*)   272, *273*
parasitoid wasp (*Venturia robusta*)   272,
    273
parasitoids   157, 165, *165*, 272, *273*
parrots, assessment of transport, release
    and establishment   68, **69**
  migratory species not suited to
    invasion   68
  species with broader niches more likely
    to establish   68
  successful invasions correlated with
    development of international
    market   68, **69**, 96
  threatened species do not figure in
    transport and release   68
patches
  with different harvest values, Peruvian
    forest   271, **272**
  patch dynamics and landscapes   263B
peat bog restoration   89–91
peregrine falcon (*Falco peregrinus*), victim
    of DDE in the food web   134, 147
periwinkle (*Littorina* sp.)   231B
Peru, forest successional mosaic   271, *272*
  old secondary forest, largest number of
    useful plants and animals   271
  value of agricultural products on cleared
    land   271
  wood and medicines most valuable forest
    commodities   271

Peruvian anchovy (*Engraulis ringens*)
    fishery
  collapse of, overfishing compounded by
    El Niño events   179–80, *180*, 184–5
  management by fixed quota   179–80
pest control   151, 245
  landscape perspective on   272–4, 287
  physical, California   142
  on St Helena   25B
  sustainable in the face of resistance
    evolution   162
  see also biological pest control
pest eradication or control
  may not be feasible   142
  of newly arrived exotic species   142
  population dynamics theory
    applied   141, 143–6B
  to reduce population to Economic Injury
    Level   143
pest management   139–71
  aim of   193–4
  biological control   154–6
  chemical pesticides   146–54, 167
  eradication or control   141–3, 167
  evolution of resistance and its
    management   162–4, 168
  integrated pest management (IPM)
    164–7, 168
  pests or pets   140–1
pesticides   253
  adverse effect on provision of high
    quality water   284
  and biodiversity reduction   283
  disrupting cellular functioning   147–50
  loss of ecosystem services through   284
  results of heavy use   2
  upsetting growth and development
    150
  widespread effects on nontarget
    organisms   153–4
pests
  changing behavior of   150–1
  defined   141
  eradication or control   141–3, 167
  evolve resistance to pesticides   162, *162*
  native or exotic imports   141
  or pets   140–1
  r-selected   142
  that seem to arise from nowhere   141
phenoxy (hormone) weed killers   150
pheromones, used to control pests   150–1
physical pest control, California   142

phytoremediation 49–50
    phytoaccumulation 49–50
    phytostabilization 50
    phytotransformation 50
Phytoseiulus persimilis, used for
        inoculation 158
picture-winged fly (*Paracantha culta*),
        adverse impact of *Rhinocyllus
        canescens* on 160
Pieris virginiensis, adversely impacted by
        garlic mustard 42
pine (*Pinus sylvestris*) 272
    flying squirrel dispersal habitat 97
pine trees *see* conifers
pink pigeon (*Columba mayeri*) 128–9,
        129
    descent of captive population 128
    inbreeding, effect on probability of
        survival to thirty years 128, *129*
    recovery program dependent on
        captive-breeding 128
    reintroduction on Mauritius 128
Pinus maximartinezii
    prone to extinction 23B
    rare and endangered 66
pioneer desert 209
pioneer species, *r*-selected, do not do well
        in *Saccharum* grassland 63
Piscidia piscipula 214
plant biomass, top-down control or bottom-
        up control 232B, 256
plant community restoration 89, 206–8
    restoration of former agricultural
        land 207
    shifting prairie mosaic 207, *207*
plant invaders, physiological and behavioral
        flexibility 67–8
plant seeds, many passively dispersed in air
        currents 83B
plantation forestry 272–3
    expansion of 272
    French Pine plantations, attacks by
        pests 272–3, *273*
plants 3
    alpine, moving higher 293B
    halophytic 50
    life histories challenge simulation
        modeling 122
    mutalists 40B
    as natural arms factories 146, 147
    Protacea species, minimum number to be
        safeguarded 280

'red list', reintroduction of 129–30, *130*
    restoration timetables for 206–8
plants, invasive
    ecosystem effects 241–2
    North America 42–3, *43*
Platte thistle (*Cirsium canescens*), density
        loss affected picture-winged fly 160
poaching 196
    effects on legal yield of Saiga
        antelope 196, *196*
polar bears (*Ursus maritimus*)
    link between birth rate and oscillating
        climate pattern *301*, 303
    under pressure from hunting and high
        pollutant levels 303
polar cod (*Boreogadus saida*), shows
        evidence of biomagnification *148*
political realities confronted 197–8
pollination 213, *214*, 215
    by native (wild bees) 103–4, *103*, 251–2,
        252
    flower-visitation webs *214*
pollutants, atmospheric 17
polluter pays principle 96
pollution 2
    by fertilizers and pesticides 4
    by human and animal waste products 3,
        247
    from intensive agriculture 245–7
Polynesian islands
    biological control gone wrong 160
    social costs of loss of *Partula varia* and
        Partula rosea 160
Ponderosa pine forest (western USA), shows
        relationship between pressure, state
        and response 252–3
Ponderosa pine (*Pinus ponderosa*) 252–3
population, carrying capacity of 112B
population determination 111B
population dynamics models 109,
        111–12B
    logistic equation 111B, 112B
population dynamics theory-1 110–13B,
        135–6
    models of population dynamics 109,
        111–12B
    population dynamics of small
        populations 113B, 135–6
    population regulation and
        determination 111B, *112B*
    uncertainty and risk of extinction
        113B

population dynamics theory-2, and pest management 143–6B
  biological control of secondary pest outbreaks 146B
  economic injury level (EIL) 143–4B
  economic threshold (ET) 144B
  evolution of resistance 146B, *146*B
  secondary pest outbreaks *145*B, 145B, 151, *152*
  target pest resurgence 144–5B, 167
population dynamics theory-3 175–8B
  fixed effort harvesting 177–8B, *178*B
  fixed quota harvesting 176–7B, *177*B
  logistic growth, underlying patterns in births and deaths 175B, *176*B
  maximum sustainable yield (MSY) 176B
  when a harvest is sustainable 175–6B
population growth, logistic 175B, *176*B
population regulation, density-dependent or density-independent 111B, *112*B, 175B
population theory 22, 24–6B, 33
  countering threat of extinction 24–5B
  harvest management, counteracting evolution towards small size 25–6B
  pest control on St Helena 25B
population viability analysis
  how good is your analysis 126–7
  for the New Zealand woodpigeon 191, *192*
  persistence in time of a population *116*, 117
  population growth rates seen as variable 117
  simple algebraic models 117–18
  simulation modeling for 119–27
populations, ecological applications at level of 24–6B
Populus deltoides × *nigra*, able to degrade TNT 50
possum (*Trichosurus vulpecula*) 223, 239
potato tuber moth (*Phthorimaea operculella*) 165–6
potato tuber moths, IPM against in New Zealand 165–6
  caterpillars protected against parasitoids and insecticides 165–6
  IPM strategy, monitoring, cultural methods and insecticides 166, *166*
prairie grassland restoration 219, 226
  Minnesota tallgrass prairie 219–20, *220*

precipitation, increased in mid and high northern latitudes 293B
predation
  of bird eggs and young 268
  can slow or stop a successional sequence 205
predator–prey spiral, anticlockwise 195, *196*
primary successions 225
  on volcanic lava flows 204B, *205*B
Principal Components Analysis, exploring relationship between plant traits and invasiveness 66, *67*
Protacea 280
Protea lacticolor
  bioclimatic model for *297*B
  prognosis *297*, 298
Protea spp. 297–8, 308–9
protected areas
  growth of 13, *13*
  suitable choice can minimize predicted losses 309, 310–11, 313
  see also marine protected areas; nature reserves; wildlife reserves
Prunus caroliniana, mycorrhizae allow seedlings to compete successfully with *Ardisia* seedlings 44
Pseudomonas putida, bacterium, used on toxic remnants of mustard gas 50–1
Puma concolor coryi (Florida panther) 24–5B, 128
Puma concolor stanleyana (Texas cougar) 128

r-selected traits 142
  in invading pine species *61*, 66
r-selection
  good invaders 300
  small opportunists 76
r/*K* concept/theory 61B, 71, 73–6
  K-species 61B, 75, 205B
  r-species 61B, 63, 142, 204B, 219, 300
rabbit calcivirus disease, in New Zealand 140, 168
rabbits, killed using 1080 163
ragwort (*Senecio jacobaea*) 142
rainbow trout (*Oncorhyncus mykiss*) 241
RAMAS-STAGE (simulation modeling package) 122–3, **122**, *126*
rats (*Rattus rattus*) 14, 223, 239
  successful invaders 65
realized niche 40B, 44, 55

recreational boaters, visiting invaded and uninvaded lakes 94, *94*

recruitment curves *176B, 177B, 305*
  and maximum sustainable yield (MSY) 304, *305*

recruitment rate, of young salmon and baby whales 185

red admiral butterfly (*Vanessa atalanta*), breeds at both ends of migration 83B

red fire ant (*Solenopsis invicta*) 15
  damage to wildlife, livestock and public health 14
  occurs in monogyne and polygyne forms 96

red-cockaded woodpecker (*Picoides borealis*), sensitivity to forest fragmentation 73

red-crested cardinal (*Paroaria coronata*) 245

red-winged blackbird (*Agelaius phoeniceus*) 208

refuge/riparian strips, for natural predators 152–3, *153*, 156–7, 284

reserve design, multipurpose 280–5, 287–8

reserve networks, designed for biodiversity conservation 277–80, 287–8
  complementarity 279, *280*
  irreplaceability 279–80, *281*

resistance, evolution of 146B, *146B*, 162–4, 168
  evolved resistance to cultural control 164

resources, limited
  interspecific competition for 38B
  intraspecific competition for 38B

restoration, of biodiverse grasslands and woodlands 13

restoration ecology 49–54, 55–6

restoration, ecosystems approach to 242–4, *245*
  carbon added to soil lowers nitrate levels 244, *245*
  exploiter mediated coexistence 242–3, *244*

restoration landscapes 274–7, 287
  reintroduction of vultures 275–6
  restoring farmed habitat 276
  willingness to pay for forest restoration 276

restoration, managing succession for 206–16, 225
  invoking enemy-interaction theory 215–16
  invoking facilitation theory 210–15
  invoking successional-niche theory 209–10
  invoking theory of competition–colonization trade-offs 209
  restoration timetable for animals 208–9
  restoration timetables for plants 206–8

restoration and species mobility 89–92
  behavior management 89
  bog restoration 89–91
  wetland forest restoration 91–2

Rhamnus alternus 211

Rhinanthus minor, hemiparasite 243

Rhinus megaphyllus, made use of trackways 98, *99*

riparian and wetland communities 257
  lost to agricultural production 248

river blindness (onchocerciasis), controlled 162

river flow, reduction in 17–18

river health 254–5
  macroinvertebrate community index (MCI) 254, *255*
  Maori Cultural Stream Health Measure (CSHM) 255, *256*
  positive relationship between pressure and health index 254
  some health indexes include more indicators 254

river restoration 53–4, 55–6
  approaches used to define minimum discharges 53–4
  restoration of natural flow regimes 45

river water quality, reduction in 17, 28

rivers
  ecosystem health of 254–5
  land use and nitrogen transport in **31**

rockroses (*Cistus*) spp., provide no nursing benefit 212

rocky shores, gaps opened by severe wave action 204B

Rodolia cardinalis, predatory ladybird beetle 155

rooks
  Corvus frugilegus, successful invader 67
  Corvus monedula, unsuccessful invader 67

royal catchfly (*Silene regia*)   122–3
  growth, fecundity and survival
    contribute to population
    performance   123
  long-lived prairie perennial with
    shrunken range   122
  management including fire
    recommended   123, *123*
  successful germination episodic   122
Rudbeckia missouriensis   216
rufous bristlebird (*Dasyornis broadbenti*),
    threatened by habit loss   46–7
rufous treecreepers (*Climacteris rufa*)
    72–3
Russian olive (*Elaeagnus augustifolia*),
    native and potential invaded
    distributional areas   42, 43, *43*
ryegrass (*Lolium multiflorum*)   224

Saccharum grassland   63–4
  species traits significant for tree seed
    success   63–4, *64*, 209
Saccharum spontaneum, invasive   63
Saiga antelope (*Saiga tartarica*)
  effects of poaching on legal yield   196,
    196
  males worth more than females   181
  simulation model for future
    management   181–2, *182*
St Helena gumwoods (*Commidendrum
    robustum*), saved by a ladybird
    beetle   155–6
St Helena, pest control on   25B
Salix koriyanagi   207
San Joaquin Valley, California, pesticide
    problems amongst cotton pests   151,
    152
sand mining
  restoration timetable for animals   208,
    208
  and successional-niche theory   210
sand shrimp (*Crangon septemspinosa*), two-
    dimension niche   *38*B, *39*B
sandwich tern (*Sterna sandvicensis*),
    maneuverable but declining
    102
scarecrows and hawks, old method of pest
    control   151
Schizachyrium scoparium   216
screwworm fly (*Cochliomyia hominovorax*),
    eradicated by biological
    trickery   143

Scutellospora   213
sea lampreys (*Petromyzon marinus*), treated
    with lampricide TFM   150
sea level, rising   18–19
sea otter (*Enhydra lutris*)
  affecting red abalone harvests   236
  locally extinct   185
seabirds
  biomagnification most marked   *148*
  development of species sensitivity index
    for (Germany)   101–2
seals   140
secondary pest outbreaks   *145*B, *145*B, 151,
    152, 167
secondary productivity   232B
secondary successions   225
  occur after partial or complete removal of
    vegetation   204B, 206B
seed dispersal   83B
seed-banks   60B, 78
  ex-swamps cf. intact swamps   91–2
seed-feeding weevil (*Rhinocyllus connicus*),
    introduced to N America to control
    Carduus thistles   160
semiochemicals   150–1
Septoria passiflorae, fungus specific to
    banana poka   156, *157*
Seraria viridis   296B
Setaria glauca   219–20
shark species   74
  fast–slow continuum   74, *74*74
  mean population growth rates (λ)   74,
    74
  various species' ability to respond to
    increased mortality   74
Sherardia arvensis, tolerant of cutting
    disturbance   *51*
Silene douglasii (var. *oraria*),
    reintroduction   129–30, *130*
  comparison of variation of
    inbreeding   129–30
  progeny of outbred, cross-pollinated
    flowers showed better
    survival   129–30, *130*
silvertop stringybark (*Eucalyptus
    laevopinea*)   98
simulation models/modeling
  future management of Saiga
    antelope   181–2, *182*
  for population viability analysis
    119–27
  RAMAS-STAGE   122–3, **122**, *126*

using susceptible-infectious-recovered (S-I-R) process   124
VORTEX   120, *121*, *126*
Sinapsis arvensis   244
Singapore island
    extinctions   7
    problems of population growth   7, *8*
size–vulnerability relationship   16
skates and rays (Rajidae), and extinctions   74, *75*
skipjack tuna   *238*
skylark (*Alauda arvensis*)
    density dependent on weed seed density   132–3, *132*
    impact of GM seed depends on effect on high-density weed patches   132–3
small populations
    genetics of   114–15B, 136
    relative importance of genetic and demographic risks   115B
small populations, population dynamics   113B
    chance events play a large role   113B
    demographic uncertainty   113B, 135
    environmental uncertainty   113B, 135
    spatial uncertainty   113B, 135
snowshoe hare (*Lepus americanus*)   221
    high densities associated with regenerating forest   222
snowy egret (*Egretta thula*)   208
social capital, increased by social learning   29–30
sociopolitical dimension, added to ecology and economics   195–8, 200
sociopolitical scenarios   30, **31**
    adapting mosaic   30, **31**
    global orchestration   30, **31**
    order from strength   30, **31**
    technogarden   30, **31**
soil mites, predatory   253, *254*
soil water balance, affected by invaders   242
    exotic grassland maintained highest water content   242, *243*
    star thistle reduced soil moisture   242, *243*
Solanum nigra   244
song sparrow (*Melospiza melodia*)   208
Sorghastrum nutans   244
South African polychaete worm (*Terebrasabella heterouncinata*), use of physical pest control   142

South African sand mine restoration   208, *208*
    successional pattern in dung beetle community   210, *211*
South American scale insect (*Orthezia insignis*)   155–6
southern damselfly (*Coenagrion mercuriale*)   87
    determination of net lifetime movement of individuals   87, *87*
    poor flight powers of adults   87
    restricted to highly fragmented habitat of ancient water meadows   87
    suggested management effort   87
Spartina spp.   208
spatial uncertainty   135
    in small population   113B
specialists and generalists   78
species
    community numbers balance between colonization and extinction   263B
    especially vulnerable when rare   185–6
    forest species, expected changes to habitable areas   305
    future distributions of   292, 312
    large/very large seed, survive well in *Saccharum* grassland   63–4
    larger, vulnerable through low reproductive rate   73–4
    maintaining a viable population   40B
    as metapopulations   263B
    named and unnamed   6
    natural and current extinction rates   6
    niche characteristics reasonably constant over evolutionary time periods   39B
    predictable sequences of appearance and disappearance   204B
    reduction in habitable area   292–3, 298
    shade-tolerant, vulnerability of   77
species conservation   *301*, 303
species conservation, managing succession for   221–4, 226
    controlling succession in an invader-dominated community   223–4
    enforcing a successional mosaic   222–3
    nursing a cultural plant back to health   224
    when early succession matters most   221–2
    when late succession matters most   223
species gain   231B

species invasions, separating into sequential
    stages 68–71
  alien plants in the Czech Republic
    68–70, 69
  California, invasions and failed
    introductions of fish in catchment
    areas 70–1
  parrots, assessment of transport, release
    and establishment 68, **69**
  see also birds, invasion success; conifers,
    invasive, in the USA
species loss 231B
species mobility
  and conservation 105
  and management of production
    landscapes 97–104, 105–6
  and restoration 89–92, 105
  why it matters 82
species pools, relationships between
    264–5B, 265B
species range shift 298, 312
  of invasive species 312
  towards poles and higher
    altitudes 293B, 294B
species restoration 52–3
species richness 4, 98, 287, 306–7
  in benzoin garden areas 217, 218
  Berlin, correlated with habit
    diversity 277, 278
  and biodiversity 287
  correlates negatively with farmers' profit
    margins 52–3, 53
  determination of 264–5B, 265B
  gamma, alpha and beta richness
    265–6B, 267B
  hotspots 277
  of Protea species 308
species traits 60–1B
  can predict invasive trees and threatened
    natives 23B
  and life cycles 77
  linking particular traits to particular
    environments 60–1B
  as predictors for effective
    restoration 61–5
  as predictors of extinction risk 71–7,
    78
  as predictors of invasion success 65–71,
    78
  and restoration 78
  significant for tree seed success in
    Saccharum grassland 63–4, 64

timing, location and distance involved in
    dispersal and migration 82
Sphagnum bogs, eastern Canada 89–91
  assessment of wind dispersal ability of
    some peatbog plants 89–91, 90
  exploited for peat then abandoned 89
  mosses, high immigration potential but
    poorly represented 91
  no persistent seed-bank, slow
    recolonization 89
  propagules capable of dispersing by water
    or animals 91
  recolonization by vascular plants well
    predicted 91
Sphyrna tiburo (fast shark) 74
spiny acacia, invasive in Australia 298,
    299
  with climate change invadable area will
    be larger 298, 299
spiny lobsters (Panulirus argus), Cuban
    fishery 194–5, 195
spiny waterflea (Bythotrephes longimanus),
    dispersal from Lake Ontario
    94–5
spottail shiner (Notropis husonius) 255
spotted alfalfa aphid (Therioaphis trifolii),
    keeping density down 144B
spotted sandpiper (Actitis macularia)
    208
spruce (Picea abies) 97
spur lupin (Lupinus arbustus) 117
squirrels 97
starfish (Leptasterias sp.) 231B
starfish (Pisaster ochraceus) 231B
  keystone species 231–2B
starling (Sturnus vulgaris), successful
    invader 65
Steller's sea cow (Hydrodamalis gigas),
    globally extinct 185
stem-mining moths (Neurostrata gunniella
    and Carmenta mimosa) 167
Stipa bungeana 296B
stoat (Mustela erminea) 223, 239
stock assessment, of harvested
    populations 186, 199
strip intercropping 284
strong interactors 231–2B
Styrax paralleloneurum, benzoin garden
    tree 217
succession 306–7
  runs from pioneer to climax community
    compositions 204B

succession and management 202–28
  managing succession for
      harvesting 216–19
  managing succession for
      restoration 206–16
  managing succession for species
      conservation 221–4
  theory of ecological succession 204–6B
  using succession to control
      invasions 219–21
succession, species traits determining
      course of 204–5B
  early-successional species (r-species),
      persistence of 204B
  late-successional species
      (K-species) 204B
succession theory 205–6B
  benzoin gardening in Sumatra 217
  nursing a community back to
      health 26B
successional mosaics 226, 263B
  first aid for butterflies 222–3
  Peruvian forest 271, **272**
  and the role of disturbance 205–6B,
      217–19, *219*
successional-niche theory 204B, 209–10,
      225
sugar beet (*Beta vulgaris*), GM modified
  adoption more likely on intensively
      managed farms 133
  impact on birds depending on cropland
      weed seeds 131–3
sulfur dioxide SO$_2$ 17
summer drought 293B
surplus yield models 176–8B, 178–86
sustainability
  ecological dimensions 22–8, 33
  economic dimensions 28–9, 33
  exploiters and preservationists 20
  in fisheries 239
  marine zoning plan for 27B
  the middle ground 20–1
  and our way of life 2
  sociopolitical dimension 29–31, 33
The sustainable biosphere initiative: an
  ecological research agenda 21
sustainable fisheries, ecological and
  economic models never
  perfect 197–8
sustainable forestry
  Australia, retention of maternity roost
      trees for bats 100

New Zealand beech forest
      proposal 190–1, **190**
sustainable future, moving towards 20–31,
      33
  ecological applications 22–8
sustainable harvest 173–4, 175–6B, 198
sustainable management 29
sweet vernal grass (*Anthoxanthum*) 49
sweetgrass (*Anthoxanthum nitens*), used in
      basketry 224
sweetgum (*Liquidamber styraciflua*),
      possible response to global
      warming 302–3
  current and predicted distributions *302*
Sydney blue gum (*Eucalyptus saligna*) 97

takahe (*Porphyrio hochstetter*), translocation
      of 23B, 48–9
  fossil evidence of former locations 48,
      48
  fundamental and realized niches
      for 48–9
  intense conservation efforts 48
  translocated island populations can
      become self-sustaining 48
tallgrass prairie restoration 219–20, *220*
  reducing soil nitrate by carbon
      addition 244, *245*
Tana River crested mangabey (*Cercocebus
      galeritus galeritus*)
  population size needed for a hundred
      years persistence 118
tangled web spider (*Nesticodes rufipes*),
      predation on broccoli 245
tapir (*Tapirus terrestris*) 271
target pest resurgence 167
  explanations for 144–5B
  pest population recovers quicker than its
      enemies 151–2
  and secondary pests 151–3, 167
Tehuacán-Cuicatlán Biosphere Reserve,
      Mexico 309–10
  cacti species under various
      scenarios **311**
  information on possible extinctions, case
      for new reserve 310
temperatures, predicted to continue
      rising 18
Tetranychus urticae, pest 158
thermal inertia 291
thick-leaved oak (*Cyclobalanopsis
      edithiae*) 215, *215*

Thlaspi caerulescens, zinc-accumulating herb 49

three-wattled bellbird (*Procnias tricarunculata*), migratory behavior in Costa Rica 84–5

Thymus serpyllum 50, 52

Thyridoptery × *ephemeraeformis*, bagworm caterpillars 157

ticks *see Ixodes*

tidal salt marshes, restoration of tidal action 208–9, 225

recovery of different aspects of ecosystem functioning 209, *209*

tobacco plant (*Nicotiana rustica*), produces nicotine 147

Tolo Harbor, Hongkong 255–6

management response, Tolo Harbor Action Plan 256

significant physical and chemical pressures 255–6

suite of ecosystem health indicators proposed 256

topshell (*Tegula* sp.) 231B

tragedy of the commons, avoidance of 173–4, 198

trees

coped with waxing and waning ice ages 302

North American, will global warming move too fast 302–3, *302*

Trema micrantha 63

Trialeurodes vaporariorum, pest 158

Trichogramma wasps, parasitize brussels sprouts 273

Trinia glauca, restricted to dry, nutrient-poor soil 50, 52

Trisetum flavescens 62

trophic cascades 232B, 240–1, *241*

control on biomass is top-down 232B

and top-down and bottom-up control of food webs 232B, 256

useful natural pest control for certain crops 245

trophic level, mean, declining in fisheries catches 239, *239*

tropical birds, range 223

mature forest specialists at risk of extinction 223

mean latitudinal range sizes 223

tropical dry forests, much converted to agriculture and pasture 213, *214*

effects of accidental fires on remaining forest 213

patches augment and link forest fragments 213

tropical forest restoration, Panama 62–5

suggestions for restoration management 64–5

tropical forests

biodiversity threatened 196–7

high species richness 265B

Panama, diversity loss 62–3

Peru, successional mosaic 271, **272**

see also tropical birds; tropical dry forests

tsunami 230

tumble mustard (*Sisymbrium altissimum*) 242

uncertainty, and risk of extinction 113B

urban settings, pest control problems in 157, *157*

Urocyon littoralis (Californian Channel Island fox)

recent extinction, translocation possible 130

relationships among subspecies 130, *131*

USA

area where wood thrush populations likely to become extinct 269, *269*

Connecticut, recovery of salt marsh vegetation 208

eastern, eradication campaign against African witchweed 142–3

eastern, pests of ornamental trees and shrubs 157, *157*

estimated annual costs associated with invaders 14, **15**

food web for a rocky shore 231B

forest blazes in California 203, 226

genetic rescue of Florida panther 24–5B

Japanese grass (*Microstegium vimineum*), invasive potential of 67–8

Minnesota, restoration of tallgrass prairie 219–20, *220*, 244, *245*

Oregon, holistic forest restoration, willingness to pay 276

Ozark glades, intrusion of fire-intolerant species 215–16

piñon-juniper woodland, butterflies in decline 222–3

problem with some introduced pine species 23B, 66, 67

southeastern, floodplains due for
    reforestation   91
study of beetle richness   266B   267B
use of organic insecticides and emergence
    of secondary pests   151, *152*
Walnut Creek, Iowa   284–5, *285*
see also California

Vanuatu, success of data-less management
    for trochus shellfishery   186
Varroa mite, arrival in South Africa
    threatens 'natural' bee colonies
    13
vernal pools, home to fairy shrimps   *310,*
    311
vertebrates
    evolved resistance to some
        pesticides   163
    as influential as insects on
        succession   215
    marine and freshwater, signs of
        population decline   6, 7
    occasional use as biological control
        agents   157–8
Vespadelus darlingtoni   99
Vespadelus pumitus   99
Vibrio cholerae, found in ballast water   93
Victoria, Lake
    introduction of Nile perch (*Lates*
        nilotica)   2, 3, 13
    loss of biodiversity   13
Virola surinamensis, survives well in
    Saccharum grassland   63–4
VORTEX (simulation modeling
    package)   120, *121, 126*
vultures, reintroduction in Valais region,
    Switzerland   275–6

Walnut Creek, Iowa
    biodiversity scenario   284, 285, 286,
        286
    present landscape   284, 285, *286*
    production scenario   284, 285, *286*
    water quality scenario   284, 285, *285,*
        286
wasp parasitoids, enemies of
    bagworms   157
water meadows, ancient, damselfly
    habitat   87
water runoff, improved   242
watermelon (*Citrullus lanatus*), and native
    bees   103

watermelon fields, northern
    California   103–4
    economic argument for preserving bee
        natural habitat   104
    pollination by native bees   103–4, *103,*
        251–2, *252*
waterways, nitrate pollution, main culprit
    farm specialization   247
weed biomass, declined with addition of
    carbon to soil   244, *245*
western yellow robins (*Eopsaltria*
    griseogularis)   72–3
weta (*Deinacrida mahoenuiensis*)   223–4
wetland construction to manage water
    quality   247–8, *248*
wetland forest restoration   91–2
    floodplains due for reforestation,
        USA   91
    major problems   91–2
wetlands   91–2, 247–8, 257
wheat, aphid pests, overwintering of
    predators   156–7
wheatgrass (*Pseudoroegnaria* sp.)   242
white skate (*Rostroraja alba*)   74
white stork (*Ciconia ciconia*)   101
white-backed vulture (*Gyps bengalensis*)
    in crisis   109
    diclofenac found in dying birds   134
    dramatic decline in numbers   133
    economic benefit of ecosystem
        services   109, 135, 136–7
    no longer enough to dispose of all
        refuse   109
    use of diclofenac banned in
        Pakistan   135
white-backed vulture, long-billed (*Gyps*
    indicus)   109
white-backed vulture, slender-billed (*Gyps*
    tenuirostris)   109
white-footed mouse (*Peromyscus*
    leucopus)   236
    transmission of Lyme disease to
        ticks   235
white-naped honeyeaters (*Melithreptus*
    lunatus), declined after logging
    72
wilderness protection   13
wildlife reserves
    and Island Biogeography Theory
        269–70
    and the Kalahari bushmen   262
willingness to pay   29, 276

wind farms 82, 100–2, 106
  expansion without detailed ecological
    information 100
  siting of turbines 105
  threats to migrating and dispersing
    birds 100
winter moths (*Operophtera brumata*) 158
wood thrush (*Hylocichla mustelina*),
    variation in fundamental rate of
    increase λ 268–9, *269*
woody plants, nitrogen fixing 241–2
  restoration using burning 242
wren (*Troglodytes troglodytes*), and a positive
    NAO index 301, *301*

yellow star thistle (*Centaurea
    solstitalis*) 14, *15*, 141, 242, *243*

yellow-fin tuna (*Thunnus albacares*) 238
  sustainable yield from 179, *180*
yield curves *177B*, *178B*
  shortcomings of 184–6

Zea mays 152–3
zebra mussel (*Dreissena polymorpha*)
  damaging invader in Great Lakes 14, *15*,
    93–4
  exerts profound ecosystem effect
    241
  predicted spread to inland lakes 93,
    93
zero-fishing zones *see* marine protected
    areas
zooplankton, northerly shift in northeast
    Atlantic *293B*, *294B*